chemistry

FIFTH EDITION

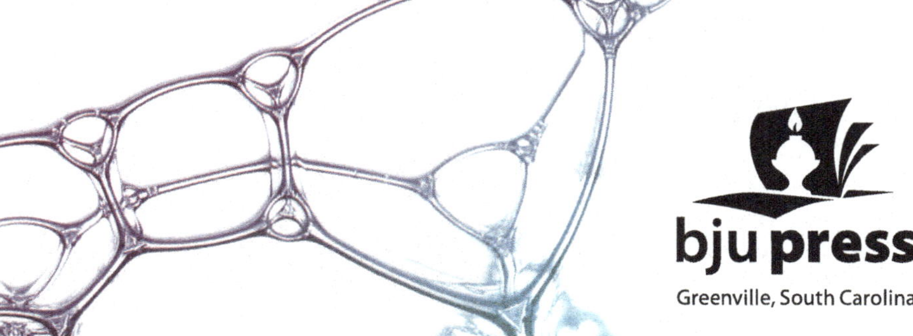

bju **press**®

Greenville, South Carolina

The writers and publisher have made every effort to ensure that the laboratory exercises in this publication are safe when conducted according to the instructions provided. We assume no responsibility for any injury or damage caused or sustained while performing activities in this book. Conventional and homeschool teachers, parents, and guardians should closely supervise students who perform the exercises in this manual.

NOTE: The fact that materials produced by other publishers may be referred to in this volume does not constitute an endorsement of the content or theological position of materials produced by such publishers. Any references and ancillary materials are listed as an aid to the student or the teacher and in an attempt to maintain the accepted academic standards of the publishing industry.

CHEMISTRY Teacher Lab Manual
Fifth Edition

Coordinating Writer
David M. Quigley, MEd

Writers
Christopher D. Coyle
Kelly Driskell

Biblical Worldview
Renton Rathbun, ThM

Academic Oversight
Jeff Heath, EdD
Rachel Santopietro, MEd

Editor
Rick Vasso, MDiv

Cover, Design, and Interior Concept Design
Sarah Lompe

Page Layout
Carrie Walker

Illustrator
Ina Stanimirova
c/o Lemonadeillustration agency

Permissions
Kathryn Jackson
Elizabeth Walker

Project Coordinators
Heather Chisholm
Chris Daniels

Photo credits appear on pages 276–77.

The text for this book is set in Adobe Minion Pro, Adobe Myriad Pro, Arial by Monotype Typography, Calibri by Monotype Typography, Helvetica, Minion Math by Typoma GmbH, Playfair Display by Claus Eggers Sørensen, Raleway, and STIX.

All trademarks are the registered and unregistered marks of their respective owners. BJU Press is in no way affiliated with these companies. No rights are granted by BJU Press to use such marks, whether by implication, estoppel, or otherwise.

The cover photo shows a close-up of a cluster of bubbles on a black background.

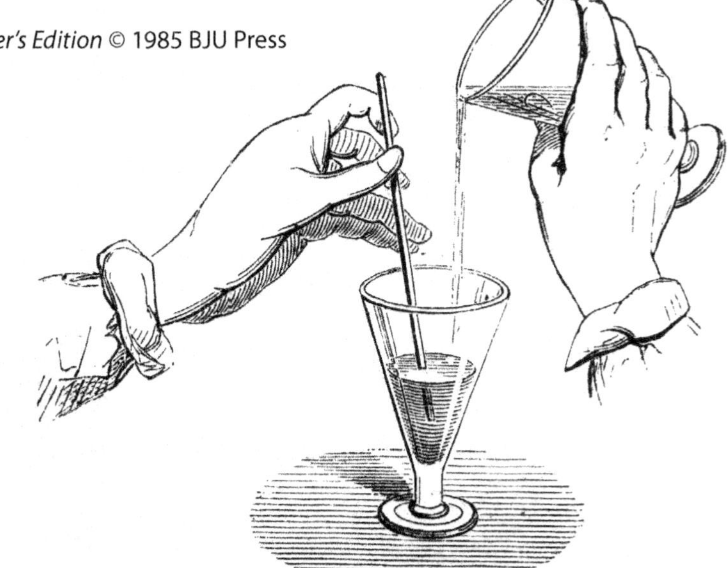

Contents

Contents

Contents

Welcome to the Laboratory!

A chemistry laboratory is anything but boring. Laboratories are places where exciting discoveries are made, old ideas are challenged, and new ideas are formed. But a laboratory can't do any of those exciting things on its own. It needs a chemist. THAT'S YOU!

Thinking like a Chemist

Who, me? Yes, you!

"But I'm just a high-school chemistry student!"

But you can be a student chemist! Even if you don't see a science career in your future, you can develop skills this year to help you work like a chemist at your own level. You'll learn to think like a scientist—being safe in the laboratory, making predictions, collecting data, and testing your ideas. Sometimes you'll have to think about how to solve a problem on your own, without any procedures spelled out in the lab activity. These lab activities will have the word *inquiring* in the subtitle. You'll learn how to collect data with the equipment that you have to work with and how to tell whether the data that you've collected is good. The equipment for these activities can come in different shapes and sizes, both high-tech and low-tech. You'll learn to use chemicals, laboratory burners, and good old-fashioned glassware. But you'll also use the internet, possibly probeware, and maybe even your smartphone.

Above all, learn to do science within the framework of a Christian worldview. Everyone has a worldview. It is the lens through which you view everything. A Christian gains his worldview from the Bible. This worldview doesn't just affect what he believes about God, Jesus, and salvation—it also affects how he views science, history, and even how he interacts with other people.

As a student scientist, keep your worldview in mind. When you do an experiment, think about how it relates to a biblical worldview. And you're not just a student chemist. You're a Christian student chemist. As a Christian student chemist, you are interpreting the world as God has already interpreted it in His Word. Christians should see science as an important gift of God, given to humans so that they can obey the Creation Mandate (Gen. 1:28) and the two great commandments (Matt. 22:37–40). When Christians do science, they bring glory to God and help their fellow humans to live in this fallen world. Therefore, some questions will specifically ask you to use your worldview to apply the information from your lab activity to a real-world problem. You will be challenged to consciously think like a Christian as you're doing science.

A Christian shouldn't see science as a pathway to truth. Rather, he should look to the Bible, the one source of absolute truth, to establish the guidelines and proper role of science in human life. Scientists working from a Christian worldview should evaluate existing scientific models in light of Scripture and challenge or revise them when they conflict. And as they create models of their own, they must test them for agreement with the revealed truths in the Bible. When we act in this way, we glorify God by using science as He intended.

But what good is all of this? Why is it important to develop the skills and mindset of a chemist? As a Christian, learn to see science as an amazing tool to glorify God and help people by obeying God's command to wisely use His creation. We should do the work of chemistry within the context of a Christian worldview and use it as God intended in ways that harmonize with His Word.

SO ...
let's get into the laboratory!

Keys to Laboratory Success

Lab activities can be both interesting and fun. But your laboratory experience largely depends on you. If you take these activities seriously, keep your brain engaged, and try to learn as much as possible, you'll come away from the laboratory feeling that your time has been well spent. On the other hand, if you treat lab activities carelessly, you probably won't learn much nor will you have a genuine sense of accomplishment. The following guidelines will help you get the most from lab activities. And they'll also help you experience the true pleasure of a brain at work.

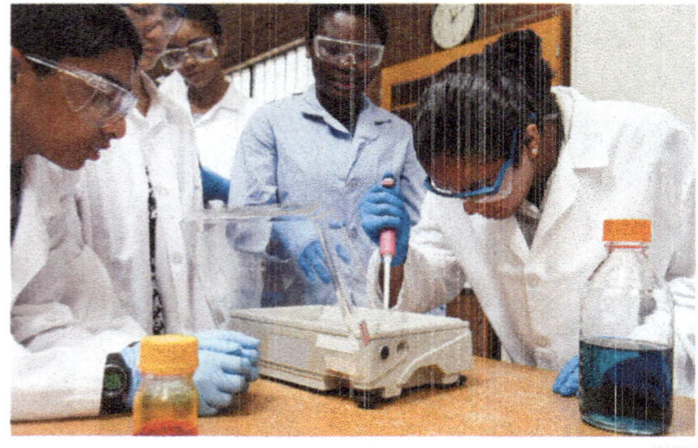

1. Read through the entire lab procedure before class. You'll know what to expect, and you'll be far less likely to make mistakes or forget a crucial step.

2. Review the parts of your textbook that apply to the lab activity. Your teacher may give you specific review guidelines.

3. Come to class well prepared. Bring your textbook, calculator, paper, and pencil. You won't necessarily need all these for every lab activity, but you'll regret it if they're sitting in your locker on a day when you *do* need them!

4. Begin the lab activity only when instructed to do so by your teacher. Read each lab procedure step carefully so that you don't make foolish mistakes. Procedure steps are identified by lettered bullets (e.g., A, B, C). If you're ever in doubt about what to do, *ask*, don't guess.

5. Lab activity questions are mixed in with the procedure steps. They're identified by numbers (e.g., 1., 2., 3.).

6. Many lab activities require you to record data in tables. Typically data tables are located at the end of the activity.

7. Record your results from measurements and calculations carefully and accurately.

8. Most lab activity questions ask you to explain your answer. Don't view this requirement as pointless busywork. Explaining something makes you think more deeply about it. It also helps you connect principles from the textbook and class discussion to what you're doing in the laboratory.

9. Keep your laboratory area tidy. At the end of class, put away your equipment and dispose of trash according to your teacher's instructions.

10. While socializing may seem like a lot more fun than the lab activity, leave that behavior outside the laboratory door. You'll be more focused, get better results, and make far fewer mistakes.

Staying Safe

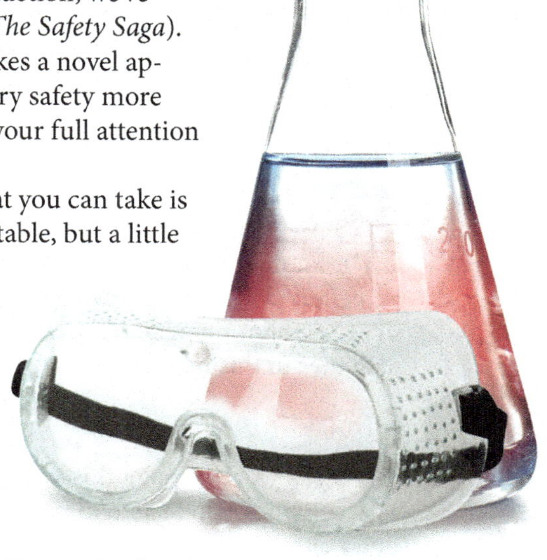

Nothing ruins a day quite so much as getting hurt in the laboratory. While the lab activities in this manual have been carefully designed to be safe, no lab activity is perfectly safe. For this reason, you should maintain constant vigilance when you're working through one. You should always use protective equipment when it's specified.

Rather than cover laboratory safety in this introduction, we've devoted an entire lab activity on the topic (Lab 1B—*The Safety Saga*). This lab activity isn't just a list of do's and don'ts. It takes a novel approach to safety, one that we hope will make laboratory safety more enjoyable and more effective. Be sure to give Lab 1B your full attention so that your class can have an accident-free year!

One of the most important safety precautions that you can take is to wear eye protection. Yes, goggles can be uncomfortable, but a little discomfort is better than losing an eye. *Always* wear your goggles anytime you are working with glassware, chemicals, or projectiles. For lab activities that require eye protection or other safety precautions, you will find the appropriate safety icons that pertain to that activity immediately below the activity's equipment list. The icons that you will see are listed below.

 **Body Protection**—Chemicals or other materials could damage your skin or clothing. You should wear a laboratory apron, chemical-resistant gloves, or both.

 **Chemical Fumes**—Chemical fumes may present a danger. Use a chemical fume hood or make sure that the area is well ventilated.

 **Corrosive Substance**—Acids and bases are corrosive substances that can cause damage to skin and eyes. Wear proper protective equipment and follow safe handling procedures.

 **Electricity**—An electrical device (e.g., hot plate, lamp, microscope) will be used. Use the device with care.

 **Extreme Temperature**—Extremely hot or cold temperatures may cause skin damage. Use proper tools to handle laboratory equipment.

 **Eye Protection**—There is a possible danger to the eyes from chemicals or other materials. Wear safety goggles.

 **Fire Hazard**—A heat source or open flame is used. Be careful to avoid skin burns and the ignition of combustible materials.

 **Gas**—Improper use of gas can result in burns, explosion, or suffocation. Be careful to check that the gas is turned off when you are finished.

 **Sharp Object**—Use care with the equipment in this lab activity to avoid cuts from sharp instruments or broken glassware.

 **Toxic Material**—A substance in the investigation could be poisonous if ingested.

Since some lab activities have specific steps with unique safety challenges, special safety notes are occasionally placed within the activity in **bold italic font** to emphasize their importance. For additional information on laboratory safety, see Appendix A.

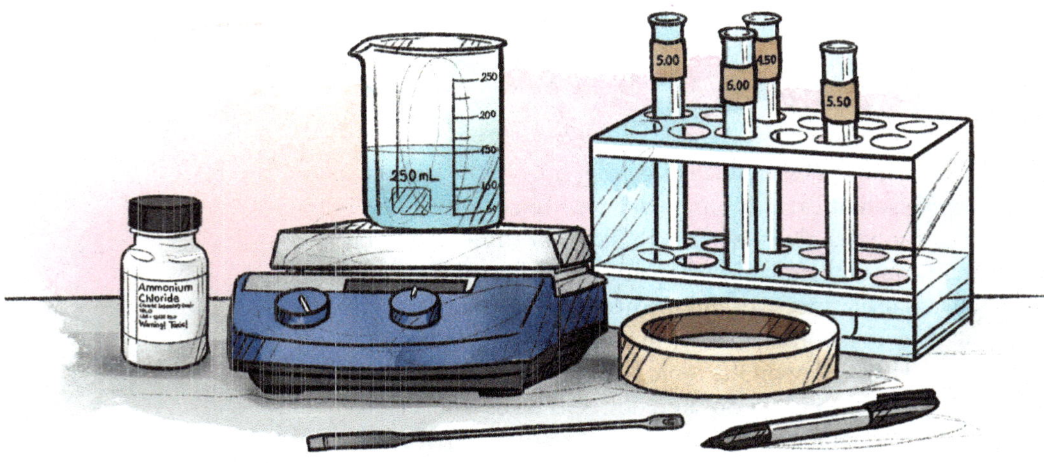

Lab Equipment and Techniques

Science involves both equipment and techniques that you may not be familiar with. To help, we have included appendixes for these. Appendix B is a visual glossary of laboratory equipment, and Appendix C explains many of the techniques that you will need for the lab activities in this manual.

Inquiry Lab Activities

As you work through this lab activity book, you will find inquiry lab activities. These activities are part of being a student scientist. Inquiry lab activities give you an opportunity to create your own activity about a given subject. You will ask questions, form hypotheses, design investigations, collect and analyze data, draw conclusions, communicate results, and often develop additional questions. These are tasks that scientists do every day.

Writing Formal Lab Reports

Communication is a key facet of science. Scientists need to communicate their findings to other scientists and ultimately to the world. Student scientists do this through writing formal lab reports, which you will write routinely if you take a college-level science course. Appendix E describes how to write a formal lab report. Use this to prepare your reports and to prepare for future science endeavors.

Let the Year Begin!

One of the pleasures that we get from writing this book is knowing that for many of you the laboratory will become a special place. We invite you to come into the laboratory part of this course expecting to learn new and fascinating things. Look forward to the challenge of getting your hands on science. Your visits to the laboratory could be life-changing for you.

WELCOME!

To The Teacher

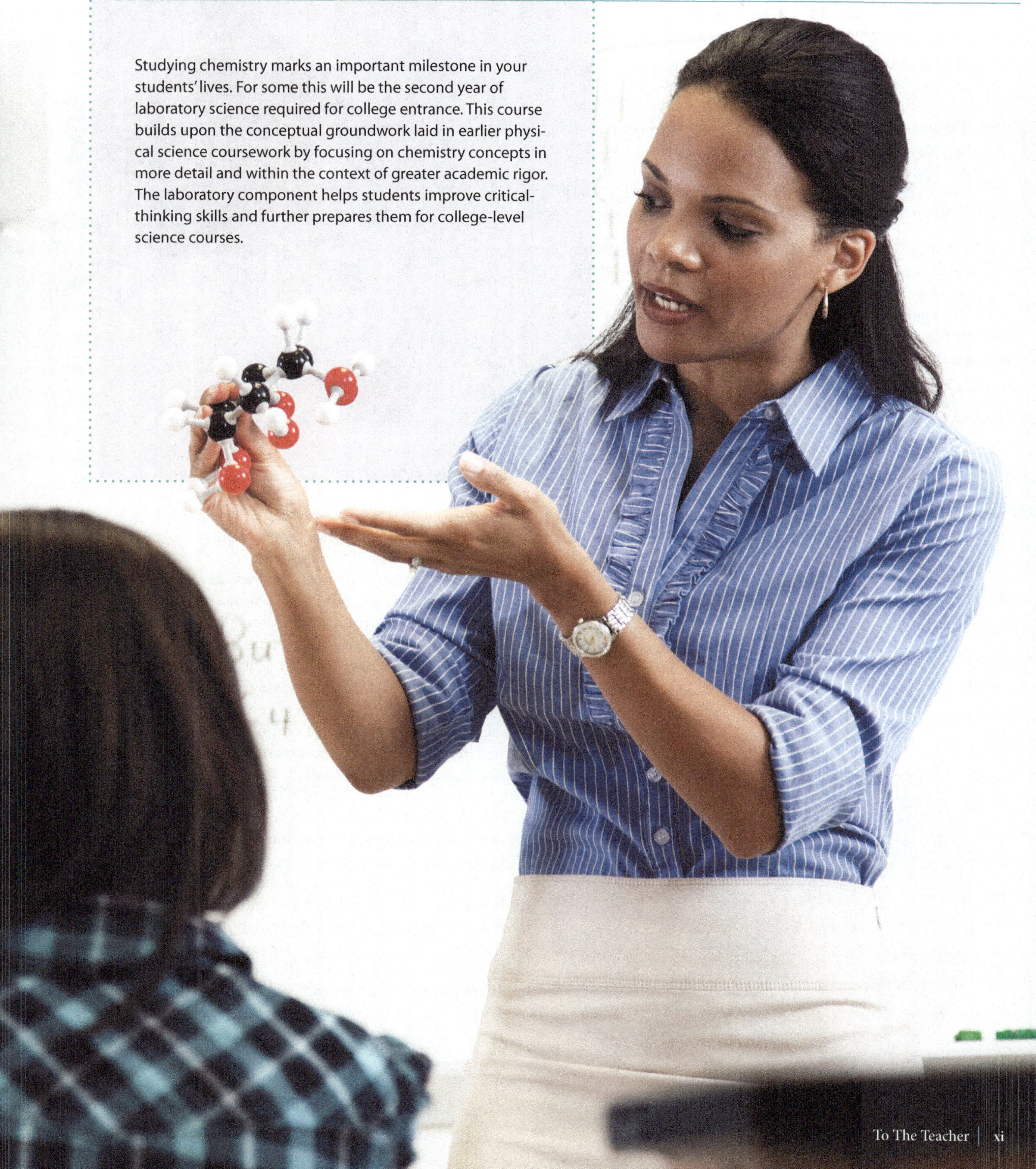

Studying chemistry marks an important milestone in your students' lives. For some this will be the second year of laboratory science required for college entrance. This course builds upon the conceptual groundwork laid in earlier physical science coursework by focusing on chemistry concepts in more detail and within the context of greater academic rigor. The laboratory component helps students improve critical-thinking skills and further prepares them for college-level science courses.

To The Teacher

As your students work their way through their lab activities this year, stress the importance of understanding the "big ideas" explored in each one. While technical skills like measuring, calculating, and graphing are important, it is far more important for them to understand the scientific concepts and principles behind each lab activity. Future science courses will make better sense and be much more exciting if students are conceptually comfortable with the subject.

This introduction introduces you to the key features of this new *Chemistry* Teacher Lab Manual 5th Edition. Read the material carefully so that you understand the direction that we're going and the reasons behind what we do. The journey will be a lot more enjoyable to you and your students if you look at the map to see where you're heading!

KEY IDEAS ABOUT LABORATORY SCIENCE

The laboratory is never very far from the science of chemistry. Over the years some of the most exciting and revolutionary ideas in chemistry began in the laboratory. And it is the place where your students put their hands on science and see it come alive. Ideas that seem abstract in class become concrete when reinforced by lab activities.

Stress this idea when you introduce the laboratory portion of your course. Help students understand that lab activities aren't "busywork"; they are rather the cement that sets key science concepts in their minds. Emphasize the fact that each lab activity has within it key ideas designed to help them understand how a particular principle works and how it fits into the area of science that they're studying.

Finally, the laboratory portion of any science course provides something that cannot be learned in any other way—the opportunity to develop *hands-on skill*. As students perform each lab activity, they will develop the critical skill of asking scientific questions and creating hypotheses. They will develop proficiency in using instruments and making measurements. They will learn to focus and refine their observational skills as they collect data and build models. This process will help them mature intellectually and socially as they take responsibility for their actions. Be sure that you communicate these key goals to your students.

MODEL MAKING

Science is all about making models. As scientists struggle to make sense of a phenomenon, they fit their data into models. If enough data no longer fits in a particular model, that model is modified or even replaced by a more workable one. It's important to remember that science is not a progression toward greater truth; it's a quest for more workable models.

Even though all models are limited, scientists can do excellent work and advance their discipline as they work within their models. Even limited models (and they are all limited to some extent) can lead to many exciting and useful discoveries. In many cases in the past, scientists' recognition of a model's weaknesses led them to create new and more workable ones.

NEW FEATURE IN THE 5TH EDITION

Some of the lab activities in the Student Lab Manual are brand new. Others, retained from the previous edition, are significantly rewritten and enhanced. An exciting new feature is inquiry lab activities. In inquiry activities students ask questions, form hypotheses, design investigations, analyze data, draw conclusions, communicate results, and often develop additional questions. Inquiry activities in the classroom allow students the opportunity to direct an investigation. After identifying the question, students form a hypothesis and design an investigation to answer it. Inquiry activities foster curiosity and require students to more effectively think critically than traditional activities in which the procedures for the investigation are spelled out explicitly.

To The Teacher

Teaching science can be a real challenge, especially if you are a new teacher or a homeschool parent. And even if you have many years of science education experience, lab activities can still contain surprises. For this reason, each activity comes with detailed teacher notes in the margins. These address a wide variety of topics ranging from equipment and material substitution suggestions to answers for those hard questions that students tend to ask.

Before performing any lab activity for the first time, read through it completely, looking at the relevant teacher notes as you go along. We did our best to anticipate your questions and point out the pitfalls that may lie in your path. We also included worked-out math, interesting extra information, and worldview commentaries. We want you to feel confident in the laboratory! To further this end, many teacher notes include icons that identify their purpose. The key below explains each one.

 **BIBLICAL WORLDVIEW**

This icon highlights biblical worldview-shaping discussion questions and activities.

 **DEMONSTRATION**

A note that includes a demonstration icon alerts you to an activity that works well as a teacher demonstration or to a demonstration that aids students as they perform the activity.

 **DIFFERENTIATED INSTRUCTION**

This icon alerts you to a procedure that may be beneficial for some students. The note may include a suggestion to alleviate a problem, for example, with a lab procedure that may be difficult for students with a disability, or it may suggest methods of instruction to help students with various learning styles.

 **DIFFICULT CONCEPT**

This icon alerts you to a concept that students may find difficult to grasp or that may be a common misconception. Information and suggestions will be given on how to clarify the material or correct the misconception.

 **EQUIPMENT**

A note with an equipment icon provides extra information about the equipment and materials needed for the lab activity. Acceptable substitutions, local sources, or preparation instructions are included.

 **HELPFUL TIP**

This icon alerts you to a note that provides you with hints and ideas for coordinating and managing the lab activity that will help students get better results or have a smoother experience.

 **MATH**

The math icon signals a note that provides guidance for calculations or other math-related tasks. Worked-out problems are often flagged by this icon.

 **RECALL**

A note with the recall icon highlights a topic from earlier chapters or courses that you may want to review to integrate and reinforce concepts.

 **SAFETY**

When you see this icon, be sure to read the accompanying note since it contains crucial information that will help you to be safe or to avoid a potential safety problem.

 **TEACHING STRATEGIES**

This icon alerts you to suggestions for ways to improve your instruction. A note so marked may include, for example, a class discussion topic, a formative assessment activity, or a prior knowledge activator.

WEB LINK

A note with this icon directs you to a resource that will help you in completing the lab activity. The note often directs you to a link on TeacherTools Online.com or suggests that you do an internet search using the keywords provided.

Lab activities always go better if you plan ahead. The Student Lab Manual contains forty-two lab activities, in most cases two for each *Chemistry* Student Edition 5th Edition chapter. You will not have time to complete them all. This gives you the freedom to choose the ones that you want your students to complete. Some require relatively little equipment, while others require more elaborate equipment. Choose lab activities that fit your educational setting. Invest some time at the beginning of the school year to decide which activities you would like to use. The *Chemistry* Teacher Edition 5th Edition includes a suggested school-year schedule in the Lesson Plan Overview that includes lab activities. This Teacher Lab Manual also includes a Lab Exam (in Appendix G).

Always read the teacher notes near the beginning of each lab activity for guidelines as to *when* you should do it. Some activities may be done at any time within the chapter, but many work far better if they are performed either before or after you've taught a particular topic.

If you've never done a particular lab activity before, run through it yourself before your students perform it. To make it easier for you to know whether your students are getting good results, most lab activities include example data showing the overall trend that their results should take. Remember, however, that their data will probably not be identical. Similarly, for calculations involving data, we usually include a sample answer that is based on the example data.

EQUIPMENT AND MATERIALS

Another facet of good preparation is ensuring that you have the necessary equipment and materials to perform each lab activity. Again, we suggest that you spend some time planning before the school year begins. Confirm that you have the required equipment and that the materials you need are on hand or available locally. Some items must be ordered. Since it's generally more economical to place a few large orders than many small ones, working out the year's equipment and materials all at once makes a lot of sense.

To make planning easier, we've included two equipment and materials lists in Appendix F of this Teacher Lab Manual. One is sorted alphabetically by item name, while the other is sorted by lab activity number. If you are teaching in a smaller school or are homeschooling, don't let the equipment guidelines discourage you. While thinking about the equipment needs for an entire year seems daunting, remember that there are many alternatives and substitutions possible. Often, everyday items or locally obtainable materials may work just as well. Suggestions for substitutions are included in the equipment and materials list, as well as in Teacher Lab Manual margin notes for each lab activity.

While most of the materials specified in this manual are reasonably safe, a few do require special care and respect. Never order a chemical unless you have viewed its Safety Data Sheet (SDS) and are comfortable with the safety guidelines. The SDS for chemicals are available through science supply companies online. Chemicals should always be stored properly. Storage requirements are included on the SDS. If you are unsure regarding safety, seek proper guidance *first*. And if you just don't feel comfortable with the requirement of a particular lab activity, choose another activity.

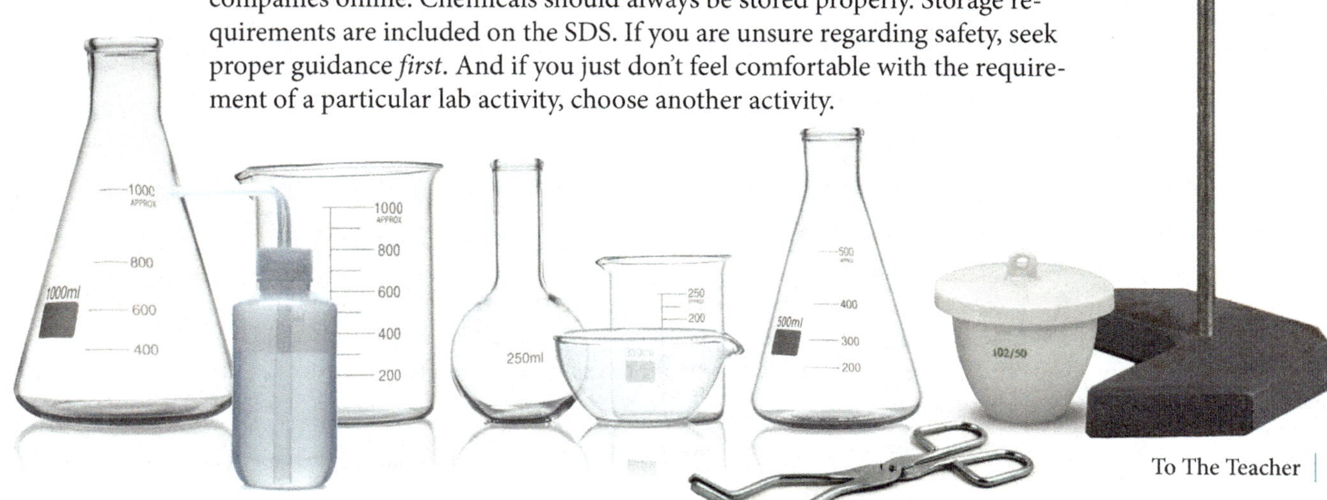

To The Teacher

GETTING THE MOST FROM A LAB ACTIVITY

Activities work best if the students arrive well prepared and ready to go. Be sure that you have covered the requisite background material in class. You may also need to review earlier concepts.

The activities in the Student Lab Manual follow a linear approach through the topic. Instead of questions being at the beginning and end of a lab activity, they are freely mixed in with the procedure steps. Questions follow a logical progression—they often check understanding of previous steps or introduce steps that follow. Arranging the material in this way encourages students to see the question-answer process as integral to the learning experience rather than something to complete as quickly as possible. To help keep students organized, procedural steps are indicated by lettered bullets (e.g., A, B, C), while questions are numbered.

Many questions specify that students explain their answers. This requirement forces them to think more deeply about the reasons behind their answers. Encourage students to write thorough explanations in complete sentences.

Unless you are homeschooling, you will probably divide your class into lab groups. Whatever size group you select, ensure that all members of the group are actively engaged in the activity. Don't allow one student to dominate a group.

MEASUREMENTS

Good science depends on quality measurements. While most students don't consider measuring the most enthralling of subjects, it's one that must be given adequate attention. Require students to make careful measurements. Also help them to recognize the limitations of their instruments. If at all possible, do the measurement activities (Labs 3A and 3B) since these address the key concepts behind measurement.

Another important point to stress is that when the procedures specify multiple measurements of a quantity, each one must be done *independently*. Encourage every student to take part in taking measurements. This will help all students improve their performance of this important skill.

FORMAL LAB REPORTS

Most college science courses require students to describe their laboratory work in the form of formal lab reports. If this is a skill that you would like your students to work on this year, refer to the information in Appendix E.

Any laboratory work includes the possibility of accidents. While it's never possible to eliminate them entirely, good preparation can greatly reduce their number. Most laboratory accidents are fairly minor, but some rapidly develop into serious situations. For this reason, you need to give some careful thought to laboratory safety before the school year begins.

First of all, it's vital that you understand your role and responsibility as the classroom teacher. You are key to developing a *think safe* mindset in your laboratory. There are many laboratory safety resources available online. Do a keyword search for "high-school laboratory safety." Flinn Scientific offers free laboratory safety training to teachers on their website.

Make sure that you have students complete the lab safety activity (Lab 1B). As you will discover, this activity takes a novel approach to laboratory safety. Instead of the usual list of rules, this exercise approaches safety from the perspective of learning to *think* safe. When the students develop the ability to anticipate problems, there are far fewer accidents. Don't be tempted to substitute a quick list of rules for this lab activity. While every laboratory needs rules, help students understand the reasons behind the rules. But rules can only do so much; a think-safe mindset will keep them safe in and out of the laboratory.

If you are a new science teacher, we strongly urge you to seek experienced help as you plan your laboratory classroom. Other teachers and reputable online resources can be very helpful. If you're teaching in a school, you should also consult your administration, school's lawyer, and the fire marshal in order to be sure that you're meeting local safety laws and guidelines.

Compliance is especially important when it comes to chemical storage and disposal, so be sure that you do your homework before bringing chemicals on the premises. All chemicals must be stored in properly labeled containers. Chemicals obtained from suppliers should have the necessary labeling on the original container. If you transfer a chemical to another container or if you prepare a diluted solution from a concentrate, properly label the new container. At a minimum the label must include the full chemical name (copied from the original label or SDS), the concentration (if applicable), the date prepared, any significant hazards (e.g., corrosive, flammable, toxic, oxidizer), and your initials.

Most countries have now adopted the Globally Harmonized System (GHS) for communicating hazard information. The GHS is a program managed by the United

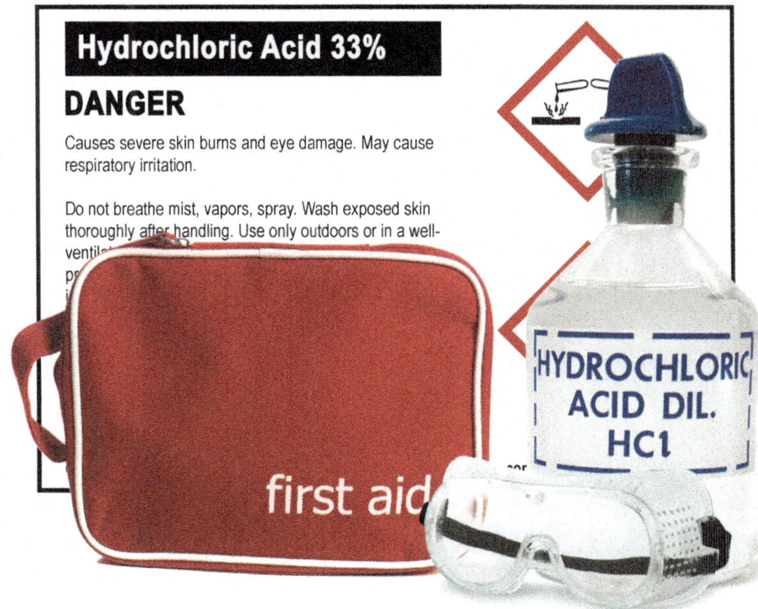

Nations to standardize hazard communication. It is your responsibility to keep informed about changes in safety and labeling standards as well as the differences between labels. For more detailed guidelines about chemical labeling and safety, visit the Occupational Safety and Health Administration (OSHA) website. You may also want to perform a general online search using keywords such as "high-school laboratory chemical safety," "HMIS labeling," "NFPA fire diamond," and "Globally Harmonized System." If you are outside the United States, be aware that safety standards as well as storage and labeling guidelines may be different. You will need to do some research appropriate to your situation.

Your laboratory should be equipped with basic safety equipment that is properly maintained and up-to-date. A fire blanket, a first aid kit, and appropriate fire extinguishers should be in every laboratory. Laboratories containing chemicals should also include an eyewash station. An emergency shower is highly desirable as well. Ensure that you have sufficient goggles for every member of the class. Research the requirements for splash protection and impact-resistant goggles for your location.

You must have an SDS for *every* chemical in your laboratory. Be sure that these SDSs are readily accessible in an emergency. Whenever you order a new chemical, it should come with an SDS. If it doesn't, download and print one from the manufacturer's or supplier's website. You should also have posters explaining the GHS, HMIS, NFPA codes, or all three, prominently displayed.

Let your year begin!

Name ___________

Date ___________

The Great Biscuit Bake-Off

Relating the Composition and Properties of Biscuits

Biscuits, or hardtack, have been known since ancient times. Because they were light, nutritious, and could keep for years if baked hard and stored dry, they were the food of choice for Roman legionnaires, British sailors, and Confederate and Union soldiers alike during the Civil War. Cooking involves chemistry, so baking biscuits is really applied chemistry. In fact, the study of chemistry may have begun with the science of food preparation. It is this chemistry of biscuit baking that you will explore today.

How do changes to the ratio of ingredients in a biscuit affect its properties?

You will be making three different biscuit recipes in a great biscuit bake-off to see which one you like the best. When you scan the list of ingredients and the procedures, you'll see that the recipes use similar ingredients and require baking at the same temperature for about the same length of time. But you will find that small variations in the recipes make a big difference in the taste and texture of the biscuits.

Salt (sodium chloride, NaCl) and sugar (sucrose, $C_{12}H_{22}O_{11}$) are chemicals that affect the taste of biscuits. Butter, milk, and shortening contain fats that affect their texture. Flour is the main ingredient—it binds the biscuit together and is what gives you energy when you enjoy one.

EQUIPMENT

- kitchen oven
- large bowl
- measuring cups
- measuring spoons
- pastry blender
- rolling pin
- biscuit cutter
- cookie sheets
- spatula
- flour
- sugar
- baking powder
- salt
- butter
- milk
- cream of tartar
- shortening
- eggs
- toppings for biscuits (optional)
- laboratory apron
- goggles

QUESTIONS

» How do I make scientific observations?

» How do the properties of foods change in response to a change in their ingredients?

» How are cooking and chemistry related?

LAB 1A OBJECTIVES

» Make scientific observations using all the senses.

» Relate the properties of foods to their ingredients.

» Explain how chemistry is related to the art of cooking.

⚠ FOOD ALLERGIES

Be aware of students that may have allergies to one or more of the ingredients used in this activity and make appropriate accommodations.

✔ EATING CHEMISTRY!

This first lab activity is important because it introduces students to the idea that chemistry affects them every day. It was created so that students can begin exploring ideas right away with minimal safety risks. Your students will explore safety ideas and laboratory safety equipment in Lab 1B.

Plan to do this lab activity in a kitchen to avoid getting chemicals in food that students will be eating because, of course, there is no reason to make biscuits if you can't eat them! Students will not need to wear goggles for this activity if you do it in a kitchen.

Consider dividing the class into three groups. Number the members of each group 1–3. Have each student with a particular number work with like-numbered students from the other groups to make a certain kind of biscuit. This allows students within a home group to compare the ingredients and processes that were used to make the different recipes and better understand the results.

Of course, biscuits aren't as good by themselves as they are with things on them! Have fun with this activity. Consider bringing in butter, jams, jellies, and maybe even some cheese, bacon, or sausage. Get students excited about your class by starting with food!

PRE-LAB CHECK

1. Name the two leavening agents used in the biscuit recipes. *(baking powder and eggs)*

2. (True or False) The study of chemistry is generally limited to the classroom and the laboratory. *(False. Chemistry is used in many areas of life, such as in food preparation.)*

3. Name some independent variables in your great biscuit experiment. *(Examples include leavening agents, the amount of milk, the type of fats. Do not accept baking time, baking temperature, or the amount of flour.)*

4. How could you use the biscuits that you make to create a survey? *(Answers will vary but should include the idea that students could survey their classmates or families on a certain quality of the biscuits, such as fluffiness, density, and crumbliness, to generate a set of data.)*

One of the most important ingredients in biscuits is the *leavening agent*, or the chemical in biscuits that makes them rise. In all three recipes, baking powder is the leavening agent. Baking powder is a mixture of chemicals including baking soda (sodium bicarbonate or sodium hydrogen carbonate, $NaHCO_3$) and some kind of acid or acid salt that chemically reacts to produce carbon dioxide gas, similar to the typical homemade volcano that uses vinegar and baking soda. Baking soda, or sodium bicarbonate, will cause food to taste bitter if there is too much of it in a baked good. Baking powder often includes other substances such as calcium dihydrogen phosphate [$Ca(H_2PO_4)_2$] and aluminum sodium bis(sulfate) [$NaAl(SO_4)_2$]. Eggs can also act as a leavening agent in baked goods when they are whipped. They also help combine the liquids in the recipe that don't normally mix together, such as oil and milk.

Procedure

RECIPE 1: BASIC BISCUITS

2 cups flour
1 teaspoon sugar
1 tablespoon baking powder
1 teaspoon salt
8 tablespoons butter, cubed
3/4 cup milk

A Preheat the oven to 425°. In a large bowl, combine the flour, sugar, baking powder, and salt.

B Using a pastry blender, cut the butter into the flour mixture until crumbly.

C Mix in milk until just moistened and turn onto a floured surface, kneading a little and rolling out to 3/4 inch thick. Cut with biscuit cutter.

D Place biscuits on a cookie sheet and bake for 10–12 minutes or until lightly browned.

RECIPE 2: BAKING POWDER BISCUITS

2 cups flour
2 tablespoons sugar
1 tablespoon baking powder
1/2 teaspoon salt
1/2 cup shortening
1 egg, beaten
2/3 cup milk

E Preheat the oven to 425°. In a large bowl, combine the flour, sugar, baking powder, and salt.

F Using a pastry blender, cut the shortening into the flour mixture until crumbly.

G Mix in milk and egg until just moistened, and turn onto a floured surface, kneading a little and rolling out to 3/4 inch thick. Cut with biscuit cutter.

H Place biscuits on a cookie sheet and bake for 10–12 minutes or until lightly browned.

RECIPE 3: DROP BISCUITS

2 cups flour
2 tablespoons sugar
1 tablespoon baking powder
1/2 teaspoon cream of tartar
1/4 teaspoon salt
1/2 cup melted butter
1 cup milk

I Preheat the oven to 425°. In a large bowl, combine the flour, sugar, baking powder, cream of tartar, and salt.

J Stir the melted butter and milk into the flour mixture until just moistened.

K Use a tablespoon to drop dollops of dough onto a cookie sheet. Bake for 10–12 minutes or until lightly browned.

Analysis

1. How do you think the biscuits that you made today differ from those used in historical times, as discussed in the introduction? You may need to do some research to answer this question.

   Historically, biscuits often didn't use a leavening agent, making them more like a cracker than a bread.

2. Taste all three types of biscuits, either plain or with the same topping on each. Which one do you like the best? Explain why.

   Answers will vary. Students should note the biscuits' taste or texture.

3. Why did Question 2 have you taste all three types of biscuits either plain or with the same topping on each?

   The purpose of the lab activity is to assess the biscuit properties, so other conditions, such as toppings, must remain constant.

4. Compare the ingredients of all three recipes. What do you notice that is different?

   See TE margin for answer.

5. Compare the textures of all three biscuits. Is there one that is different from the others?

   See TE margin for answer.

6. Cut apart a biscuit. Look at the holes in it. Where do you think the holes come from?

   See TE margin for answer.

   The chemical reaction that occurs in the process of leavening is shown below.

   $$NaAl(SO_4)_2\,(s) + 3\ NaHCO_3\,(s) \longrightarrow Al(OH)_3\,(s) + 2\ Na_2SO_4\,(s) + 3\ CO_2\,(g)$$

   The first two chemicals—aluminum sodium bis(sulfate) and sodium hydrogen carbonate—are found in baking powder.

7. What do you think would happen if you added more baking soda (sodium hydrogen carbonate) to the biscuits without changing any other ingredients?

   Students should conclude that it is possible that more carbon dioxide gas would be released, causing the biscuit to rise more.

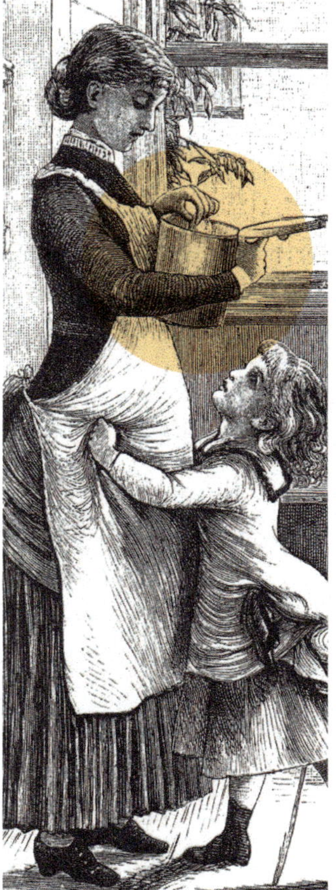

Question 4 Answer

Students should point out that the amount of flour is the same in all three recipes. All three recipes contain milk, though in different amounts. The second recipe used shortening instead of butter. The third recipe required melting the fat prior to mixing.

Question 5 Answer

Answers will vary. Students may notice a difference in the dropped biscuits since they are not kneaded and shaped with a cutter. The cream of tartar may also create a different texture in the biscuits.

Question 6 Answer

Students should conclude that the holes in the biscuits are pockets that were formed by a gas (carbon dioxide) in the chemical reactions activated by the leavening agents.

GLUTEN AND KNEADING

Gluten is a protein that forms when two other chemicals form bonds. This reaction can be sped up by kneading and by adding dough conditioners such as ascorbic acid (vitamin C). Gluten makes bread chewy and holds it together.

8. What do you think would happen if you added more baking soda (sodium hydrogen carbonate) than the aluminum sodium bis(sulfate) could react with?

   The sodium hydrogen carbonate would stay unreacted in the biscuit, affecting the taste. Since sodium hydrogen carbonate has a bitter taste, the biscuits would taste bitter.

One of the three recipes includes an ingredient called cream of tartar, which is the chemical potassium hydrogen tartrate, or potassium bitartrate ($KHC_4H_4O_6$). It is used in addition to the baking powder in your recipes and acts in a manner similar to aluminum sodium bis(sulfate).

9. Look at the reaction below Question 6 on page 3. If cream of tartar is an acid salt like aluminum sodium bis(sulfate), what is it likely to react with to form carbon dioxide gas? Where is this chemical coming from?

   See TE margin for answer.

10. Suggest some other things that could affect the texture and taste of biscuits other than its ingredients.

    See TE margin for answer.

11. The milk in these recipes adds moisture to the biscuits because of its high water content. If there were a fourth recipe that had no milk in it, what do you think the texture would be like? Besides milk or water, do you think there is another ingredient that you could increase or decrease to account for the missing moisture?

    It would likely be a powder and not a biscuit unless egg were added. Adding more oil would help but would not completely solve the problem. Buttermilk is often added to biscuit recipes.

12. One of your classmates is allergic to milk. How could you modify one of these recipes so that he could eat these biscuits?

    Answers will vary. Students will probably suggest alternative milks such as coconut milk, soy milk, rice milk, or almond milk.

13. How does food science involve chemistry?

    See TE margin for answer.

1B LAB

EQUIPMENT

- digital camera, smartphone, or camcorder
- laboratory safety equipment
- chemical bottle with GHS label
- SDS

The Safety Saga

Thinking Safe in the Laboratory

Explosions! Fire! Flashes of color! These are what you want to see in the chemistry laboratory. But, we *don't* want them to happen in an uncontrolled way. Your teacher, administrator, and parents want you to explore chemistry safely. But safety is no fun, right? Wrong! *Getting hurt* is no fun. In this lab activity, you'll have some fun learning to "Think Safe."

How can I prevent accidents and injuries in the laboratory?

QUESTIONS

- » What are safe and unsafe behaviors in the chemistry laboratory?
- » What should I do if my lab partner gets injured?
- » Where is the safety equipment in the chemistry laboratory?
- » What information can I find on a GHS safety label?
- » What information can I find on a Safety Data Sheet (SDS)?

LAB 1B OBJECTIVES

- » Identify unsafe behavior in the chemistry laboratory.
- » Describe safe behavior in the chemistry laboratory.
- » Describe how to treat injuries in the chemistry laboratory.
- » Locate safety equipment in the chemistry laboratory.
- » Find information on a GHS safety label.
- » Find information on a Safety Data Sheet (SDS).

✔ ASSEMBLING THE SAFETY SAGA

You may choose to have groups of students pick visual content to create rather than have each group do all of them. If you have a "bring your own device" policy, it should be no problem to get equipment to make videos or take pictures. However, it will be essential to show all the created content to the class so that they can be educated about how to think safe in the laboratory in all of these different areas.

This exercise could be a lot of fun, though, as students think about teaching and performing for their classmates. You will probably need to show the visuals during a later lab period. Students can share their video or picture files with you so that you can pull your "safety saga" together for the next class day.

The table below shows safety equipment that is often used in chemistry laboratories. Be sure that you can quickly and easily locate each piece of safety equipment in your laboratory.

Equipment	Purpose
eyewash station	removes chemicals that may splash into the eyes
shower	removes chemicals that may splash onto clothing; also, may be used to extinguish clothing fires
fire blanket	extinguishes clothing fires; extinguishes fires on the laboratory bench
fire extinguisher	extinguishes fires in the laboratory (different types of extinguishers for different kinds of fires)
first aid kit	provides basic medical supplies for treating common injuries

You will be taking some pictures and possibly creating some videos to teach the other students in your class safe laboratory behavior—a "Chemistry Safety Saga." To help you know what to do, see Appendix A on Laboratory Safety and First-Aid Rules and Appendix C on Laboratory Techniques to give you some hints on how to create your videos.

Procedure

THE SAFETY SAGA

Create videos or take pictures that teach your classmates the ten safe laboratory skills and behaviors listed below. Be sure to answer the questions! If you demonstrate safety equipment in your video or pictures, don't actually activate them. just pretend to. Don't forget the information in the appendixes! Have fun!

A What should I wear in the laboratory?

B How should I smell chemicals in the laboratory?

C How should I mix acid and water when called for in a lab activity?

D What are some examples of unsafe behavior in the laboratory? (Remember to make this video or picture in a safe way!)

E What should I do if I break glassware?

F How should I treat minor heat burns in the laboratory?

G What should I do if I spill a chemical on my skin?

H What should I do if I get a chemical in my eye?

I What should I do if there is an accidental fire in the laboratory?

J What should I do if my clothes catch on fire in the laboratory?

Name _______________________

Now that you have completed your safety saga, consider the labels below. They are some of the symbols used to communicate hazards with certain chemicals. The labels are part of a system recently introduced called the GHS, or the Globally Harmonized System (see example below). The idea is that they can communicate safety hazards to anyone, anytime, anywhere, even if someone can't read English. Try to determine what these labels mean. Your teacher will confirm your guesses later.

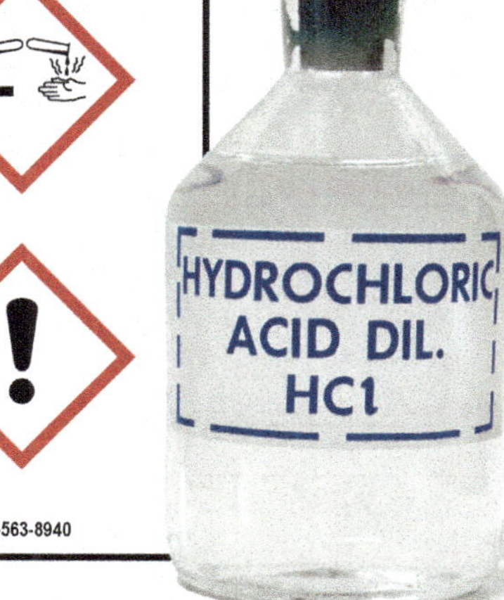

CHANGES IN CHEMICAL SAFETY

SDS sheets have replaced MSDS sheets. HMIS labels have likewise been replaced by GHS labels. While changes like these in the field of chemistry take several years to implement, all US employers were required to switch to these new formats by June 2015.

For more information on these new standards, do a keyword search for "globally harmonized system" and "chemical safety data sheet."

1. explosive

2. corrosive

3. toxic (health hazard)

4. flammable

5. toxic (to the environment)

6. gas under pressure

ANSWERS TO SDS QUESTIONS

Consider using the same SDSs each year. Then make notes below or on the reduced student page that reflect the safety sheets that you have selected.

Now look at the Safety Data Sheet (SDS) document that your teacher has given you. Laboratories keep these in an accessible notebook so that people can easily consult them. They are also available in digital form. Answer the following questions about your SDS.

7. Give the product name for the chemical whose SDS you have.

   Answers will vary.

8. List any synonyms for this substance (a maximum of three).

   Answers will vary.

9. What is the percentage composition of this substance?

   Answers will vary.

10. What are the potential effect(s) that this chemical will have for each of the following types of exposure?

    eye contact

    Answers will vary.

    skin contact

    Answers will vary.

    inhalation

    Answers will vary.

    ingestion

    Answers will vary.

11. What first aid is recommended for each kind of exposure?

    eye contact

    Answers will vary.

    skin contact

    Answers will vary.

    inhalation

    Answers will vary.

    ingestion

    Answers will vary.

12. What personal protection is recommended?

    Answers will vary.

13. Describe each of the following physical characteristics for the chemical. Write n/a if no information is given.

    melting point

    Answers will vary.

    boiling point

    Answers will vary.

    solubility in water

    Answers will vary.

    color

    Answers will vary.

14. Is your chemical stable?

    Answers will vary.

15. Is it designated as being incompatible with any specific substances? If so, with what?

    Answers will vary.

I have read the Laboratory Safety and First-Aid Rules in Appendix A and have located all the safety equipment in the laboratory.

Signature ____ **Date** ________________

POST-LAB CHECK

Use these questions to assess students' understanding of safety procedures. Don't rush—it's important to satisfy yourself that students are "thinking safe" before moving on.

1. What is the best thing to do if you spill a chemical right beside your laboratory burner? *(Turn off the burner and then clean up the spill in accordance with guidance from the SDS and your teacher.)*

2. What is your primary source of information about the chemicals that you are using in the laboratory? *(While the teacher will provide much information, the primary source of information is the SDS.)*

3. What is the first thing to do if a person swallows a chemical? *(Find out the specific substance ingested and contact the poison control center in your area immediately. Students may also suggest referencing information on the SDS.)*

4. Your lab partner splashes a tiny bit of acetic acid on your hand. How do you handle this situation? *(Since it is an acid, you need to rinse your hand in plenty of water in accordance with the SDS.)*

Name ___________________

Date ___________________

Needle in a Haystack

Separating Mixtures

In 1898 chemist-physicist Marie Curie was up against a tremendous challenge. She was trying to isolate two unknown radioactive elements in an ore called *uraninite*, which is a mixture of about thirty different substances. The more abundant of the two unknown elements represents just 0.000 000 1% of the uraninite's total mass! The second element represents even less. It was like looking for a needle in a haystack.

After four years, Curie had isolated 0.1 g of radium from a 1000 kg sample of uraninite. In this lab activity, you will do something similar. Remembering that a mixture is a physical combination of substances, you will use the physical properties of three substances to isolate them from the mixture. Let's tackle that haystack!

How can I separate a mixture even when the components are similar?

QUESTIONS
» How can I separate a mixture?
» How can I analyze the composition of a mixture?
» How can I assess my ability to separate the components of a mixture?

EQUIPMENT

- laboratory balance
- hot plate
- Erlenmeyer flask, 250 mL
- spatula
- weighing dish
- beaker, 150 mL
- bar magnet
- filtering funnel
- clay triangle
- ring stand and ring
- graduated cylinder, 10 mL
- wire gauze
- beaker, 250 mL
- watch glasses, 150 mm (2)
- filter paper
- plastic sandwich bag
- sand, salt, iron mixture, 2–3 g
- goggles

Needle in a Haystack | 9

LAB 2A OBJECTIVES

» Identify physical properties that can be used to separate a mixture.

» Use laboratory procedures to separate a mixture into its components.

» Analyze the composition of a mixture.

» Assess your ability to separate the components of a mixture.

MAKING IT AN INQUIRY LAB ACTIVITY

The lab activity as written is designed to give students practice with some basic laboratory techniques that will be used throughout the year. If you feel that your students already have a good handle on the techniques used in this activity, you could do it as an inquiry lab activity by giving them the mixture and the task of separating the mixture. Inquiry lab activities leave it up to students to determine the procedures.

Note: If you do this activity as an inquiry lab activity, don't allow students to look at the Student Lab Manual because that will limit their ability to direct the inquiry. Also, expect the activity to take longer as students work through how to accomplish the task.

You may also want to consider having students submit a lab report. See Appendix E for writing formal lab reports.

THE CURIES AND THEIR DISCOVERIES

Marie and Pierre Curie's discoveries of polonium and radium marked the beginning of an exciting era for physics and chemistry, but their work has been somewhat romanticized in many science histories. While they did work for several years under primitive and rigorous conditions, they had some assistance from an industrial chemical company, which did the initial processing of the large quantities of uraninite (pitchblende) ore. They also received a number of financial grants, including one from the Institute of France.

Polonium and radium are decay products of uranium, the main component of uraninite. After the Curies' successful isolation of radium chloride in 1902, Marie Curie went on to isolate pure radium metal in 1910. She never succeeded in isolating pure polonium,

---

PRE-LAB CHECK

1. How will you separate the iron from the mixture? *(by using a magnet)*

2. What solvent will you use in this lab activity to dissolve the salt in the mixture? *(hot tap water)*

3. Should you measure the amount of solvent that you will add to the mixture? Why or why not? *(No. The water is vaporized eventually, so the amount is not important.)*

4. How will you separate the sand from the solution? *(by filtering the sand from the solution)*

5. How will you separate the solute from the solvent? *(by vaporizing the water, leaving the salt behind)*

which exists in even smaller quantities than radium. It also has a much shorter half-life—138 days for polonium-210, compared with 1601 years for radium-226.

 MAKING THE MIXTURE

To make the mixture, combine 20 g of salt with 20 g of sand and 10 g of iron filings. This will give you the correct percent compositions of salt (40%), sand (40%), and iron (20%).

 WHY MASSING EQUIPMENT?

Do not tell the students why they are determining the mass of the equipment. This topic will be addressed by a question near the end of the activity.

 SUGGESTIONS FOR STEP B

Laboratory Balances

You may want to spend a few minutes reviewing the proper use of a laboratory balance.

Providing Samples

To save time, you may instead choose to make available tubes that contain the approximate amounts so that each group will have one. You will have to modify the procedures to allow students to determine the initial mass of the mixture.

PROPERTIES

Students may come up with other physical properties, but the property of solubility is the most obvious and the most direct means to separate the mixture.

LABORATORY TECHNIQUES

Remind students that Appendix C contains many helpful tips on laboratory techniques, including how to use a laboratory balance and how to set up gravity filtration.

Procedure

A. Measure the masses of the flask, filter paper, and plastic bag. Record your results in Table 1. Refer to Appendix C on laboratory techniques for help in measuring mass.

B. Using the laboratory balance, spatula, and weighing dish, measure out between 2–3 g of the sand, salt, and iron mixture. Record the mass in Table 1. Transfer the mixture to the 150 mL beaker.

1. Is it possible to separate this mixture by hand or by using a sieve or strainer? Explain.

   Yes. But the size of the particles and similar appearance of the sand and salt makes hand separation impractical. A sieve or strainer will not work because the particles are essentially the same size.

2. Identify physical properties of the three components that are significantly different and might be useful for separating the mixture.

   Answers will vary. The key distinguishing properties are magnetism and solubility. Iron is magnetic, while sand and salt are not. Salt readily dissolves in water, while sand does not.

SEPARATING THE IRON

C. Insert the bar magnet into the plastic bag and then stir the mixture in the beaker with the magnet until all the iron filings are removed from the mixture.

D. Carefully invert the bag, capturing the iron filings. Remove the magnet. Measure the mass of the bag and iron and record your data in Table 1.

3. What is the purpose of the plastic bag in Step C?

   The bag allows us to remove the iron from the magnet after separation.

SEPARATING THE SAND

E. Fold a piece of filter paper as shown in Appendix C. Shape the filter paper into a cone and insert it into the funnel. Moisten the paper with water and press it against the funnel wall.

F. Set the funnel in the clay triangle as it rests on the ring stand's ring. Lower the ring until the tip of the funnel stem touches the inside rim of the flask as shown in the image at left. (When the funnel stem touches another surface, it drains better.)

G. Pour the remaining mixture from the beaker into the filter paper. You may need to tap the beaker with a pen or pencil to dislodge any particles that adhere to the sides or bottom.

H Using the graduated cylinder, slowly pour four portions of hot tap water (approximately 5 mL each) over the mixture. The liquid—called the *filtrate*—will collect in the flask. Allow each 5 mL portion to run through the mixture before adding the next. Pour the final 5 mL of water around the upper edge of the filter paper.

4. Is the water in the flask still plain water? Explain.

   No. It dissolved the salt as it flowed through the mixture, effectively
   separating the salt from the sand. The liquid is now salt water.

5. Why do you think that you used hot water for this step? Would you have expected different results had you used cold water instead? Explain.

   Salt dissolves more quickly in hot water than in cold water. If cold
   water had been used, it would likely have left some of the salt behind.

6. Identify physical properties of salt and water that are significantly different and that might be useful for separating this new mixture. What about the water and sand mixture?

   Answers will vary. Water is a liquid, while the salt is a dissolved solid.
   The water can evaporate and leave the solid salt behind. A similar
   situation will apply to the water and sand. The water can evaporate
   and leave the solid sand behind as well.

SEPARATING THE SALT

I Place the flask containing the salt solution on the hot plate, turn the hot plate on medium-high, and bring the liquid to a gentle boil. Allow the salt water to simmer until most of the water has been boiled off and crystals begin to appear in the dish. ***Do not heat so strongly that splattering occurs.***

J Turn the hot plate to low and continue evaporating the water. When no water is visible, turn off the hot plate and allow the residual heat to evaporate the remaining water. ***Do not heat dry glassware.***

K While you are waiting for the salt solution to evaporate, carefully spread the filter paper containing the sand onto a watch glass. Set the watch glass at the base of the burner. By the time your salt solution has evaporated, the sand and filter paper should also be dry. (If they are not dry by the time the solution has evaporated, leave them overnight and do Step M the next day.)

L Once the flask is cool, measure the mass of the flask and salt and record your data in Table 1.

M When the sand and filter paper are dry, measure the mass and record your data in Table 1.

7. Why did you have to take such care in evaporating the water from the saltwater mixture? Why not just heat the water vigorously until it all boiled away?

   See TE margin for answer.

   ____

   ____

   ____

   ____

⚠ HEATING HAZARD

Emphasize the point that the liquid must be boiled gently. If it splatters, it could burn them. Also, splattering will cause the loss of some of the salt, resulting in inaccurate data.

This is also a good time to remind students that hot laboratory equipment usually doesn't look hot. It can cause significant burns despite its appearance.

Question 7 Answer

Vigorous heating could splatter salt water out of the flask dish. Salt would be lost, affecting the final results and reducing the overall accuracy of the experiment. Some students may mention the danger of overheating glassware, which could result in breakage.

MASS OF IRON, SALT, AND SAND

Some students may be unsure of how to determine the masses of the iron, salt, and sand. Do not give them the answer directly. Instead, encourage them to remember the first step of the experiment and think about how the information that they obtained can help them with their present problem.

EXAMPLE CALCULATIONS

Throughout the Teacher Lab Manual, worked-out solutions to problems are based on sample data given in the tables. While student's numerical values will likely differ, their processes should be similar. The worked-out solution for Question 8 is for iron, but we also provide the answers for salt and sand.

Analysis

N. Calculate the masses of the iron, salt, and sand recovered. Record your answer in Table 2.

O. Calculate the total mass recovered and record your answer in Table 2.

When dealing with combinations of materials, it is common to calculate the percent composition. The formula below can be used to calculate the percentage of each of the three components.

$$\% \text{ composition} = \left(\frac{\text{component mass}}{\text{total mass}}\right)100\%$$

8. Use the space below to calculate the percent composition of the iron. Record your answer in Table 2. Repeat for salt and sand.

Answers will vary.

What we know: $m_{iron} = 0.48$ g, $m_{mixture} = 2.56$ g

Unknown: $\%_{iron}$

Write the formula and solve for the unknown:

$$\%_{iron} = \left(\frac{m_{iron}}{m_{mixture}}\right)100\%$$

Evaluate:

$$\%_{iron} = \left(\frac{0.48 \text{ g Fe}}{2.56 \text{ g}}\right)100\%$$

$$= 19\% \text{ Fe}$$

$$\%_{salt} = 38\%$$

$$\%_{sand} = 43\%$$

9. Obtain the original percent composition from your teacher. How do your percent composition calculations compare? Explain any discrepancies.

   Answers will vary. Any variation in the iron value is most likely due to leaving iron in the sand-salt mixture or sand and salt particles being trapped between iron filings. Incomplete dissolving of the salt would increase the sand percentage and decrease the salt percentage.

10. On the basis of your experimental percent compositions and the percent composition of the original mixture, how well do you think you did at separating the mixture? Explain.

    See TE margin for answer.

   Scientists will many times need to know how efficiently they have accomplished a task. When working in chemistry, we are often interested in how much of the original sample we ended up with. We can use percent recovered to check our efficiency.

$$\% \text{ recovered} = \left(\frac{m_{\text{recovered}}}{m_{\text{initial}}}\right)100\%$$

11. Use the space below to calculate the percent recovered and record your data in Table 2.

    Answers will vary.

    What we know: $m_{\text{initial}} = 2.77$ g, $m_{\text{recovered}} = 2.56$ g

    Unknown: $\%_{\text{recovered}}$

    Write the formula and solve for the unknown:

    $$\%_{\text{recovered}} = \left(\frac{m_{\text{recovered}}}{m_{\text{initial}}}\right)100\%$$

    Evaluate:

    $$\%_{\text{recovered}} = \left(\frac{2.56 \text{ g}}{2.77 \text{ g}}\right)100\%$$

    $$= 92.4\%$$

12. Did you recover all of the original sample? If not, explain possible reasons for this discrepancy. How well do you think you did at recovering the sample?

    See TE margin for answer.

13. Why was it necessary to find the mass of the flask, filter paper, and plastic bag at the beginning of the experiment?

    Doing so allowed for subtracting the mass of the equipment when determining the mass of the various products.

Question 10 Answer

Answers will vary. Percent compositions should be fairly close. Assess students' answers on the basis of their reasoning more than their actual assessments.

Question 12 Answer

Answers will vary. If students recover over 100%, it is most likely due to incompletely drying the sand, salt, or both. Percents recovered below 100% are often due to lost material, such as sand falling off the filter paper or salt splattering out of the flask.

A NEEDLE IN A HAYSTACK

As a final fun bonus question, ask students to think of a good method for separating a needle from a haystack because of the significantly different physical properties between the hay and the needle. Probably the most direct answer will be to use a powerful magnet since the needle is ferromagnetic, while the hay is not.

TABLE 1: *Data*

Object	Mass (g)
flask	46.24
filter paper	1.26
plastic bag	1.76
mixture, initial	2.77
plastic bag and iron	2.24
flask and salt	47.21
filter paper and sand	2.37

TABLE 2: *Results*

Object	Mass (g)	Percent Recovered	Percent Composition
iron, recovered	0.48		19
salt, recovered	0.97		38
sand, recovered	1.11		43
mixture, recovered	2.56	92.4	

2B LAB

Zebroids, Wolphins, and Ligers, Oh My!

Classifying Matter

Did you know that zebras and horses can produce offspring together? So can whales and dolphins, and tigers and lions. These animal hybrids definitely look different from either of their parents! People have learned how to breed animals and plants such as horses and wheat to increase their productivity and use for mankind. Chemists do something similar when they form new substances for specific purposes.

Does physically or chemically combining elements change their properties?

Today we'll explore how two chemicals change when they come together to form a completely new substance with its own custom set of physical and chemical properties. In this lab activity, you will observe the properties of iron filings, sulfur powder, a mixture of iron filings and sulfur powder, and the compound iron(II) sulfide. You will see under what conditions the physical and chemical properties change.

EQUIPMENT

- laboratory burner and lighter
- magnifying glass
- bar magnet
- deflagration spoon
- spatula
- graduated cylinder, 10 mL
- test tubes (3)
- test tube rack
- weighing paper (8 pcs.)
- iron (Fe) filings
- sulfur (S) powder or granules
- iron(II) sulfide (FeS)
- iron and sulfur mixture
- hydrochloric acid (HCl), 6 M, 7.5 mL
- goggles
- laboratory apron
- nitrile gloves

QUESTION

» What properties are helpful in classifying mixtures?

LAB 2B OBJECTIVES

» Classify mixtures and compounds on the basis of observed differences.

» Analyze observations.

 ACIDS

Review with your students the hazards associated with acids. Acids are corrosive, and 6 M hydrochloric acid is a highly concentrated acid. Review safety procedures for handling acids as well as handling spilled acids.

As always, refer to the safety data sheet (SDS) before using chemicals.

MAKING IRON(II) SULFIDE

You may want to create the iron sulfide that students will analyze in the laboratory while they observe. This will show them the chemical change that results in a substance with completely different properties than its comprising elements.

To form iron sulfide, put a 7:4 mixture by mass of iron powder to sulfur powder into a test tube. Put a piece of steel wool into the mouth of the test tube. Place the test tube over a laboratory burner and heat until the mixture begins to glow. This process may release some toxic gases, so do this in a vented hood area.

To remove the iron sulfide, break the test tube by immersing the hot test tube in water or use a tool to break the glass. ***Use caution in this process.*** Remove the iron sulfide and crush it with a mortar and pestle. You and your students may find that the iron sulfide is mildly magnetic but much less so than the iron filings.

PRE-LAB CHECK

1. What two elements will be evaluated in this lab exercise? *(iron and sulfur)*

2. What mixture will be evaluated? *(a mixture of iron filings and sulfur powder)*

3. What compound will be evaluated in this exercise? *[iron(II) sulfide]*

4. How will you distinguish between a mixture and a compound? *(Answers will vary. Students should mention that the mixture and the compound should have different observable properties, such as their physical appearances, their behavior in the presence of a magnetic field, solubility in water, and chemical changes in the presence of the other element. Physical methods can be used to separate the elements of the mixture but not of the compound.)*

Procedure

APPEARANCE

A Using the spatula, obtain pea-sized samples of iron, sulfur, the iron-sulfur mixture, and iron(II) sulfide (FeS), and place them on separate pieces of weighing paper.

B Describe the physical appearance of each sample, noting any differences between them. Use a magnifying glass to aid your inspection. Record your observations in Table 1.

You've seen how matter can be classified either as a pure substance or as a mixture. Pure substances consist of only a single substance: an element or a compound. Elements consist of only one type of atom, while compounds are chemical combinations of two or more elements in fixed ratios. Mixtures are physical combinations of two or more substances in a changeable ratio.

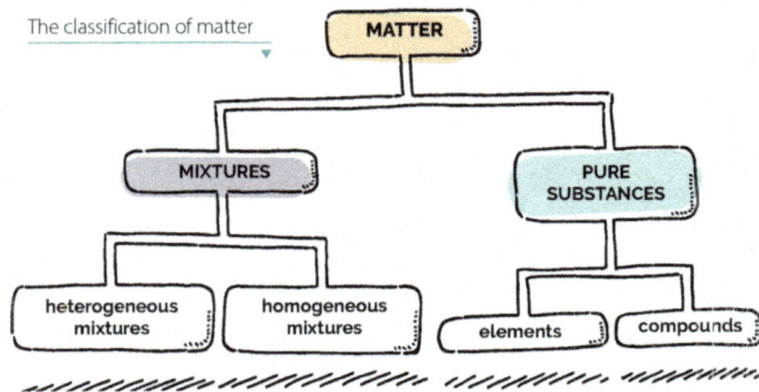

1. From your observation of its physical appearance, determine whether the iron-sulfur mixture is homogeneous or heterogeneous. How can you tell?

   It is heterogeneous. Particles of iron and sulfur powder can be seen mixed together; each retains its own physical appearance. On the basis of a magnified inspection, the mixture is distinctly nonuniform.

2. Would you make the same observation for the compound iron(II) sulfide? Why or why not?

   See TE margin for answer.

Question 2 Answer

No. The heterogeneous mixture classification applies only to mixtures and not to compounds, which are pure substances and always homogeneous. Its properties are different from those of the elements of which it is made.

Name _______________________

MAGNETISM

C Observe the magnetism of the substances by passing a magnet under each sample. *Keep the weighing paper between the samples and the magnet so that magnetic substances will not stick to the magnet.* Record your observations in Table 1.

3. What happened to the magnetic properties of iron and sulfur when they were combined in a mixture? What happened when they were chemically combined in iron(II) sulfide?

See TE margin for answer.

____

____

____

COMBUSTIBILITY

D Place approximately half of your iron sample on a deflagration spoon and insert it into the hottest part of a laboratory burner flame for about thirty seconds *under a fume hood.* Look for evidence that the iron burned. Record your observations in Table 1. If it burns, allow it to burn completely. Repeat for the other three substances.

4. After the chemicals were burned there was less of them. Was that matter destroyed? Explain.

See TE margin for answer.

____

____

____

E If any substance remains after the combustion test, place it on a clean piece of weighing paper and test it with a magnet. Record your observations in Table 1.

REACTIVITY

F Working under the fume hood, pour 7.5 mL of 6 M HCl into your graduated cylinder. Then divide it into nearly equal parts among the three test tubes.

G Drop the remainder of your iron sample into one test tube and observe whether a chemical change occurs (e.g., a color change occurs, a new substance forms, or bubbles form). Repeat for the sulfur and the iron-sulfur mixture. *Do not test the iron(II) sulfide in the acid.* Record your observations in Table 1.

5. What evidence was there to indicate that the iron was reacting with the hydrochloric acid?

Bubbles formed on the iron filings. The bubbles indicated that a gas was being formed.

____

____

H Pour the remaining acid solutions (HCl) into the designated acid waste container located in the fume hood.

✓ MAGNET TROUBLE

To avoid getting the chemicals stuck to the magnet in Step C, consider having students place the magnet in a disposable plastic bag.

Question 3 Answer

Iron and sulfur in the mixture had the same magnetic properties as isolated elements. In the compound iron(II) sulfide, however, it was expected that iron and sulfur would combine to make a compound with different magnetic properties, and this was observed.

Question 4 Answer

According to the law of conservation of matter, we know that matter cannot be created or destroyed. Therefore, the material wasn't destroyed but was changed into something different. Most likely it formed a gas that left the room through the fume hood.

Answers will vary. The student should mention that the mixture and the compound should have different observable properties, such as their physical appearances, their behavior in the presence of a magnetic field, water solubility, and chemical changes in the presence of other substances. Physical methods should be able to separate the elements of the mixture but not of the compound.

ANOTHER SEPARATING TECHNIQUE

Not discussed in the lab activity is using the inorganic solvent carbon disulfide (CS_2), which will dissolve the sulfur but not the iron. This method will *not* break down iron(II) sulfide and separate it into its elements.

Question 8 Answer

No. The properties of compounds are usually totally different from the properties of their individual elements. Students may conclude that the iron(II) sulfide melting point would be somewhere in between these melting points. (The actual melting point is approximately 1194 °C.)

 SAMPLE DATA

The data provided is sample data. Student data may be different.

Analysis

6. How will you distinguish between a mixture and a compound?

    See TE margin for answer.

7. How can an iron-sulfur mixture be separated into its elements? Can this method extract elements from iron(II) sulfide?

    A magnet can pull the iron from the mixture but not from the

    compound. Sulfur in the mixture can be removed by burning.

8. Iron melts at 1538 °C and sulfur melts at 115 °C. Does this information help you determine the melting point of iron(II) sulfide? Explain.

    See TE margin for answer.

TABLE 1

	Elements		Mixture	Compound
	Iron Filings (Fe)	**Sulfur Powder (S)**	**Iron and Sulfur Mixture**	**Iron(II) Sulfide (FeS)**
What does each sample look like?	granular or filings; gray or rust-brown color	fine, bright yellow powder	distinctly visible particles of iron mixed in and covered with sulfur powder	crystalline solid; may be white if pure or black if it contains impurities
Did the sample exhibit magnetic behavior?	yes	no	Yes. The iron was still magnetic.	slightly
Did the sample burn?	no	yes	yes	no
If a residue remained, was it magnetic?	yes	N/A	yes	slightly
Did the sample react with HCl?	yes	no	yes	Yes. It reacts to form a gas (hydrogen sulfide).

°3A LAB

Name ______________________
Date ______________________

Metric Unicorns

Exploring the Metric System

Have you ever heard of mutchkins, yojanas, grzywnas, and virgates? They aren't fantastical creatures like ogres, fairies, and unicorns from some enchanted land. They're actually units of measurement! These strange-sounding units of measurement might help us appreciate the standard that the SI system gives us for measuring anything, whether length, luminosity, or electric current. By using a common system, scientists all over the world can exchange information easily and without confusion.

More than likely, you've been measuring things since elementary school, but have you ever really thought about what lies behind measurement? Let's explore the subject by getting out of our comfort zone. We'll do this by inventing a brand-new unit of measure and seeing what we discover.

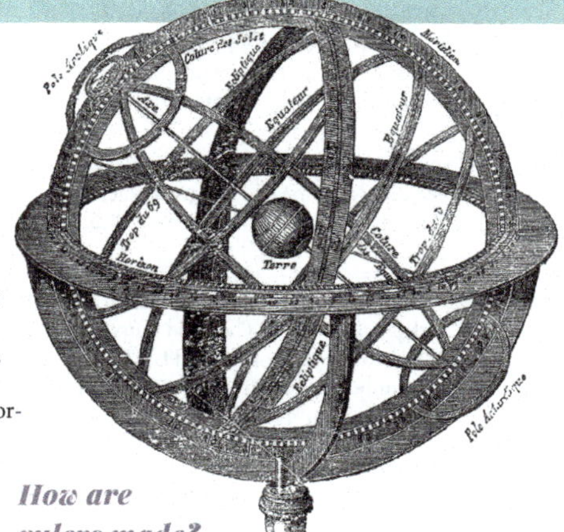

How are rulers made?

Procedure

All measurement units are based on a reference known as a *standard*. Some standards are physical objects, while others are phenomena. To keep things easy, we'll use a physical object—your textbook—as a standard of length.

WORKING WITH THE UNIT

A Create a ruler with a length of one book by cutting a piece of paper tape equal to the length of your textbook's spine.

1. What is the relationship of your paper ruler to the standard?

   The ruler is a copy of the standard and ideally should have an equal length.

2. Identify two objects in the room that could be measured with your new ruler.

   Answers will vary. *Examples:* Their lab manual, the length of their shoe, and perhaps their forearms are about a book long.

3. What is a significant limitation of your ruler?

   See TE margin for answer.

QUESTIONS

» What goes into creating a new unit of measurement?

» How are metric prefixes helpful?

» How do I determine conversion factors?

» How do I use unit conversion factors?

EQUIPMENT

- paper tape
- textbook
- scissors
- metric ruler
- meter stick

UNITS FOR WHAT?

If your students are curious about these unusual-sounding units of measurement, have them do a keyword search on the internet.

EQUIPMENT OPTIONS

Textbook

The Student Edition textbook can be replaced with any object as the standard for the new unit of measure. Alternatively, you could allow students to pick whatever object they want for the standard.

Paper Tape

The best source for paper tape is adding machine tape. An excellent substitute for the paper tape is duct tape. Duct tape can be kept for a longer period of time if you want the students to use their rulers later in the year. Tear off pieces of duct tape and fold the tape in half widthwise, sticking it to itself so that it is no longer sticky.

PHYSICAL CONSTANTS VERSUS ARTIFACTS

Depending on how much you discussed the topic of standards, you may want to review the fact that the SI fundamental units are all based on physical constants. The kilogram was the last fundamental unit to switch from being based on a physical artifact. This lab activity focuses on how units were created on the basis of artifacts.

Question 3 Answer

Since hardly any objects will be exactly 1 book long, measurements will always be an approximation. Objects that are a fraction of a book long will require estimating.

PRE-LAB CHECK

1. In this lab activity, what is your standard for the measurement of length? *(The standard is the textbook.)*

2. How can you create larger and smaller units that are based on your standard? *(Students will multiply their standard of length to scale up and subdivide to scale down.)*

3. What is a conversion factor? *(a ratio of equal measurements used to convert from one unit to another)*

Question 6 Answer

Since the ruler was originally marked at 0 and 1 book, you could measure to 0.1 books. After graduation, you could measure to 0.001 books.

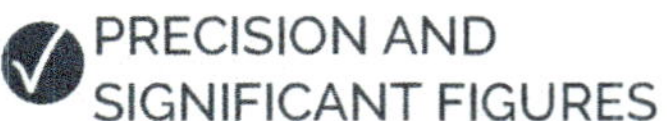

PRECISION AND SIGNIFICANT FIGURES

Don't feel that you need to have students worry about significant figures since we have not covered them at this point in Chapter 3. Lab 3B explores significant figures.

Question 7 Answer

Since the divisions are eyeballed, they are probably not very consistent and therefore not equal. Commercially produced metric rulers do have this problem but to a lesser degree. The graduations on commercially produced rulers are made by machines that have the ability to space them more evenly, but there is still uncertainty in those markings.

To make a measuring tool like a ruler more useful, it must be divided into fractions of a unit. This process is called *graduation*. The more graduations a ruler has, the more *precise* it becomes.

B Fold your ruler in half to find the middle. Unfold the ruler, draw a pencil line down the fold, and label it *5*. Label the left edge of the ruler *0* and the right edge *10*.

C Add four equally spaced lines to the left of the fold labeled *1*, *2*, *3*, and *4*. Then add four equally spaced lines to the right of the fold labeled *6*, *7*, *8*, and *9*. Use only your eyes to space the lines. Do the best that you can!

D Finally, divide the first division (between *0* and *1*) into ten small divisions by drawing nine equally spaced tick marks along the edge of the paper.

4. Using the standard metric prefixes, what would you call the large divisions? the small ones?

   The large divisions are decibooks (0.1 book), while the small ones are centibooks (0.01 book).

5. What was the smallest marking of the ruler originally? After adding tick marks, what was the smallest marking on the ruler between 0 and 0.1?

   The ruler originally only had marks at 0 book and 1 book. After graduation the ruler had marks every 0.01 book between 0 and 0.1.

6. With what precision should you measure using the ruler as originally marked? After graduation, with what precision could you measure between 0 and 0.1?

   See TE margin for answer.

Recall that when making measurements, you read all of certain digits and then estimate the next whole decimal place. For example, if you are using a ruler marked to 0.01 cm, then you can make measurements to the thousandths place.

E Using your improved ruler, measure the width of your hand and record the value in Table 1. Come up with a symbol for your base unit of books and enter it in the column header of Table 1. Also enter the two additional prefixed units that your ruler allows for.

F Express your measurement to the same precision but using the two prefix-scaled units. Record your values in Table 1.

7. What is a serious limitation to your improved ruler? Does a commercially produced metric ruler have this limitation? Explain.

   See TE margin for answer.

We've measured things smaller than a book with your new unit. Now let's measure things much bigger than a book.

G Create a tape measure with a length of ten books by cutting a piece of paper tape ten times the length of your textbook's spine. Mark and number 1-book divisions on the tape.

8. Using a standard metric prefix, what is the name of this new unit of length? Enter the new prefixed unit that your tape measure allows for.

   a dekabook (dabk, 10 books)

H Using the tape measure, measure the length of your classroom. Record the value in Table 1.

I Express your measurement to the same precision but in the new unit and record the value in Table 1.

CONVERTING UNITS

Sometimes it's necessary to change between two different units. But in order to perform this task, you need to know the relationship between the two units. A *conversion factor* expresses this relationship, either as a fraction or as an equation. For example, 1 m = 3.28 ft. Let's create a conversion factor to convert between centimeters and books.

J Using the metric ruler, measure your 1-book ruler to the nearest 0.1 cm.

9. Create a conversion factor that relates books to centimeters.

   See TE margin for answer.

10. Using your conversion factor from Question 9, convert the measurement of your hand in books to centimeters. Use the space below to show your work. Record this value in Table 1.

    Answers will vary. *Example*: 0.28 bk $\left(\dfrac{27.56 \text{ cm}}{1 \text{ bk}}\right)$ = 7.7 cm

K Using the meter stick, measure your dekabook tape.

11. Create a conversion factor that relates dekabooks to meters. Write your conversion factor in either fractional or equation form.

    See TE margin for answer.

12. Using your conversion factor from Question 11, convert the measurement of your classroom in dekabooks to meters. Use the space below to show your work. Record this value in Table 1.

    Answers will vary. *Example*: 2.71 dabk $\left(\dfrac{2.756 \text{ m}}{1 \text{ dabk}}\right)$ = 7.47 m

Analysis

13. Using what you've learned from your textbook, state why your textbook is a good standard for measurement.

    Standards must be consistent objects that can be easily replicated.

    The textbook is relatively consistent dimensionally and is easily

    replicated. For everyday measurement it would actually be an

    acceptable unit.

Question 9 Answer

Answers will vary within the uncertainty of metric rulers (typically ± 0.05 cm).

Example:

1 bk = 27.56 cm or 27.56 cm = 1 bk

Or as a ratio:

$\dfrac{1 \text{ bk}}{27.56 \text{ cm}}$ or $\dfrac{27.56 \text{ cm}}{1 \text{ bk}}$

(Any centimeter measurements from 27.51 cm to 27.61 cm are acceptable.)

Question 11 Answer

Answers will vary within the uncertainty of the meter stick (typically ± 0.05 cm).

Example:

1 dabk = 2.756 m or 2.756 m = 1 dabk

Or as a ratio:

$\dfrac{1 \text{ dabk}}{2.756 \text{ m}}$ or $\dfrac{2.756 \text{ m}}{1 \text{ dabk}}$

(Any meter measurements from 2.751 m to 2.761 m are acceptable.)

14. Again, using what you've learned from your textbook, state why your textbook is *not* a good standard for measurement.

The textbook is not manufactured to the very tight tolerances of actual measurement standards like the standard kilogram. It would therefore not be suitable for precision measurement.

15. Why do we generally avoid creating customized measuring units such as the one that you created in this lab activity?

See TE margin for answer.

16. Discuss how standardized measuring systems are an example of good dominion that is pleasing to God.

Standardized measuring systems enable people to collaborate effectively when they exercise dominion over the earth. Construction, science, and medicine all benefit from a common measuring system. Furthermore, a common measuring system reduces the number of potentially dangerous mistakes that can occur when converting between multiple measuring systems.

Question 15 Answer

Since measurement works best when everyone follows the same system, customized measuring systems are limited in usefulness. The metric system works well in science because it eliminates the confusion that existed in the past due to different national measuring systems.

 SAMPLE DATA

The sample data in Table 1 is typical. Students' values may be somewhat different. The key to the measurements is that they are estimated to one additional decimal place beyond what is marked on the instrument.

TABLE 1

Object	Width (bk)	Width (dbk)	Width (cbk)	Width (dabk)	Width (cm)	Width (m)
hand	0.28	2.8	28		7.7	
room	27.1			2.71		7.47

3B LAB

Name _______________

Date _______________

You Are My Density

Inquiring into Measurement

Intensive properties, such as density, are ones that remain constant regardless of the amount of material in a sample. Therefore, density is useful for identifying substances. Recall that density is defined as the mass per unit of volume ($\rho = m/V$). In the metric system, density has the units g/mL or g/cm³ for liquids and solids and g/L for gases.

How do different methods affect the accuracy of experimental methods?

In this inquiry lab activity, you will develop procedures to determine the density of two objects. Both objects are mixtures of elements. From your density data, you will then estimate the percent composition of each object.

Procedure

PLANNING/WRITING SCIENTIFIC QUESTIONS

A Obtain a penny and a coil of solder from your teacher.

B Brainstorm with your lab group about how you can determine the density of each object.

C Brainstorm with your lab group about how you can use your density data to determine the elements that make up each object.

D Write specific questions related to calculating the density of each object that you could answer by collecting data.

DESIGNING SCIENTIFIC INVESTIGATIONS

E Write procedures to collect data that will allow you to answer the questions that you wrote in Step D above.

F Have your teacher approve your procedures.

CONDUCTING SCIENTIFIC INVESTIGATIONS

G Following the procedures that you have written, collect the data to answer the questions that you wrote.

QUESTIONS

» How do I design a lab activity?

» How can I use density to help identify the composition of an object?

» How accurately can I measure?

EQUIPMENT

- penny
- coil of solder
- goggles

LAB 3B OBJECTIVES

» Devise a process to determine the density of an object.

» Evaluate the composition of a material on the basis of its density.

» Evaluate measurements using percent error.

✓ INQUIRY LABS

This is the first inquiry lab activity in the Student Lab Manual. Inquiry labs give students a question to answer or a problem to solve by collecting empirical data. The students have freedom to approach the task in different ways. A Teacher Guide immediately follows the reduced Student Lab Manual pages in this volume. In the guide you will find a sample procedure that students may come up with. While you can use that sample procedure as a guided discovery lab activity, your students will grow tremendously through the process of designing their own activity.

EQUIPMENT NOTES

Pennies

Due to changes in the composition of pennies, this activity will work best if you give the students only pre-1982 pennies or only post-1982 pennies. Obviously, post-1982 pennies will be easier to obtain in large quantities.

Solder

A great solder choice is 59/41 Sn/Cu, but you may use other solder. You may elect to avoid using lead solder. If you use lead solder, remind students to wash their hands at the end of the lab activity, as they should always do anyway.

H From Table 1, select some possible metals that your objects may contain. Other physical properties can help you make a determination.

I Assuming that the solder is made of copper and one other metal in a 41/59 ratio, use your density data to determine the other metal.

J Assuming that a penny is made of only copper and zinc, use your density data to determine the proportions of copper and zinc.

K Look up the actual proportions of zinc and copper in pennies. Make a claim about the accuracy of your process. Support your claim with evidence from your prediction and test.

TABLE 1

Metal	Density (g/cm³)
aluminum	2.75
copper	8.96
gold	19.30
iron	7.87
lead	11.34
magnesium	1.74
nickel	8.90
silver	10.49
tin	7.26
zinc	7.14

Teacher Guide

INQUIRING INTO MEASUREMENT

MANAGING INQUIRY LABS

An inquiry lab activity allows students a degree of freedom, but the teacher has to guide the process to achieve the desired educational outcomes. As teachers start using inquiry lab activities, they often struggle with this process of guiding students. The best method is to think through the goals that you want students to achieve and then use questions to guide them to reach those goals. You want to balance students' freedom to investigate the question in the manner they choose with the need to meet your educational goals.

Procedure

PLANNING/WRITING SCIENTIFIC QUESTIONS

Each lab group starts by getting a penny and a coil of solder. They will then brainstorm how to determine the density of these objects and then how to use that information to determine the composition of those objects.

Each lab group should write a series of questions related to those tasks that they can answer by measurement in the laboratory. Help students write specific, testable questions. For example, "How can I find the density?" is not a good question. But "How can I measure the mass and volume of each object?" is more specific and testable. Also make sure that students are thinking about the challenges of determining the volume of the irregular coil of solder. You may want the groups to submit their questions for your review.

DESIGNING SCIENTIFIC INVESTIGATIONS

Once the groups have written their questions, they need to write specific procedures to answer them. This is a new task for many students, so you may want to do part of this task as a class discussion or review their procedures prior to their starting to collect data.

Help students think through how they will factor out extraneous information. For example, a student may decide to measure the mass of a substance in a cup. Often students won't think through the need to know the mass of the empty cup in order to subtract that mass from the total.

CONDUCTING SCIENTIFIC INVESTIGATIONS

Once the groups have good procedures, allow them to collect data.

DEVELOPING MODELS

Students will analyze their data to determine the composition of the objects.

Students will use their density information to determine the other metal in the solder. The question and answer below assume that you used 59/41 Sn/Cu solder. If you use a different solder, adjust the percentages for them.

Students will calculate the proportions of copper and zinc in a penny. This is a system of equations, which some students may struggle with. This is a great opportunity to emphasize that the skills they learn in math class are needed in science also.

EQUIPMENT

- laboratory balance
- pennies (15)
- coil of solder
- graduated cylinder, 25 mL
- goggles

SCIENTIFIC ARGUMENTATION

Students will analyze their process by comparing the proportions that they calculated with the accepted values. Most students will use percent error to evaluate their performance.

Sample Procedure

COLLECTING DATA

A   Measure the mass of fifteen pennies and the coil of solder and record your data in Table 1.

B   Fill the graduated cylinder with approximately 15 mL of water. Record the actual value in Table 1.

C   Carefully submerge the pennies in the water. Record the volume in Table 1.

D   Repeat Steps B and C for the solder.

E   Calculate the volume of the pennies and the solder. Record your data in Table 1.

F   Calculate the density of the pennies and the solder. Record your data in Table 1.

DEVELOPING MODELS

1. From Table 1, select some possible metals that your objects may contain. Other physical properties can help you make a determination.

   On the basis of densities, tin or zinc would be likely candidates for pennies. Of course, the color of a penny would indicate that it is made of copper. On the basis of densities, iron or nickel are good candidates for solder; on the basis of color, nickel, tin, or silver are good candidates for solder.

2. Assuming that the solder is made of copper and one other metal in a 41/59 ratio, use your density data to determine the other metal.

   Answers will vary. The other metal is most likely tin.

   What we know: $\%_{copper} = 41\%$, $\%_{unknown} = 59\%$, $\rho_{mixture} = 8.03$ g/mL, $\rho_{copper} = 8.96$ g/cm^3

   Unknown: $\rho_{unknown}$

Write the formula and solve for the unknown.

$$\rho_{mixture} = (\%_{copper})(\rho_{copper}) + (\%_{unknown})(\rho_{unknown})$$

$$\rho_{mixture} - (\%_{copper})(\rho_{copper}) = \cancel{(\%_{copper})(\rho_{copper})} + (\%_{unknown})(\rho_{unknown}) - \cancel{(\%_{copper})(\rho_{copper})}$$

$$\frac{\rho_{mixture} - (\%_{copper})(\rho_{copper})}{(\%_{unknown})} = \frac{\cancel{(\%_{unknown})}(\rho_{unknown})}{\cancel{(\%_{unknown})}}$$

$$\rho_{unknown} = \frac{\rho_{mixture} - (\%_{copper})(\rho_{copper})}{(\%_{unknown})}$$

Evaluate.

$$\rho_{unknown} = \frac{8.03 \text{ g/mL} - (0.41)(8.96 \text{ g/mL})}{(0.59)}$$

$$= \frac{8.03 \text{ g/mL} - 3.6736 \text{ g/mL}}{(0.590)}$$

$$= \frac{4.3564 \text{ g/mL}}{(0.590)}$$

$$= 7.38 \text{ g/mL}$$

3. Assuming that a penny is made of only copper and zinc, use your density data to determine the proportions of copper and zinc.

Answers will vary. A penny is 94.6% zinc and 5.4% copper.

What we know: $\rho_{mixture} = 7.21$ g/mL (Copper and zinc make up 100% of the mixture.)

Unknown: $\%_{copper}$, $\%_{zinc}$

This is a system of two equations.

$$\%_{copper} + \%_{zinc} = 100\%$$

$$\rho_{mixture} = (\%_{copper})(\rho_{copper}) + (\%_{zinc})(\rho_{zinc})$$

Solve the first equation for $\%_{copper}$ and then substitute into the second equation.

$$\%_{copper} + \%_{zinc} = 100\%$$

$$\%_{copper} + \cancel{\%_{zinc}} - \cancel{\%_{zinc}} = 100\% - \%_{zinc}$$

$$\%_{copper} = \frac{100\%}{100} - \%_{zinc}$$

$$\%_{copper} = 1.00 - \%_{zinc}$$

$$\rho_{mixture} = \%_{copper}\,\rho_{copper} + \%_{zinc}\,\rho_{zinc}$$

$$\rho_{mixture} = (1.00 - \%_{zinc})\,\rho_{copper} + \%_{zinc}\,\rho_{zinc}$$

(continued)

Distribute ρ_{copper}, then subtract ρ_{copper}.

$$\rho_{mixture} - \rho_{copper} = \cancel{\rho_{copper}} - \%_{zinc}\,\rho_{copper} + \%_{zinc}\,\rho_{zinc} \cancel{- \rho_{copper}}$$

Factor out $\%_{zinc}$, then divide both sides by $(\rho_{zinc} - \rho_{copper})$.

$$\frac{\rho_{mixture} - \rho_{copper}}{\rho_{zinc} - \rho_{copper}} = \frac{\%_{zinc}\cancel{(\rho_{zinc} - \rho_{copper})}}{\cancel{(\rho_{zinc} - \rho_{copper})}}$$

$$\%_{zinc} = \frac{\rho_{mixture} - \rho_{copper}}{\rho_{zinc} - \rho_{copper}}$$

Evaluate.

$$(\%_{zinc}) = \frac{7.21 \text{ g/mL} - 8.96 \text{ g/mL}}{7.11 \text{ g/mL} - 8.96 \text{ g/mL}}$$

$$= \frac{-1.75 \cancel{\text{ g/mL}}}{-1.85 \cancel{\text{ g/mL}}} = 0.946 = 94.6\%$$

ASSESSING METHODS

4. Look up the actual proportions of zinc and copper in pennies. Make a claim about the accuracy of your process. Support your claim with evidence from your prediction and test.

The actual percentages for copper and zinc are 2.5% copper and 97.5% zinc. Our process was very accurate. The actual mixture in a penny would result in a 7.11 g/mL density. Our density (7.21 g/mL) represents about a 1.4% error.

TABLE 1

Object	Mass (g)	Volume of Water (mL)	Volume of Water and Object (mL)	Object Volume (mL)	Density (g/mL)
pennies (15)	38.57	15.12	20.47	5.35	7.21
solder	42.72	14.88	20.20	5.32	8.03

» Calculate the weighted average of a mixture of isotopes.

» Calculate the percentage of each isotope in a mixture.

4 LAB

Name ___________________

Date ___________________

EQUIPMENT

- laboratory balance
- chocolate-covered candies of two different varieties (50 total)

All that Glitters Is Not Copper-63

Investigating Mixtures of Isotopes

Pennies are made of copper, right? Well, sort of. Pennies minted before 1982 were made of 95% copper. But starting in that year, because of the rising cost of copper, the US Mint started making pennies from 97.5% zinc with a thin copper coating. That means that not all pennies in circulation today have the same composition, even though they are all pennies.

Copper atoms are not all alike either. They exist as one of two different but common and stable isotopes—copper-63 and copper-65. On the periodic table, the atomic mass of copper is not listed as either 63 or 65, but 63.55. Why is this? Copper, like most naturally occurring elements, is a mixtures of isotopes. The atomic masses listed on the periodic table are *weighted averages* of the various isotopes of each element.

Why are the masses on the periodic table not whole numbers?

QUESTIONS

» How do I determine the weighted average mass of a mixture of isotopes?

» How do I determine the percent composition of a mixture?

ISOTOPE MIXTURES

Give each student or lab group approximately 50 M&M® candies in a mixture of plain and peanut (be aware of food allergies) or mini and regular. You may also use another brand of similar candy. The intent is that students will recognize that the candies represent isotopes of the same element. If you want to model the ratio of isotopes in smelted copper, make the mixture an approximate ratio of 2.5:1 (copper-63:copper-65). Copper-63 accounts for about 70% of the isotopes found in natural copper. This will allow students to compare their calculated values to the actual values.

M&M's are not, of course, perfectly analogous to real copper isotopes since we can mass individual candies but not individual copper atoms.

INQUIRY LAB

This lab activity can be modified to be done as an inquiry activity by allowing students to formulate a set of procedures for determining weighted average mass and percent composition. To see an example of a similar activity done in this fashion, see page 63 of BJU Press *Physical Science* 6th Edition.

PRE-LAB CHECK

1. What are isotopes? *(Isotopes are atoms with the same number of protons but with a different number of neutrons.)*

2. How many "isotopes" will you have in your mixture in this experiment? *(two)*

3. Which isotope does each candy represent? Explain. *(The larger candy represents copper-65, while the smaller candy represents copper-63. The larger candy has the higher mass just as copper-65 has a higher mass number.)*

4. Why are most of the atomic masses on the periodic table not whole numbers? *(Most naturally occurring elements consist of mixtures of isotopes; their atomic masses are weighted averages.)*

5. Are all isotopes stable? Explain. *(No. Some isotopes are unstable and radioactive.)*

We calculate average atomic mass by dividing the total mass of all the atoms by the total number of atoms.

$$\text{average atomic mass} = \frac{m_{\text{total}}}{n_{\text{total}}}$$

To determine the total mass, we sum the masses of each isotope (kind of atom). The formula for this calculation is shown below.

$$\text{total mass} = \sum m_{\text{isotope}}\, n_{\text{atoms}}$$

The Greek letter Σ indicates "to take the sum of." If we were taking the sum of the masses of only two isotopes, the total mass would be calculated as shown below.

$$m_{\text{total}} = \sum m_{\text{isotope}}\, n_{\text{atoms}} = \overbrace{m_{\text{isotope 1}}\, n_{\text{atoms of isotope 1}}}^{\text{Isotope 1}} + \overbrace{m_{\text{isotope 2}}\, n_{\text{atoms of isotope 2}}}^{\text{Isotope 2}}$$

So the formula for calculating the average atomic mass is

$$\text{average atomic mass} = \frac{\sum m_{\text{isotope}}\, n_{\text{atoms}}}{\sum n_{\text{atoms}}}.$$

In this investigation, you will use a mixture of two varieties of chocolate-covered candies to represent two different isotopes in 1 mol of the "element" emenemium.

Procedure

A. Obtain a mixture of fifty candies from your teacher.

B. Find the mass of five large candies and five small candies. Remember that we always report one estimated digit in our measurements. Record these masses in Table 1.

C. Calculate the average mass for each type of candy. Record these values in Table 1. Don't forget significant figures!

1. A student from a different class mentioned that when they divided the mass of their five large candies (10.67 g) by five to calculate the average, they got 2.134. The lab partners couldn't agree whether that should be reported as 2.134 g or 2.13 g. How do you reply?

   See TE margin for answer.

   ____

   ____

   ____

   ____

D. Count the total number of both large and small candies in your mixture and record your data in Table 1.

SETTING THE RATIO OF CANDIES

The sample calculations assume that you are choosing to follow the ratios of Cu-63 to Cu-65. If you simply dump two large bags of candies together, the ratios will probably be different.

Question 1 Answer

The goal of using significant figures is to report the precision of measured data. Therefore, the rules for significant figures aim to preserve the precision of the measurements. The original measurement was precise to the hundredth of a gram, and so the average value can be precise only to a hundredth of a gram. Therefore, the answer should be 2.13 g.

Analysis

Now you will calculate the weighted average of the masses of candies in the mixture as instructed below. Refer to Example 4-3 on page 89 of your textbook. Be sure to follow the rules for significant figures.

E Do not measure the mass of your mixture. Calculate the total mass of your sample of candies and record your data in Table 2.

Answers will vary.

What we know: m_L = 2.02 g/candy, n_L = 15 candies, m_S = 1.49 g/candy, n_S = 35 candies

Unknown: m_{total}

Write the formula and solve for the unknown:

$$m_{total} = \sum m_{isotope}\, n_{atoms} = \overbrace{m_L\, n_L}^{\text{large candies}} + \overbrace{m_S\, n_S}^{\text{small candies}}$$

Evaluate:

$$m_{total} = \overbrace{m_L\, n_L}^{\text{large candies}} + \overbrace{m_S\, n_S}^{\text{small candies}}$$

$$= \left(2.02\,\frac{g}{\text{candy}}\right)(15\ \text{candies}) + \left(1.49\,\frac{g}{\text{candy}}\right)(35\ \text{candies})$$

$$= 30.30\ g + 52.15\ g$$

$$= 82.45\ g$$

F Measure the total mass of your mixture of candies and record your data in Table 2.

G Calculate the percent error between your calculated total mass and the actual total mass of your mixture and record your calculation in Table 2.

Answers will vary.

What we know: m_C = 82.45 g, m_M = 82.31 g

Unknown: $\%_{error}$

Write the formula and solve for the unknown:

$$\%_{error} = \left(\frac{m_C - m_M}{m_M}\right)100\%$$

Evaluate:

$$\%_{error} = \left(\frac{82.45\ g - 82.31\ g}{82.31\ g}\right)100\%$$

$$= \left(\frac{0.14\ \cancel{g}}{82.31\ \cancel{g}}\right)100\%$$

$$= 0.17\%$$

The mass of a peanut M&M is about 2.02 g, and that of a regular M&M is about 1.49 g. The sample data assumes that you are using these two types of candy. Remind students that they are never to eat in the laboratory, even if it is M&Ms. You may want to have a clean supply that you will share with the students at the end of class as a reward for a lab activity well done.

H Calculate the weighted average mass of a candy piece from your calculated total mass of your sample and record your data in Table 2.

Answers will vary.

What we know: m_{total} = 82.45 g, n_L = 15 candies, n_S = 35 candies

Unknown: $m_{average}$

Write the formula and solve for the unknown:

$$m_{average} = \frac{m_{total}}{n_L + n_S} \text{ or } m_{average} = \frac{m_L n_L + m_S n_S}{n_L + n_S}$$

Evaluate:

$$m_{average} = \frac{82.45 \text{ g}}{15 \text{ candies} + 35 \text{ candies}}$$

$$= \frac{82.45 \text{ g}}{50 \text{ candies}}$$

$$= 1.65 \frac{\text{g}}{\text{candy}}$$

$$\text{or} \quad m_{average} = \frac{\left(2.02 \frac{\text{g}}{\text{candy}}\right)(15 \text{ candies}) + \left(1.49 \frac{\text{g}}{\text{candy}}\right)(35 \text{ candies})}{15 \text{ candies} + 35 \text{ candies}}$$

$$= \frac{30.30 \text{ g} + 52.15 \text{ g}}{15 \text{ candies} + 35 \text{ candies}}$$

$$= \frac{82.45 \text{ g}}{50 \text{ candies}}$$

$$= 1.65 \text{ g/candy}$$

2. How does this calculation relate to average atomic masses for elements?

The process is similar to that used by chemists to determine the average atomic mass of elements found on the periodic table.

Name ___________________

1. Using the average mass of each candy and the average candy mass for the entire mixture, calculate the percentage of each "isotope" in the candy mixture.

Record your data in Table 1.

Answers will vary.

What we know: $m_L = 2.02$ g/candy, $m_S = 1.49$ g/candy, $m_{average} = 1.65$ g/candy

Unknown: $\%_L$, $\%_S$

This is a system of two equations with two unknowns.

Equation 1: $\%_L + \%_S = 100\%$

Equation 2: $m_{average} = \%_L m_L + \%_S m_S$

Solve Equation 1 for $\%_L$ and then substitute into Equation 2.

$$\%_L + \%_S = 100\%$$

$$\%_L + \%_S - \%_S = 100\% - \%_S$$

$$\%_L = \frac{100\%}{100} - \%_S$$

$$\%_L = 1.00 - \%_S$$

$$m_{average} = \%_L m_L + \%_S m_S$$

$$= (1.00 - \%_S)\, m_L + \%_S m_S$$

Distribute m_L, then subtract m_L.

$$m_{average} - m_L = m_L - \%_S m_L + \%_S m_S - m_L$$

Factor out $\%_S$, then divide both sides by $(m_S - m_L)$.

$$\frac{m_{average} - m_L}{m_S - m_L} = \frac{\%_S (m_S - m_L)}{(m_S - m_L)}$$

$$\%_S = \frac{m_{average} - m_L}{m_S - m_L}$$

Evaluate:

$$\%_S = \frac{1.65\text{ g} - 2.02\text{ g}}{1.49\text{ g} - 2.02\text{ g}}$$

$$= \frac{-0.37\text{ g}}{-0.53\text{ g}}$$

$$= 0.70 = 70\%$$

The small candies are 70% of the mixture and the other 30% is large candies.

J. Now verify the percent calculations from Step I using the numbers of candies ("isotopes") that are used. Record your data in Table 1.

Answers will vary. What we know: $n_L = 15$ candies, $n_S = 35$ candies

Unknown: $\%_L$, $\%_S$

Write the formula and solve for the unknown:

$$\%_L = \left(\frac{n_L}{n_L + n_S}\right)100$$

$$\%_S = \left(\frac{n_S}{n_L + n_S}\right)100$$

Evaluate:

$$\%_L = \left(\frac{15\text{ candies}}{15\text{ candies} + 35\text{ candies}}\right)100\%$$

$$= \left(\frac{15\text{ candies}}{50\text{ candies}}\right)100\% = 30\%$$

$$\%_S = \left(\frac{35\text{ candies}}{15\text{ candies} + 35\text{ candies}}\right)100\%$$

$$= \left(\frac{35\text{ candies}}{50\text{ candies}}\right)100\% = 70\%$$

3. Account for any differences between the answers for Steps I and J.

   The percentages should be close. Any small differences are due to using rounded values in calculations.

4. Now explain why the atomic mass on the periodic table for copper is not a whole number. Which isotope is more common in natural copper?

   See TE margin for answer.

   Copper has twenty-seven other isotopes, all of which are radioactive, or *radioisotopes*. You'll learn more about radioisotopes in Chapter 22. These isotopes are unstable, short-lived, and relatively rare.

5. How would radioisotopes of copper affect the atomic mass of copper on the periodic table?

   They would have relatively little effect since they make up such a small fraction of the isotopes in natural copper.

6. The radioisotopes of copper are very useful in medicine for making images of the body and for treating cancers. Where do you think scientists get these radioisotopes?

   Students should conclude that these radioisotopes need to be isolated in nature or created artificially.

Question 4 Answer

Students should recognize that the value on the periodic table is a weighted average of the individual atomic masses of copper isotopes in natural copper. They should conclude that the more common isotope is copper-63.

7. How can the chemistry of copper be used to fulfill God's commands to love and serve others?

Using copper for conducting electricity, imaging, treating cancers, and engaging in similar purposes shows love to God's image-bearers by improving and perhaps even saving their lives.

TABLE 1

Object	Mass of Five (g)	Average Mass (g/candy)	Number in Mixture	%$_{composition}$ from Mass	%$_{composition}$ from Number of Candies
Large Candies	10.12	2.02	15	30	30
Small Candies	7.43	1.49	35	70	70

TABLE 2

Object	Total Mass (calculated) (g)	Total Mass (measured) (g)	% Error	Average Atomic Mass (g)
Mixture	82.45	82.31	0.17	1.65

5A LAB

Bulls-Eye!

Modeling an Atomic Orbital

We commonly see an atom depicted as a nucleus surrounded by a number of electrons traveling in circular orbits. But as physicists learned more about atoms, they quickly deduced that electrons do not travel in neatly defined orbits. Instead, they are found in indistinct regions called *orbitals*. What does an orbital look like? And what is the difference between an orbital in one level and another found in a higher energy level?

Procedure

ATOMS WITH LOW-ENERGY ORBITALS FILLED

Select one lab partner to be the "dropper" and the other to be the "catcher."

A Place a target sheet flat on a hard floor.

B The dropper should hold the marble over the target sheet at waist level, aim at the X, and release the marble.

C The catcher should catch the marble after it bounces the first time, before it can strike the paper a second time.

D Repeat Steps B and C ninety-nine additional times. Use a scratch piece of paper to keep a tally of how many times the marble is dropped.

E Using a compass, draw a circle on the target sheet such that ninety of the hits are inside the circle, leaving ten of the hits outside the circle.

1. If the X on the paper represents the nucleus of an atom, what do you think the impression left by each marble drop represents?

   Each impression represents a "snapshot" of where an electron might

   be found in an atom.

QUESTIONS

» How can I model electron orbitals?

» How is the energy of an electron related to the size of its orbital?

Why can't we know exactly where electrons are located?

EQUIPMENT

- carbonless paper targets (2)
- marble or ball bearing
- drawing compass
- meter stick

LAB 5A OBJECTIVES

» Create a model of an electron orbital.

» Relate orbitals to electron energies.

MATERIALS

The targets should be printed on two-part carbonless paper, available from print shops. The target should consist of a small X in the center of the top sheet. You can either have the targets printed at the shop or draw them yourself. If carbonless paper is not available, you can use two sheets of computer paper with carbon paper between them for each trial.

Any small marble or ball bearing should work for this activity. As always, test the procedure first to see that acceptable results can be obtained.

This lab activity should be done in groups of two. One student drops the marble while the other serves as the catcher.

PRE-LAB CHECK

1. What is this lab activity attempting to model? *(an atomic orbital)*

2. Why do we use carbon *(or carbonless)* paper in this activity? *(to record where the marble strikes each time)*

3. What will the compass be used for? *(to determine the area covered by marble strikes)*

4. How will you evaluate the relationship between atomic radius and drop height? *(by creating a scatterplot)*

F Repeat Steps A through E using the second target. This time, though, drop the marble from eye level rather than waist level.

2. What does each marble represent in this second set of drops?

   The marbles dropped from eye level represent electrons with more

   energy than those dropped from waist level.

Analysis

G Measure the radius of each circle and record your measurements in Table 1.

3. How do the radii of the two targets compare?

   Answers will vary, but the high-energy radius should be significantly

   larger.

4. What does the size of the circle represent in this particular atomic model?

   The size of the circle represents the diameter of the atom according to

   the probable location of its electrons.

5. Does the variation in target size make sense? Explain.

   Answers will vary. Students should compare the dropping heights

   (more potential energy) with the size of the circle.

6. Why didn't all the dropped marbles land in the same spot?

   There is a slight variation from drop to drop in how and where the

   marble is dropped.

7. Compare the distribution of "electrons" between the two targets. How does the distribution change with the height of the dropped marbles? How does this model the arrangement of electrons in an actual atom?

   The marbles dropped from a greater height produced a wider

   distribution of strikes. This models the tendency in real atoms of

   higher-energy electrons to exist in larger orbitals.

H Create a scatterplot of the atomic radius versus the drop height. Include a curve of best fit that includes the origin as a data point.

8. When you include (0, 0) as a point on your graph, is the relationship linear? Explain.

   Answers will vary. Most students will find that the relationship is not

   linear.

9. Your target model is, of course, a two-dimensional representation of a three-dimensional structure. What would the circle drawn on each target look like in a three-dimensional model?

   It would be spherical.

10. Describe the shapes of the orbitals depicted in your textbook.

    The orbitals depicted in the Student Edition are varied. Some are spherical-shaped, while others look like barbells or even donuts.

11. On the basis of your descriptions from Questions 9 and 10, do you think that the circular orbital depicted in this lab activity is a good model for all the electrons in an atom?

    Yes. Some atoms contain only spherical orbitals, so this lab activity models those very well. But even in atoms with more complex orbitals, the combination of all the orbitals results in a nearly spherical distribution of electrons for the entire atom.

Going Further

12. In what way(s) does using a dropped marble aimed at a target do a good job of modeling the locations of electrons? In what way(s) does it fall short?

    See TE margin for answer.

TABLE 1

Target	Drop Height (m)	Atomic Radius (cm)
1	1.0	
2	2.0	

Question 12 Answer

The act of dropping from a height introduces an element of randomness to the locations produced on the paper. Higher drops, corresponding to electrons with greater energy, produce a wider distribution of results. However, the type of constraint limits the randomness in each system. The constraint on the modeled system's randomness is the act of repeatedly aiming at the same spot on the paper, whereas the constraint in an actual atom is the electrostatic attraction that exists between oppositely charged subatomic particles.

5B LAB

Seeing Light in a New Way

Exploring Spectroscopy

News flash! Our solar system is not the only one out there. As of 2020, scientists have confirmed the existence of more than 4100 other planets that are orbiting stars other than our sun. How did scientists know that they were exoplanets and not something else, like a distant galaxy or star?

Light from stars gives us information. Some of this light is visible, but some of it is in forms that we can't see, such as gamma rays, x-rays, microwaves, and radio waves. Visible light can tell us the elements that make up stars and the speed that stars are moving relative to Earth. Scientists usually find an exoplanet when it passes in front of a star, dimming the light for a short period of time. To get information from a star, astronomers use a tool called a *spectroscope*. Although any form of electromagnetic radiation could theoretically be used in spectroscopy, scientists often work with visible light because it is easiest to observe.

Visible light, like other kinds of radiation in the electromagnetic spectrum, can be either emitted or absorbed by atoms when their electrons move to a lower energy level or jump to a higher energy level. This means that scientists can distinguish two kinds of visible light spectra—emission spectra and absorption spectra. Emission spectra show the wavelengths of light that something puts out by showing lines of light on a black background. See page 95 of your textbook for a picture of an emission spectra. Absorption spectra show the wavelengths of light that something absorbs by showing lines of black on the background of a continuous spectrum. Both absorption and emission spectra are line spectra.

In this lab activity, you'll learn just how informative light can be. You'll observe the emission spectra of several salts using a simple diffraction grating spectroscope. After sketching these spectra, you'll use your sketches to reveal the secret identity of an unknown salt.

EQUIPMENT

- diffraction grating spectroscope
- laboratory burner and lighter
- beaker, 100 mL
- incandescent light source
- colored pencils
- chloride salts
- nitrate salts
- presoaked wooden splints
- LED light source
- fluorescent light source
- goggles
- laboratory apron

QUESTIONS

» How do elements produce visible light?

» Why do different elements produce different colors of light?

» How can we use a diffraction grating to study light?

» What is the relationship between the color of light and the electrons in an atom?

Seeing Light in a New Way | 37

» Explain how elements can produce visible light.

» Perform a flame test on different compounds.

» Sketch a spectrum on the basis of observations made using a simple diffraction grating spectroscope.

» Relate the color of light that an element emits to the shift of electrons into different energy levels.

PREASSESSMENT

If your students completed a course in physical science using BJUPress *Physical Science* 6th Edition materials, they may recall doing spectroscopy in Lab 3A of that course. Consider preassessing student understanding of this topic. If students demonstrate sufficient subject matter competency, you may choose to skip this activity.

PRE-LAB CHECK

1. What region of the electromagnetic spectrum is commonly used in spectroscopy? Why is this? *(Visible light is used because it is the easiest to observe.)*

2. What instrument will you use in this experiment to get information from light? *(a diffraction grating spectrometer)*

3. Why will you burn the mineral salts in this procedure? *(to see the emission spectrum produced by each salt)*

4. List the light sources that you will study. *(incandescent lights, open flame tests, fluorescent light, and LEDs)*

⚗ PREPARING THE SPLINTS

Make sure that you soak the wooden splints in water overnight. This extends the length of time that the salt-coated splints can be held in the flame before the splints begin to burn.

Nichrome or platinum wires can be substituted for wooden splints, but they are more susceptible to corrosion. They have the benefit of not igniting as the wooden splints do. To use wires for the flame tests, dissolve the salts in 6 M HCl and test. Between tests, rinse wires with HCl and dry in the burner flame.

If you choose to use wires and HCl, make sure to prepare your students to handle the acid safely, being careful to avoid spills and acid burns.

SPECTROSCOPY

A diffraction grating is a transparent material containing many closely spaced grooves or lines that diffract light to produce a spectrum. See pages 95–98 of the Student Edition for more information on spectroscopy.

LABORATORY BURNERS

In this lab activity, it's important that students have their laboratory burners adjusted properly. For tips, see Appendix C, Using a Laboratory Burner.

⁉ CONTINUOUS SPECTRUM

The acronym ROYGBIV is often used to refer to a continuous spectrum of visible light. This is a helpful way for students to remember the basic colors in a continuous spectrum, but be careful to avoid propagating the faulty idea that the colors in a continuous spectrum are discrete frequencies with well-defined boundaries.

The boundaries between colors will be indistinct, but for your reference, students should show colors at about the following wavelengths:

Red: 750–620 nm

Orange: 620–590 nm

Yellow: 590–570 nm

Green: 570–495 nm

Blue: 495–450 nm

Indigo/Violet: 450–380 nm

Procedure

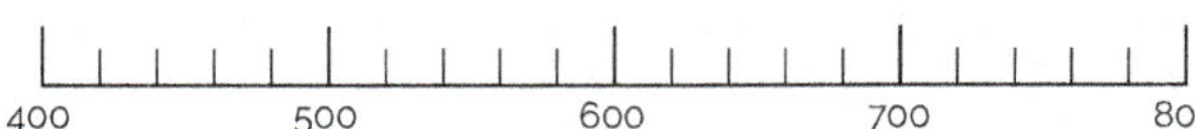

Incandescent light bulbs produce light when electricity runs through a fine tungsten wire called a *filament*. This filament begins to glow, producing visible light as electrons are excited by an increase in temperature and jump to higher energy levels. As electrons return to lower energy levels, they release a variety of visible light wavelengths.

A Observe the spectrum of an incandescent light by looking at it through your spectroscope. Note the calibration marks (400–800 nm) in the spectrum below. Use colored pencils to sketch the spectrum or label the colors that you observe. Add the letters ROYGBIV (representing red, orange, yellow, green, blue, indigo, and violet) at the appropriate locations.

1. What kind of spectrum does an incandescent light bulb emit? What color is this kind of light?

   An incandescent light bulb emits a continuous spectrum. It produces white light.

400		500		600		700		800

B Obtain three pairs of samples of certain nitrate and chloride salts that your teacher has chosen. The salt samples should be in pairs according to their metal ions (e.g., sodium nitrate and sodium chloride).

C Fill the beaker with water and keep it close by.

D Light your laboratory burner. Adjust the burner until it produces a blue or colorless flame.

E The wooden splints that you will be using have been soaked in water overnight. Dip a presoaked splint into either of the first pair of salt samples and then insert the salt-encrusted tip into the hottest part of the burner's flame (see below). Record the flame color that is produced by the salt in Table 1. If the splint catches fire, simply douse it in the beaker of water that you prepared in Step C.

F Repeat Step E for the other salt in the pair.

G Repeat Steps E and F for the remaining salt pairs.

2. Which element in these compounds do you think is responsible for the color change? How can you tell?

   See TE margin for answer.

3. What do you think is happening to the electrons in these atoms that causes them to emit visible light? Where do the colors come from?

   Electrons are absorbing energy from the burner, moving them to higher energy levels. As the electrons return to lower energy levels, they emit light in the visible spectrum. The colors come from the different wavelengths of electromagnetic energy that the electrons emit as they move to lower energy levels.

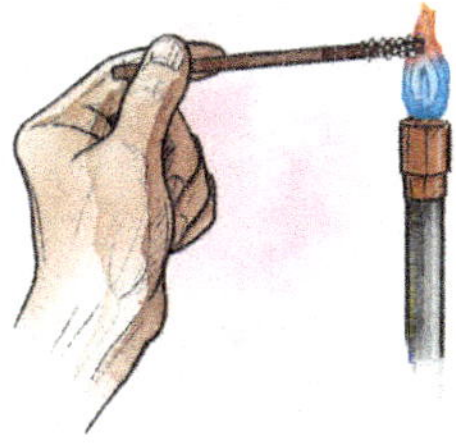

38 | Lab 5B

✋ SEEING THE SPECTRUM

If you have students with color vision deficiency, they may need some accommodation, although the spectroscope will help quantify the colors that they are seeing.

Question 2 Answer

Students should assume from testing multiple pairs that the metal cations in these salts are responsible for the color changes because the anions remain unchanged and the cations are the only variable that changes directly with color.

⚗ SELECTING SAMPLES AND GETTING THEM TO STUDENTS

The following pairs of salts work well for this procedure:

sodium nitrate and sodium chloride

strontium nitrate and strontium chloride

lithium nitrate and lithium chloride

potassium nitrate and potassium chloride.

Name ___________________________

ELEMENTS AND SPECTRA

H Now that you have found which element in these compounds is responsible for the color change in the flame test, obtain other chloride salts that your teacher has provided.

I Dip a presoaked splint and have your lab partner look through the spectroscope at the flame. Put the tip of the splint in the flame long enough for the salt to burn as you did before, noting its color before the wooden splint begins to burn. Record your observation in Table 1.

J Have your lab partner observe the spectrum that forms. It will probably be different than when looking at an incandescent light bulb! Use the spectrum graphs on pages 42–43 to sketch the emission spectra of your flame tests. Either label or color the bright lines with colored pencils to identify them. Switch roles with your lab partner so that each of you observes the burning splints while using the spectroscope. If you need to see a flame test again, use a fresh presoaked splint and a fresh sample of salt.

K Repeat Steps I and J for the remaining salts, using a new splint for each test.

4. How does a spectroscope help you identify elements in a flame test?

See TE margin for answer.

____

____

5. What did you notice when you compared the colors of the flame and the line spectra?

See TE margin for answer.

____

____

____

MYSTERY SALT

L Now to solve a mystery! Your teacher will give you two unknown salts. Do a flame test of each salt, and sketch its emission spectra in the appropriate spectral graphs on page 43.

6. What was the identity of the chloride salt Unknown 1?

Answers will vary depending on which salt you give students.

7. What was the identity of the chloride salt Unknown 2?

Answers will vary depending on which salt you give students.

It is necessary at a minimum to do at least two pairs of these so that students can come away with the idea that the metal ion is the element that imparts color to the flame.

Use labeled watch glasses for the salts to allow students easy access to the samples. They may put a small amount in a small beaker to dip their splints or they may choose to get samples on weighing paper or on their own watch glasses. Do not have students dip directly from the stock containers. If students accidentally spill salt into a laboratory burner, advise them to let it burn completely before proceeding to the next test, or you may choose to turn the burner off, allow it to cool, then tap the spilled particles out of the inverted burner. In the interest of safety, it is best that you do this task rather than the student.

 EMISSION SPECTRA

For examples of different emission spectra from flame tests, visit TeacherToolsOnline.com. Under Additional Resources, select Chapter 5 Web Links. Then click on *Element Emission Spectra*. You can also conduct an internet search using the keywords "element emission spectra."

 CHLORIDE SALTS

You may choose to have students observe a variety of chloride salts, including calcium chloride ($CaCl_2$), copper(II) chloride ($CuCl_2$), manganese(II) chloride ($MnCl_2$), potassium chloride (KCl), lithium chloride (LiCl), sodium chloride (NaCl), and strontium chloride ($SrCl_2$). Observe as many as time and budget allow.

These chemicals may be obtained from science supply companies. Large quantities are not needed. Sodium and potassium chloride can both be obtained from any grocery store, where they are sold as table salt and salt substitute, respectively. Calcium chloride is also available at some grocery stores in the pickling supplies section and from home centers where it is sold as a sidewalk deicer.

Question 4 Answer

Many chemicals give off similar colors when they are heated, and spectroscopes help us tell them apart in ways that are less subjective. The emission lines are unique, so they give a more specific identification of an element.

Question 5 Answer

Students should notice that what appears to be a single color in a simple flame test is usually a series of colored lines in a spectroscope. The individual colors seen in a spectrum are not apparent to the unaided eye.

 UNKNOWN SALT

You may choose to have all groups identify the same two mystery salts or select a different pair for each group. Place each one in an unmarked watch glass at the students' stations.

OPTIONAL MATERIAL

The material on artificial lights and spectra is optional. Use this material as time and student interest permit.

Analysis

8. Since each element produces a characteristic spectrum, what can you conclude about the location of the electrons?

   The electrons are located in definite energy levels.

9. Suppose that you had used the same wooden splint to burn all the salts in succession. What difficulty could this have introduced?

   Contamination of each salt with the previous one would have ruined the results and made identification uncertain.

Look at the absorption spectrum from a star shown at left. The black lines display wavelengths of light that are absorbed by the gases in the star.

10. Why is this information useful?

    It can help us identify the kinds of elements in a star.

11. If emission lines are created in a spectrum when an electron falls from an excited state to a less energetic state, how would you explain absorption lines?

    Students should reason that stellar gases absorb energy from a spectrum when their electrons absorb energy.

Going Further

ARTIFICIAL LIGHTS AND SPECTRA

In this lab activity, you have been looking at emission spectra of different elements. These bright lines are emitted when atoms receive energy that causes one or more of the atom's valence electrons to jump from their original positions to higher energy orbitals and almost immediately fall back to lower orbitals, emitting visible light. The wavelengths (colors) of the light depend on the energy differences between the atom's various orbitals. As a result, atoms of each element generate characteristic lines of colors. You also looked at the emission spectra of an incandescent bulb, which produced a continuous spectrum rather than lines.

LED (light-emitting diode) bulbs make light by the movement of electrons through *semiconductors*, materials that contain elements such as silicon and antimony. They don't have filaments like incandescent bulbs do. They produce light just like the elements in your flame test: the electrons are excited to higher energy levels and then return to lower ones.

M Observe the emission spectrum of an LED bulb by looking at it with your spectroscope. Sketch the spectrum in the appropriate spectral graph on page 43.

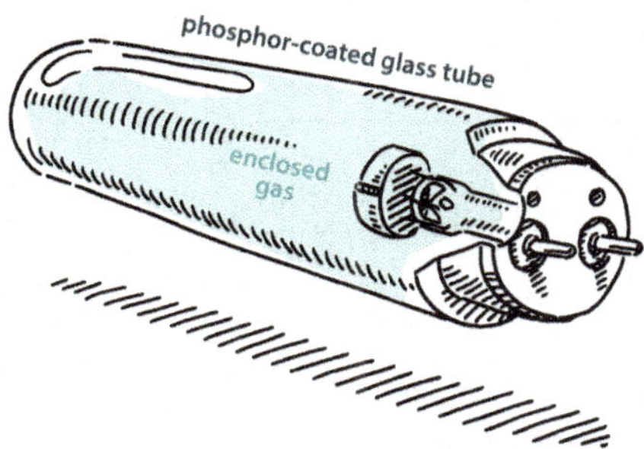

Fluorescent bulbs make light in a different way. They contain a gas that produces ultraviolet light when electricity excites electrons. The reason that the electrons don't produce visible light is because the jump between energy levels is larger and releases ultraviolet light, which humans can't see. The inside of the bulb is coated with *phosphors*.

Phosphors are chemicals that glow when they are exposed to ultraviolet light, emitting visible light as their electrons move between energy levels. So fluorescent bulbs convert electricity to visible light in a multistep process.

N Observe the emission spectrum of a fluorescent bulb by looking at it with your spectroscope. Sketch the spectrum in the appropriate spectral graph on page 43.

12. You may have noticed that the spectra of some artificial lights were different than those of the incandescent light. Compare and contrast both the color of the light and the spectra.

Students may notice that LED and fluorescent bulbs appear bluer than incandescent bulbs, which appear more yellow. A spectroscope reveals that these bulbs do not produce a continuous spectrum.

13. What type of lamp would produce light in which different-colored objects look the most natural? Why?

Incandescent lamps produce the most natural-looking light since they emit almost all visible wavelengths and therefore render colors the most accurately.

14. Can you think of any settings in which very good lighting is essential for work and productivity?

Answers will vary. Students may name art, photography, and museums with artifacts that need natural light to properly observe them.

Salt name:

Color of the flame:

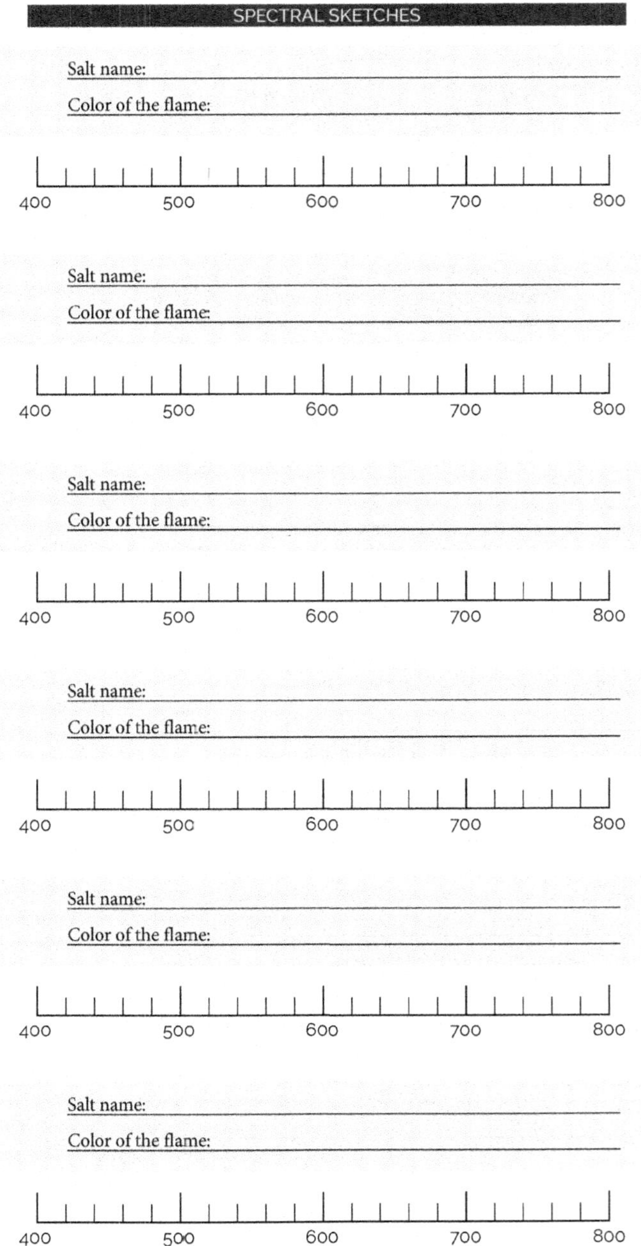

Salt name:

Color of the flame:

Salt name:

Color of the flame:

Salt name:

Color of the flame:

Salt name:

Color of the flame:

Salt name:

Color of the flame:

Salt name:

Color of the flame:

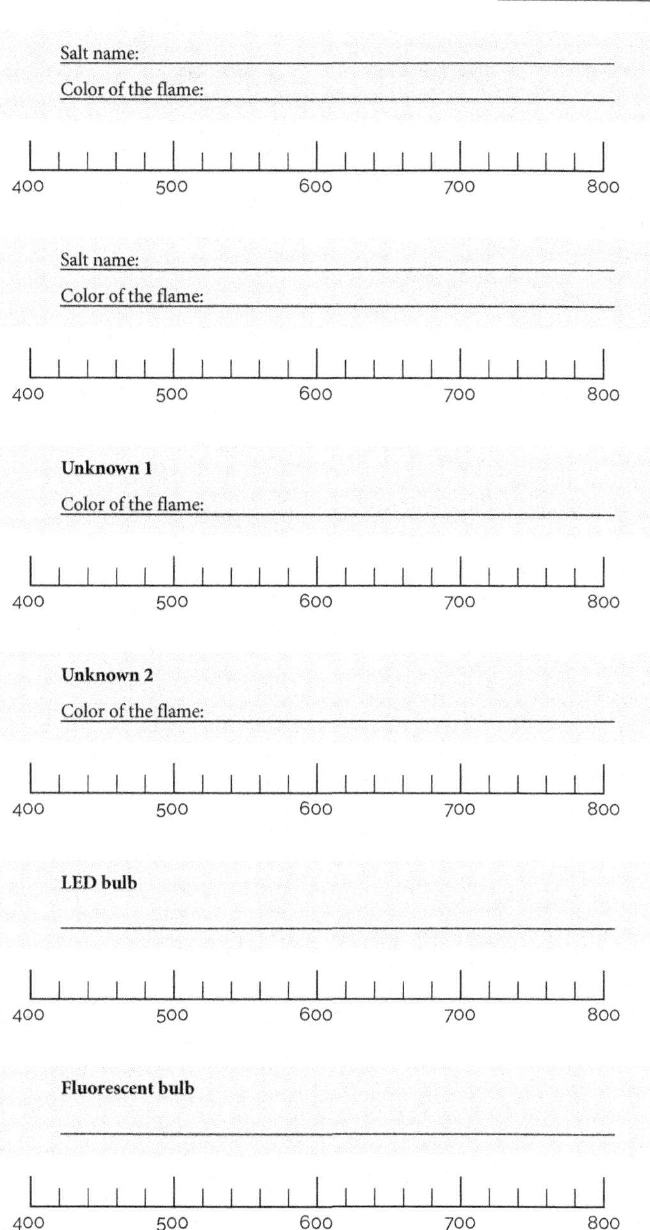

400	500	600	700	800

Salt name:

Color of the flame:

400	500	600	700	800

Unknown 1

Color of the flame:

400	500	600	700	800

Unknown 2

Color of the flame:

400	500	600	700	800

LED bulb

400	500	600	700	800

Fluorescent bulb

400	500	600	700	800

TABLE 1

Chloride Salt	Color Observed	Nitrate Salt	Color Observed
sodium chloride	yellow	sodium nitrate	yellow
potassium chloride	lilac	potassium nitrate	lilac
lithium chloride	red	lithium nitrate	red
strontium chloride	red	strontium nitrate	red

6A LAB

Exposed to the Elements

Inquiring into Properties of Elements

Throughout history, people have made bridges out of different materials. We have made them from wood, stone, steel, and concrete. Why have we used these materials for building bridges? Each one has properties that make it good for constructing a bridge. Engineers try to match the properties of materials with the physical requirements of the object that is being built.

How do we classify elements?

There are currently 118 known elements. Each element has different physical and chemical properties. As scientists discovered elements, they recognized that they would need a way to organize them. Some scientists made lists and others made tables. As the periodic table of the elements was developed, scientists were surprised to discover that individual elements had properties that were similar to other elements located nearby on the table.

In this lab activity, you will observe the properties of various elements.

Procedure

RESEARCH

1. What are the general characteristics of metals, nonmetals, and metalloids? Use your textbook or other resources as needed.

See TE margin for answer.

EQUIPMENT

- small plastic cups (7) with samples
- goggles
- laboratory apron
- nitrile gloves

QUESTIONS

» How do metals, nonmetals, and metalloids compare?

» How can I use the characteristics of metals, nonmetals, or metalloids to classify them?

» Characterize metals, nonmetals, and metalloids on the basis of their properties.

» Classify elements as metals, nonmetals, or metalloids on the basis of collected data.

SCHEDULING

This lab activity can be done either as an introduction to Section 6.1 or as a follow-up.

INQUIRY LAB ACTIVITY

This activity is written as an inquiry lab activity, in which the students develop their own procedures for answering a series of questions. You can also provide students the sample procedure from the Teacher Guide and have students do the activity as a guided discovery lab activity.

SAMPLES

There are a number of samples that you can use in this lab activity. The ones that you choose will depend on chemicals currently in your stock and and on available funds if purchasing other samples. Science supply stores sell element sample kits. Many have a good selection of elements for the testable samples.

If you are going to put together your own collection of elements, we recommend the following: aluminum, carbon, copper, iron, magnesium, nickel, silicon, sulfur, tin, and zinc. As a minimum, include carbon, sulfur, and silicon so that you have some nonmetals and metalloids.

Question 1 Answer

Metals tend to be solids at room temperature. They typically have a metallic luster and are malleable and ductile. They are also typically good conductors of electricity and thermal energy. Many metals react with acids. Nonmetals are typically gases, liquids, or dull, brittle solids at room temperature. They are typically poor conductors of thermal energy and electricity. Metalloids are typically solids at room temperature and have properties between metals and nonmetals.

A. Brainstorm with your lab group about how you could test the seven samples provided by your teacher to enable you to classify each element as either a metal, nonmetal, or metalloid.

B. Write specific questions related to the samples and the characteristics of metals, nonmetals, and metalloids that you could answer by collecting data.

C. Write procedures to collect data that will enable you to answer the questions that you wrote in Step B above.

D. Have your teacher approve your procedures.

E. Following the procedures that you have written, collect data to answer the questions that you wrote.

F. Organize your data in a table. Include in the table your classification of each element as a metal, nonmetal, or metalloid.

G. Obtain the identification of the elements from your teacher. Consult a periodic table to see whether your classification matches the organization of the periodic table.

H. Make a claim about the accuracy of your classification of the elements that you tested. Support your claim with evidence from your prediction and test.

Teacher Guide

INQUIRING INTO PROPERTIES OF ELEMENTS

MANAGING INQUIRY LABS

An inquiry lab activity allows students a degree of freedom, but the teacher has to guide the process to achieve the desired educational outcomes. As teachers start using inquiry lab activities, they often struggle with this process of guiding students. The best method is to think through the goals that you want students to achieve and then use questions to guide them to reach those goals. You want to balance students' freedom to investigate the question in the manner they choose with the need to meet your educational goals.

Procedure

PLANNING/WRITING SCIENTIFIC QUESTIONS

Each lab group starts with seven cups containing the sample elements that you have chosen. After reviewing the characteristics of metals, nonmetals, and metalloids, students should brainstorm how they could test the seven elements in order to classify them as metals, nonmetals, or metalloids.

Each lab group should write a series of questions related to the samples and the characteristics that they identified that can be answered by collecting data. Help students write specific, testable questions. For example, "How can I tell whether something is electrically conductive?" is not a good question. "How can I measure the conductivity of each element?" might work. You may want the groups to submit their questions for your review.

DESIGNING SCIENTIFIC INVESTIGATIONS

Once the groups have written their questions, they need to write specific procedures to answer them. This is a new task for many students, so you may want to do part of this task as a class discussion.

Always have students get your approval on laboratory procedures prior to their starting to collect data.

CONDUCTING SCIENTIFIC INVESTIGATIONS

Once the groups have good procedures, allow them to collect data. If students are going to use acids, review safety procedures for acids. ***Never add water to an acid! Always wear appropriate personal protective equipment!***

DEVELOPING MODELS

Students will classify their elements as metals, nonmetals, or metalloids.

SCIENTIFIC ARGUMENTATION

After identifying the seven elements, students should compare their classifications with the periodic table. They will state a claim about how well they were able to classify the elements. They must support their claim with specific evidence from their data.

Teacher Guide

Sample Procedure

EQUIPMENT

- conductivity tester
- small plastic cups (7)
- wash bottle (filled with distilled water)
- spatula
- weighing paper

- small hammer
- test tube rack
- test tubes (14)
- graduated cylinder, 10mL
- element samples (7)
- distilled water, 35 mL

- 1 M hydrochloric acid, 35 mL
- goggles
- laboratory apron
- nitrile gloves

Conductivity Tester

You can purchase conductivity testers for a reasonable cost. Hardware stores also sell circuit testers that should also work as a conductivity tester. You can also make a simple conductivity tester with a 9 V battery, a small lamp, and some wire. To learn how to make a conductivity tester, do an internet search using the keywords "how to make an electrical conductivity tester."

Hammer

Your school's earth science teacher may have small rock hammers, which would work well.

Preparing Acid

You can purchase 1 M HCl solution already prepared. To make your own solution, add approximately 70 mL of distilled water to a 100 mL volumetric flask. Add 8.3 mL of 12 M HCl to the volumetric flask and swirl gently. Add distilled water to the graduation mark. If you need more solution, you can scale these instructions.

Have sodium carbonate (soda ash) available to neutralize any acid spills.

QUALITATIVE OBSERVATIONS

A Observe the samples in the cups. Record your observations in Table 1. Note the state of matter and appearance (e.g., color and luster) of each sample.

CONDUCTIVITY

B Using the conductivity tester, test the first sample in one of the plastic cups. Record your observations in Table 1.

C Using the wash bottle, clean the conductivity tester. Dry the tester.

D Repeat Steps B and C for each of the other six elements in your cups.

BRITTLE OR MALLEABLE

E Using the spatula, move a small sample of the element from the first cup to a piece of weighing paper.

F Lightly tap on the sample to determine whether it is brittle or malleable. Record your observations in Table 1.

G Repeat Steps E and F for the other six elements.

H Dispose of the tested material appropriately.

REACTIVITY WITH WATER

I Label seven of the test tubes from 1 to 7.

J Using the spatula, transfer a small sample of each element from the cups to the appropriate test tube.

K Using the graduated cylinder, add 5 mL of distilled water to each test tube. Watch for indications of a chemical reaction (formation of a gas [bubbles forming], transfer of energy [heating up or cooling down], or change in color). Record your observations in Table 1.

Teacher Guide

L Dispose of the tested material appropriately.

REACTIVITY WITH HYDROCHLORIC ACID

M Label the other seven test tubes from 1 to 7.

N Using the spatula, transfer a small sample of each element from the cups to the appropriate test tube.

O Using the graduated cylinder, add 5 mL of 1 M hydrochloric acid to each test tube. Watch for indications of a chemical reaction (formation of a gas [bubbles forming], transfer of energy [heating up or cooling down], or change in color). Record your observations in Table 1.

P Dispose of the tested material appropriately.

Reactivity with Acids Cautions

Check the SDSs for the sample elements that you have chosen for their reactivity with acid. Typically the reactions will release hydrogen gas, though other products are possible.

Make sure that there are no ignition sources in the room during this portion of the activity. In cases where the elements are reactive with hydrochloric acid, one of the products is hydrogen gas, which is flammable. The amount and concentration of the acid that you are using will keep the volume of gas small.

TABLE 1

Cup	Element	Appearance	Conductive?	Brittle?	Malleable?	Reactive with Water?	Reactive with Acid?	Metal, Nonmetal, or Metalloid?
1	sulfur	solid, yellow chips; light, dull luster; powdery surface	no	yes	no	no	no	nonmetal
2	aluminum	solid, silver gray; light, metallic luster; smooth strip	yes	no	yes	no	yes	metal
3	carbon	solid, black; light, dull luster; rough surface, chunks	no	yes	no	no	no	nonmetal
4	copper	solid, reddish-brown; heavy, metallic luster; smooth surface, like BBs	yes	no	yes	no	no	metal
5	iron	solid, black; heavy, dull luster; rough surface, small pieces	yes	no	yes	no	yes	metal
6	silicon	solid, dark gray; light, metallic luster; smooth surface	no	yes	no	no	no	metalloid
7	magnesium	dark gray; light, dull luster; smooth strip	yes	no	yes	no	yes	metal

Teacher Guide

		Other Possible Samples						
Cup	Element	Appearance	Conductive?	Brittle?	Malleable?	Reactive with Water?	Reactive with Acid?	Metal, Nonmetal, or Metalloid?
	nickel	silver-gray; heavy, metallic luster; smooth surface, chunks	yes	no	yes	no	yes	metal
	tin	silver-gray; heavy, metallic luster; smooth surface, chunks	yes	no	yes	no	yes	metal
	zinc	silver-gray; heavy, dull luster; rough surface, chunks	yes	no	yes	no	yes	metal

6B LAB

Name _______________________
Date _______________________

An Elemental Merry-Go-Round

Exploring Periodic Trends

Periodic tables are not "one size fits all." They can use spirals, steps, circles, and even amoeba shapes to communicate information about how the properties of elements change with their atomic number. The regular repetition of the properties of elements is described by the *periodic law*.

Regardless of the shape of a periodic table, elements with similar properties are located in similar positions in the table. In the traditional periodic table, the elements with similar properties are in the vertical columns. For example, since lithium, sodium, potassium, rubidium, and cesium have similar reactivity, they are placed in a vertical column. We call vertical columns *groups*, or *families*, and horizontal rows *periods*, or *series*. Properties vary as you move across a period or down a group.

So just how periodic is the periodic table? In this lab activity, you will graph the trends of four different properties: atomic radius, electronegativity, electron affinity, and ionization energy. You will then use these graphs to predict the properties of other elements. You will also explore how atomic radii increase across a period and down a group of the periodic table.

What does the periodic table tell us about chemical and physical properties?

The main idea of this spiral periodic table is to show the periodic nature of the elements. Makes you think of a merry-go-round!

QUESTIONS

» How do atomic radii and atomic numbers vary along the periodic table?

» How can I predict the atomic radius of an element on the basis of periodic patterns?

» How is the periodic table a type of model?

EQUIPMENT

- computer with spreadsheet software

PERIODICITY

LAB 6B OBJECTIVES

» Determine periodic patterns by graphing atomic radii and atomic numbers.

» Predict atomic radii on the basis of periodic patterns.

» Relate the function and organization of the periodic table to the modeling nature of science.

✔ SCHEDULING

This lab activity is a great introduction to periodic trends. Consider scheduling it before covering Section 6.2. Students can see how the data shows the periodic trends.

If time is limited, consider having students graph only two of the trends. In this case it will be best to have them graph atomic radius and electronegativity values.

GRAPHING: COMPUTER OR BY HAND?

The lab activity is designed to use computers with spreadsheet software. Students can plot by hand, but the graphing will be tedious and will take time away from investigating the trends. There are many free spreadsheet programs available online. If the graphing must be done by hand, consider having students plot only through element number 54. You will have to modify expected answers to the questions if they do this.

If students need help on graphing, refer them to Appendix D in the Student Lab Manual.

PRE-LAB CHECK

1. Define *atomic radius. (The atomic radius of an element is the distance from its nucleus to its outermost energy level.)*

2. The sizes of atoms and ions and the forces between the nucleus and electrons are direct offshoots of electron configurations. Would you say that the electron configurations are periodic? Why or why not? *(Yes. The periodic table is arranged by atomic number and electron configuration. The number and placement of electrons follow a pattern related to the orbitals and the number of energy levels of a neutral atom.)*

3. State the periodic law. *(Many of the properties of the elements are periodic functions of their atomic number.)*

4. What is a periodic function? *(A periodic function passes through cycles with high and low values at regular intervals.)*

5. What periodic property will you be graphing? *(atomic radii)*

✔ PREDICTING RADII (QUESTION 5)

Students should use the elements to the left and right of Mg, Fe, and Pb to make their predictions in Question 5. This method is most successful when the elements in the question line up in the middle of a period and not at the end because there is a jump in the values at the end of a cycle. For example, if the students attempt to predict the radius of Na from the values before it (O, F, and Ne) and the values after it (Mg, Al, and Si), they will make wrong predictions. In this case it is best to look at the vertical trend, that is, the one within the group.

Procedure

ATOMIC RADIUS

Atomic radius is the average distance from the nucleus to the outermost electrons in an atom. Atomic radii range from a billionth to a trillionth of a meter. You can imagine how difficult it would be to measure these values. There are a number of techniques that scientists use to measure the radius of an atom. The atomic radii values in Table 1 are *covalent radii*. The covalent radius is one of the measures that scientists use for atomic radius.

A Using a spreadsheet program and the data from Table 1, create a scatterplot of the atomic radii given. Plot the atomic number on the *x*-axis and the atomic radii on the *y*-axis.

1. What general trends do you notice on the graph of atomic radii?

   There are five segments on the graph that start with a high radius and then decrease.

B Label the major peaks on the graph with the symbol of the appropriate element.

2. Which elements occupy the peaks in your graph of atomic radii?

   lithium, sodium, potassium, rubidium, and cesium

3. Where are these elements on the periodic table?

   in the first column

4. Are there any areas of your graph that seem to contradict the pattern?

   Students should notice that the pattern is broken when the transition metals enter the mix.

5. Using this graph, predict the radii of magnesium (12), iron (26), and lead (82).

   magnesium, 150 pm; iron, 140 pm; lead, 180 pm

6. Using your graph of atomic radii, indicate the general trend across a period and down a column of the periodic table.

   Across a period, the atomic radius generally decreases. Down a column, the atomic radius increases.

ELECTRONEGATIVITY

Electronegativity is a chemical property indicating the attraction of an atom for electrons that it shares with other atoms when it bonds. The electronegativity values in Table 1 are based on the Pauling scale.

C Create a scatterplot of the electronegativity values. Plot the atomic number on the *x*-axis and the electronegativity on the *y*-axis.

7. What general trends do you notice on the graph of electronegativity? How does it compare with the atomic radius graph?

   Five segments on the graph start with low electronegativity values and then increase. The graph looks like the inverse of the atomic radius graph.

D Label the major peaks on your electronegativity graph with the symbol of the appropriate element.

8. Which elements occupy the first four peaks of the cycles shown in your graph?

fluorine, chlorine, bromine, and iodine

9. Where are these elements on the periodic table?

in Group 17 (second to last column)

10. Using this graph, predict the electronegativity values for magnesium (12), iron (26), and selenium (34).

magnesium, 1.31; iron, 1.83; selenium, 2.55

11. Using your graph of electronegativity, indicate the general trend across a period and down a column of the periodic table.

Across a period, electronegativity values increase. Down a column, they decrease.

ELECTRON AFFINITY

Electron affinity is a measure of the amount of energy released when an electron is added to a neutral atom to make it a negatively charged anion. The electron affinity values in Table 1 are measured in kJ/mol.

E Create a third scatterplot with atomic number on the x-axis and electron affinity on the y-axis.

12. What general trends do you notice on the graph of electron affinity?

The data in this case seems much more random than in the first two graphs. Some students may notice that the first four highest values are the first four elements in Group 17.

FIRST IONIZATION ENERGY

The first ionization energy of an atom is the minimum amount of energy needed to remove an electron from a neutral atom. The first ionization energy values in Table 1 are given in electronvolts (eV).

F Make a fourth scatterplot showing atomic number on the x-axis and first ionization energy on the y-axis.

13. What general trends do you notice on the graph of first ionization energy? What graph does it look similar to?

This graph looks very similar to the graph of electronegativity, with five segments on the graph that start with low electronegativity values and then increase.

14. Do the peaks of this graph match those in the graph that you created in Step C? Explain.

Yes. The peaks are the same elements.

15. In the last three segments, there should be a series of four or five elements that significantly deviate from the pattern. Is there a pattern to the location of these elements on the periodic table?

The three deviations start at gallium (31), indium (49), and thallium (81). These elements are all in Group 13, which resumes the p sublevels on the periodic table following the transition metals.

16. Would it be useful to organize the elements in an alphabetized table? Why or why not?

It would be useful if we were simply trying to organize or find information relating to elements, but to order them alphabetically would obscure the periodicity that the periodic table demonstrates.

17. How is the periodic table an example of a model in science? (See pages 1–2 of your textbook.)

Students should recognize that the periodic table is a representation of something that we observe in the natural world—the properties of elements—in an effort to understand, describe, and relate these properties for the purpose of better using this information.

Going Further

18. Predict a relationship between atomic radius and electronegativity if you were to graph atomic radii on the x-axis and electronegativity values on the y-axis. Explain the basis of your prediction.

See TE margin for answer.

G Test your prediction in Question 18 by creating a scatterplot showing the atomic radii on the x-axis and electronegativity values on the y-axis.

19. Does your scatterplot support your prediction in Question 18? Explain.

Answers will vary. Make sure that students' answers connect the scatterplot to their predictions.

Question 18 Answer

Since the patterns on the scatterplots for atomic radii and electronegativity values seem to be inverses, it is likely that the relationship between them would be inverse as well. When an atom is smaller, its shared electrons would be closer to its nucleus and the protons in its nucleus would have a greater attraction for those electrons.

Name _______________________

TABLE 1

Element	Atomic Number	Atomic Radius (pm)	Electronegativity	Electron Affinity (kJ/mol)	First Ionization Energy (eV)
Li	3	145	0.98	60	5.4
Be	4	105	1.57	−50	9.3
B	5	85	2.04	27	8.3
C	6	70	2.55	122	11.3
N	7	65	3.04	−7	14.5
O	8	60	3.44	141	13.6
F	9	50	3.98	328	17.4
Na	11	180	0.93	53	5.1
Mg	12			−40	7.6
Al	13	125	1.61	42	6.0
Si	14	110	1.90	134	8.2
P	15	100	2.19	72	10.5
S	16	100	2.58	200	10.4
Cl	17	100	3.16	349	13.0
K	19	220	0.82	48	4.3
Ca	20	180	1.0	2	6.1
Sc	21	160	1.36	18	6.6
Ti	22	140	1.54	7	6.8
V	23	135	1.63	51	6.7
Cr	24	140	1.66	65	6.8
Mn	25	140	1.55	−50	7.4
Fe	26			15	7.9
Co	27	135	1.88	64	7.9
Ni	28	135	1.91	112	7.6
Cu	29	135	1.90	119	7.7
Zn	30	135	1.65	−60	9.4
Ga	31	130	1.81	41	6.0
Ge	32	125	2.01	119	7.9
Se	34	115		195	9.8
Br	35	115	2.96	325	11.8
Rb	37	235	0.82	47	4.2
Sr	38	200	0.95	5	5.7
Y	39	180	1.22	30	6.2
Zr	40	155	1.33	42	6.6
Nb	41	145	1.6	89	6.8
Mo	42	145	2.16	72	7.1
Tc	43	135	1.9	53	7.3
Ru	44	130	2.2	101	7.4
Rh	45	135	2.28	110	7.5
Pd	46	140	2.20	54	8.3
Ag	47	160	1.93	126	7.6

DATA TABLES

The data in this table is easily found online and can be copied into a spreadsheet program.

THE RELATIVITY OF ATOMIC RADII

We cannot assume that a given atom has a definite size under all conditions any more than we can assume that the electrons in an atom are in definite orbits. Since the radius of an isolated individual atom cannot be measured, the atomic radii listed in sources often vary according to the method used and the conditions under which they were measured.

The data points in this table are calculated values from a paper published in 1967 by Enrico Clement. They were obtained by different methods. For metals, the radius is one-half the distance between the atoms' centers in metal crystals, where metal atoms are packed like stacks of spheres. For nonmetals, covalent radii in molecules of elements were used, measured as half the distance between the atoms' centers in the bonded molecule. Since the noble gases do not form compounds and are monatomic, their radii were based on van der Waals radii, similar to the stacked spheres. This can account for the failure of noble gases to follow the general trend. Atomic radii in bonded molecules, where electrons are shared between nuclei, are expected to be smaller than those in unbonded atoms.

Despite the difficulties in obtaining the data, the general trend and the periodic nature of this property of atomic radii should be clear when the students create their graphs.

Element	Atomic Number	Atomic Radius (pm)	Electronegativity	Electron Affinity (kJ/mol)	First Ionization Energy (eV)
Cd	48	155	1.69	−70	9.0
In	49	155	1.78	37	5.8
Sn	50	145	1.96	107	7.3
Sb	51	145	2.05	101	8.6
Te	52	140	2.1	190	9.0
I	53	140	2.66	295	10.5
Cs	55	260	0.79	46	3.9
Ba	56	215	0.89	14	5.2
La	57	195	1.10	53	5.6
Ce	58	185	1.12	55	5.5
Pr	59	185	1.13	93	5.5
Nd	60	185	1.14	185	5.5
Pm	61	185	1.13	12	5.6
Sm	62	185	1.17	16	5.6
Eu	63	185	1.2	11	5.7
Gd	64	180	1.20	13	6.2
Tb	65	175	1.1	112	5.9
Ho	67	175	1.23	33	6.0
Er	68	175	1.24	30	6.1
Tm	69	175	1.25	99	6.2
Yb	70	175	1.1	−2	6.3
Lu	71	175	1.27	23	5.4
Hf	72	155	1.3	17	6.8
Ta	73	145	1.5	31	7.5
W	74	135	2.36	79	7.9
Re	75	135	1.9	6	7.8
Os	76	130	2.2	104	8.4
Ir	77	135	2.20	151	9.0
Pt	78	135	2.28	205	9.0
Au	79	135	2.54	223	9.2
Hg	80	150	2.0	−50	10.4
Tl	81	190	1.62	36	6.1
Pb	82		1.87	34	7.4
Bi	83	160	2.02	91	7.3
Po	84	190	2.0	183	8.4
At	85	115	2.18	78	9.3

Name ______________________

Date ______________________

» Compare single, double, and triple covalent bonds.

» Analyze the polarity of covalently bonded molecules.

The Name's Bond—Covalent Bond

Modeling Covalent Bonds

Building models of molecules may not be new to you. You may have built such models using toothpicks and colored marshmallows, gumdrops, or foam balls. These models can be useful, but certain aspects of molecular structure can't be modeled well by stiff toothpicks. You'll see that in this lab activity and again in Chapter 8. Molecular modeling sets that are purposefully designed for the task can tell us more about what happens when atoms get together to share electrons. In this lab activity, you'll explore multiple covalent bonds and polarity using one such product—a modeling set made by MolyMod.

Can physical models accurately represent what happens when atoms make covalent bonds?

QUESTIONS

» How do single, double, and triple covalent bonds compare?

» What affects the polarity of covalently bonded molecules?

EQUIPMENT

• MolyMod® modeling set

Procedure

EXPLORING SINGLE, DOUBLE, AND TRIPLE BONDS

MolyMod sets use different-colored balls to represent different elements: black (carbon), red (oxygen), and white (hydrogen). There are also different kinds of connectors for joining atoms together—rigid connectors for single bonds and flexible connectors for multiple bonds.

A Compare the MolyMod carbon, oxygen, and hydrogen atoms.

1. Other than color, describe the differences that you observe between the MolyMod "atoms" of each element.

See TE margin for answer.

____

____

____

2. Thinking in terms of atomic structure, what do you suppose the differences that you described in Question 1 represent?

See TE margin for answer.

____

____

____

SCHEDULING

Lab 7A can be done as an introductory lab activity prior to teaching the material in Section 7.2.

MOLECULAR MODELING SETS

Although this laboratory activity can be done with any molecular modeling set, we particularly recommend sets made by MolyMod. Their parts are durable, assemble easily, hold securely, and have the holes for connectors molded in such a way that modeled molecules will have the correct bond angles. The Inorganic/Organic Student Set (MMS-009) is suitable for this activity and can be purchased from various online sources.

Question 1 Answer

Students should observe that the hydrogen atoms are smaller than other atoms in the set. They should also note that each atom has a different number of holes for placing connectors: one for hydrogen, two for oxygen, and four for carbon.

Question 2 Answer

The smaller size of the hydrogen atoms represents the fact that real hydrogen atoms are smaller than most other atoms. The holes on each atom represent its "missing" electrons—the electrons that an atom of that element needs to fill its valence shell.

PRE-LAB CHECK

1. What information can molecular modeling sets provide about molecules? *(Molecular modeling sets show us the kinds of atoms and bonds in a molecule, along with their arrangements.)*

2. Which ball color repesents carbon? hydrogen? oxygen? *(black; white; red)*

3. What do the rigid connectors in a MolyMod set represent? What do the flexible connectors represent? *(single bonds; portions of double or triple bonds)*

4. (True or False) The polarity of a molecule cannot be determined by observing a molecular model. *(False)*

5. Where and how is ozone formed in Earth's atmosphere? *(in the upper atmosphere through bombardment by ultraviolet light)*

3. Observe the placement of the holes in the oxygen and carbon atoms. What significance will the fixed locations of the holes have for building molecules, and how is this different from, for example, toothpick and marshmallow models?

See TE margin for answer.

4. What do the gray connectors in your model represent?

shared pairs of electrons

The reasoning behind the locations of the holes in the various MolyMod atoms will be explored in the Chapter 8 material on bond geometry. For now it suffices to know that bonds between real atoms do tend to form in the positions suggested by the MolyMod atoms.

A SIMPLE SINGLE BOND

B Make a model of molecular hydrogen using two hydrogen atoms and a rigid single-bond connector. Leave this model intact for use later in the activity.

5. How many ways are there to build a model of molecular hydrogen using the indicated parts?

one

6. Is this an accurate representation of what happens when hydrogen bonds with itself? Explain.

See TE margin for answer.

COMPARING SINGLE AND MULTIPLE BONDS

C Build a model of ethane (C_2H_6) using the appropriate atoms and single-bond connectors. This model and those that you build in Steps D and F should be left intact for the discussion that follows.

D Build another ethane molecule. Once you finish it, examine it and consider what you would need to do to replace the carbon-carbon bond with a double bond.

7. What do you need to do to your ethane model to form a double bond in it?

   A hydrogen atom needs to be removed from each carbon so that the double bond can form between the carbon atoms.

E In your second ethane model, replace the single bond with a double bond on the basis of your answer to Question 7. You will need to use the flexible connectors between the carbon atoms. The resulting molecule is *ethene*.

F Now repeat Steps D and E to make a third molecule that contains a triple bond rather than a double bond. This is a molecule of *ethyne*.

Now compare your three models—ethane, ethene, and ethyne—paying particular attention to the bonds.

8. On the basis of bond strength, rank the three compounds from least to most strong and explain why you ranked them in that order.

   See TE margin for answer.

THINKING ABOUT POLARITY

Now let's consider how physical models such as those built using Moly-Mod sets can help us visualize polarity in molecules.

9. What causes polarity in a covalent molecule?

   See TE margin for answer.

G Take another look at your model of molecular hydrogen.

10. Is H_2 polar or nonpolar? How does the model show this?

    Molecular hydrogen is nonpolar. Since the two atoms in the model are both hydrogen, there can be no unequal sharing of electrons.

11. Now look at your model of ethane. Are its carbon-hydrogen bonds polar? Explain.

    Yes. Each is polar because there is unequal sharing of electrons between the central carbon atom and the four hydrogen atoms.

Question 8 Answer

ethane, ethene, ethyne; Each connector represents a shared pair of electrons. The creation of multiple bonds increases the number of electrons between the two nuclei, increasing the amount of electrostatic attraction between the two atoms. Therefore, there is greater force holding together the atoms with a triple bond than with either a double or single bond.

Question 9 Answer

Polarity is the result of the unequal sharing of electrons that occurs between atoms having different electronegativities. The unequal sharing can cause an uneven distribution of electrical charge throughout a molecule.

12. Look carefully at your models of ethane, ethene, and ethyne. Are any of them polar molecules? Explain.

   See TE margin for answer.

H Now remove two of the hydrogen atoms from one carbon atom in your ethane model and replace them with a double-bonded oxygen atom. The resulting molecule is *ethanal*, a compound classified as an *aldehyde*. (You will learn about aldehydes in Chapter 20.)

13. Is ethanal polar? Explain.

   See TE margin for answer.

14. Would *saturating* your ethanal molecule—by opening one of the pair of bonds in the double bond and adding a hydrogen atom to the oxygen— make the resulting molecule (ethanol) nonpolar? Explain.

   No. The oxygen atom would still pull shared electrons toward itself,

   creating a region of negative charge on the molecule.

Going Further

Ozone, O_3, is a compound formed in the earth's upper atmosphere when ordinary oxygen, O_2, is bombarded by ultraviolet light. The region of the atmosphere where the greatest concentration of ozone occurs is known as the *ozone layer*. The ozone layer helps protect the earth from the harmful effects of UV light by blocking most of the incoming UV rays before they reach the planet's surface.

I Create a model of ozone by connecting three oxygen atoms together using single bonds.

15. Sketch your ozone molecule in the space provided.

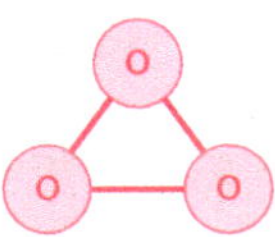

Name ___________________

Conduct an internet search using the keyword "ozone" to find information about the molecular structure of ozone.

16. Is the structure that you sketched in Question 15 correct? Explain.

See TE margin for answer.

____

____

____

17. How does your answer to Question 16 illustrate a shortcoming with the MolyMod system?

See TE margin for answer.

____

____

____

____

____

18. If the MolyMod system has this particular limitation, why is it still used for modeling? Can you think of any other models that are still used as educational tools despite their limited ability to portray reality? (*Hint*: Think back to Chapter 5.)

See TE margin for answer.

____

____

____

____

____

____

____

Question 16 Answer

No. Ozone is an example of a resonating hybrid structure in which each of the outer oxygen atoms forms one and a half bonds with the central atom. It is impossible to make this structure with the regular MolyMod oxygen atoms since each can make bonds in only whole-number quantities, that is, one or two, not one and a half.

Question 17 Answer

The MolyMod system is limited in the kinds of bonding that it can accurately illustrate. In most instances, oxygen makes two covalent bonds. The half-bonds in ozone result in an unstable molecule that breaks down rapidly. Even though many MolyMod kits include an oxygen atom that can make four bonds, thus allowing for the construction of an ozone model, they still do not accurately portray what is happening with the bonding in ozone.

Question 18 Answer

Despite its limited usefulness in depicting some aspects of chemical bonding, such as resonance structures, the MolyMod system is still very useful for modeling many of the kinds of bonds and compounds that chemistry students routinely encounter in their studies. This is similar to the limitations inherent in the Bohr model of atoms. Though the Bohr model was quickly superseded by the quantum model, the latter is a far more difficult model to comprehend. In many instances, simplified Bohr model representations are sufficient to model what is happening at the level of atomic structure. In each instance, the workability of the modeling system is a key in determining its usefulness.

Name __________________

Date __________________

Bulletproof Chemistry

Relating Chemical Bonds and Physical Properties

A woman approached a teller one April morning in 2012 at a BB&T bank in Smyrna, Georgia. In the slot under the wall of glass, she passed to the teller a typed demand note. When the teller moved too slowly, the woman pulled a gun and demanded money. Happily, the glass was bulletproof. Eventually, the would-be bank robber left without a dime, foiled by chemistry!

Chemical bonds are responsible for the physical properties of substances such as bulletproof glass. We can use these properties to keep our homes warm, cook our food, and make great running shoes.

How can we use physical properties to identify bond types in substances?

Ionic, covalent, and metallic bonds play a big role in determining the physical properties of substances. If you can observe the physical properties of a substance, you can often determine its bond type. In the table on the next page, notice the properties that each bond type typically produces in substances.

In this lab activity, you will examine the melting point, solubility, and conductivity of several substances in order to determine the types of bonds they contain. These bonds and the properties they create could make the difference in the way your next grilled cheese sandwich turns out or how warm your house stays on a cold day.

QUESTIONS

» How do the properties of compounds relate to their chemical bonds?

» Is there a relationship between the bond types in a compound and its melting point?

» Is there a relationship between the bond types in a compound and its solubility and conductivity?

» How can I identify an unknown substance on the basis of empirical evidence?

EQUIPMENT

- laboratory burner and lighter
- conductivity tester
- ring stand and ring
- wire gauze
- evaporating dish
- test tubes (6)
- test tube rack
- wash bottle with distilled water
- weighing paper
- unknown substances (3)
- acetone

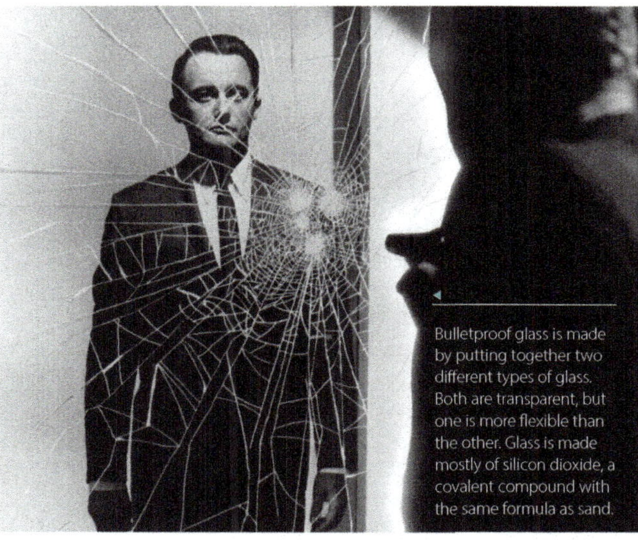

Bulletproof glass is made by putting together two different types of glass. Both are transparent, but one is more flexible than the other. Glass is made mostly of silicon dioxide, a covalent compound with the same formula as sand.

Bulletproof Chemistry | 59

LAB 7B OBJECTIVES

» Relate physical properties of compounds to their chemical bonds.

» Compare the melting points of substances on the basis of bond types.

» Compare the solubility and conductivity of substances on the basis of bond type.

» Identify unknown substances on the basis of empirical evidence.

EQUIPMENT NOTES

Use the following substances for the three unknowns and label them by number only: (1) a salt such as table salt ($NaCl$), salt substitute (KCl), or any other safe ionic compound; (2) iron filings (Fe), granular zinc (Zn), granular copper (Cu), or any finely divided, safe metal; and (3) paraffin wax or beeswax. Quantities the size of a pea should be sufficient for each group. Sodium chloride, potassium chloride, and paraffin wax (used for sealing canning jars) are available from grocery stores. Beeswax is available at craft and hobby stores.

Substitutions

1. A Pyrex custard dish can be used instead of the evaporating dish.

2. Small baby food jars can replace the test tubes since they will not be heated.

3. Although some nail polish removers contain acetone, they do not work well for this experiment. Acetone can be purchased in most home improvement stores in the paint department.

Assembling a Conductivity Tester

Assemble the conductivity tester according to the diagram on page 61. Strip the two ends of the insulated wires. Connect one end of one of the wires to the positive pole of a 9 volt battery, and then connect one end of the other wire to the negative pole of the battery. Put electrical tape across the wires at the poles to keep them connected.

On the positive terminal of the battery, wire in a 1 kΩ resistor. After the resistor, add more wire with the ends stripped, and wire in a light-emitting diode (LED). Make sure that you get the LED wired in correctly since they do have a polarity. You can identify the polarity of a new LED by looking at the wires. The positive end of the LED is longer.

PRE-LAB CHECK

1. What kind of atom forms metallic bonds? *(a metal atom)*

2. What kind of atom forms ionic bonds? *(an ion, often a metal with a nonmetal)*

3. What kind of atom forms covalent bonds? *(a nonmetal)*

4. What causes the physical properties of substances? *(the bond type and chemical properties of the substance)*

5. If a substance has a high melting point and is insoluble in an organic solvent, what type of bond does it contain? *(either ionic or metallic)*

6. If a substance has a high melting point and is soluble in water, what type of bond does it contain? *(ionic)*

7. How will a conductivity tester indicate that an electrical current is flowing? *(The light will glow.)*

You may just use trial and error to determine how your LED should be wired in. To the end of the LED, add more insulated wire with both ends stripped. Touch the stripped ends of your wires to see whether the LED illuminates to confirm that everything is wired correctly.

Cover all wire junctions with electrical tape. Secure the loose wires, LED, and resistor to the battery with electrical tape to prevent the electrical components from being pulled apart during handling.

Characteristics \ Bond Type	Ionic Bond	Covalent Bond	Metallic Bond
Electrons	transferred	shared between two atoms	free among all atoms
Smallest Unit	formula unit	molecule	atom
Melting Point	high	relatively low	relatively high
Solubility	often soluble in polar solvents but insoluble in nonpolar solvents	soluble in polar solvents if polar covalent; soluble in nonpolar solvents if nonpolar covalent	insoluble in both polar and nonpolar solvents
Conductivity	conducts electricity when melted or dissolved	usually does not conduct electricity	good conductor of electricity

Procedure

MELTING POINTS

A Obtain small samples of the three unknown substances provided by your teacher. Put each sample on a separate piece of weighing paper.

B Set up an apparatus as shown below.

C Place a small amount (about the size of an un-cooked grain of rice) of Unknown 1 in an evapo-rating dish. Set the dish on the wire gauze and gently heat the contents.

D If the unknown does not readily melt, heat it strongly for a minute or two. Describe how easily the substance melts in Table 1.

E Repeat Steps C and D for Unknowns 2 and 3, and record your observations in Table 1.

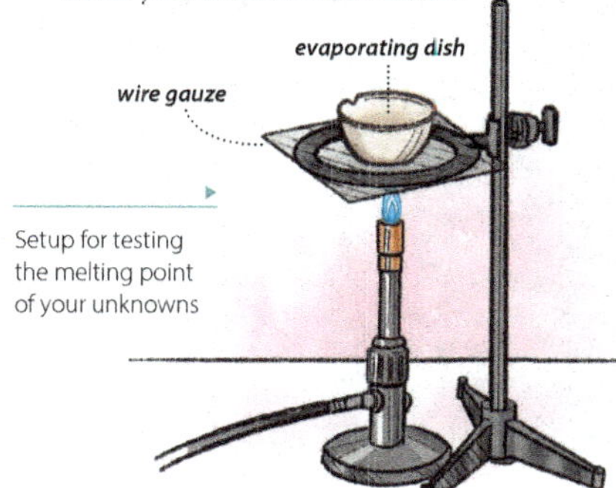

1. Using the table in the introduction, rank the types of bonds by increasing bond strength on the basis of melting points.

covalent, metallic, ionic

2. Ionic compounds and metallic compounds have higher melting points than covalent compounds. Why do you think this is so?

See TE margin for answer.

3. On the basis of your observations, make a hypothesis about which unknown has metallic bonds, which has ionic bonds, and which has covalent bonds before you proceed with more testing.

Students should conclude that Unknowns 1 and 2 are either metallic or ionic and that Unknown 3 is covalent since it has a low melting point.

SOLUBILITY

F Try to dissolve Unknown 1 by placing a small amount of the substance (about the size of an uncooked grain of rice) in a test tube and adding distilled water to fill the test tube one-quarter full. Record the relative solubility in Table 1. You'll need this sample for the conductivity test, so don't discard it! You may want to label your test tube.

G Repeat Step F for relative solubility with Unknowns 2 and 3. Record your observations in Table 1.

4. Is water a polar or a nonpolar liquid?

   polar

One general rule about dissolving substances is that "like dissolves like," meaning that polar liquids tend to dissolve polar substances, and that nonpolar liquids tend to dissolve nonpolar substances. This will be discussed in more detail in Chapter 14.

5. Would water be more likely to dissolve ionic or covalent compounds? Explain.

   See TE margin for answer.

   ____

   ____

   ____

Before you continue with the second part of the procedure, **make sure that no one in the laboratory has a laboratory burner lit.** You will be using acetone, a nonpolar liquid used in fingernail polish remover, and it is extremely flammable! You will not need a laboratory burner for the rest of this activity.

H Repeat Step F for each unknown using acetone as the solvent. Record your observations in Table 1.

6. Acetone is a nonpolar liquid. Is it more likely to dissolve ionic or covalent compounds? Explain.

   Acetone is more likely to dissolve nonpolar

   covalent compounds since it is nonpolar.

   Acetone should not dissolve ionic compounds.

   ____

7. On the basis of your observations, make a hypothesis about which unknown is metallic, which is ionic, and which is covalent before you proceed with more testing.

   See TE margin for answer.

   ____

   ____

   ____

CONDUCTIVITY

Now you will test to see whether your unknowns allow electricity to pass through them easily, a property known as *conductivity*.

I Lower the two electrodes of the conductivity tester (see right) into the aqueous solution for Unknown 1. If the sample is conductive, the light will illuminate. Record your observations in Table 1.

J Using the wash bottle, rinse the electrodes with distilled water and then dry the electrodes.

K Repeat Steps I and J for the remaining five test tubes.

L Test the conductivity of the remaining small amounts of each unknown solid by touching the electrodes to each one. Again, rinse and dry the probes between each test. Record your observations in Table 1.

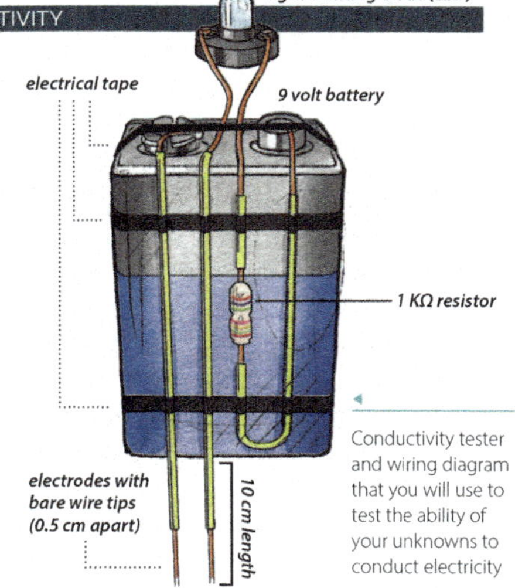

Conductivity tester and wiring diagram that you will use to test the ability of your unknowns to conduct electricity

---

⚠ BURNERS OFF!

Make sure that all students have laboratory burners turned off before they begin to work with the highly flammable acetone. Burners should remain off for the remainder of the lab activity. You may even consider collecting the burners and lighters before students start this next part. Also make sure that the room is well ventilated.

ⓝ RINSE AND REPEAT

Ask students, "When scientists clean glassware in the laboratory, they often wash it with water and then rinse it with acetone afterward. Why do you think they do this?" *(Rinsing with both a polar and a nonpolar solvent helps ensure that both kinds of substances are removed from the glassware.)*

Question 5 Answer

Water is more likely to dissolve ionic compounds because it is highly polar, attracting the ions in ionic compounds. The solubility of a covalently bonded substance in water depends on the polarity of the substance.

Question 7 Answer

Students should confirm that Unknown 3 is a covalent compound. Unknown 1 is probably ionic since it dissolved in water. That leaves Unknown 2 as a metallic compound.

⚗ WASTE DISPOSAL

If you use sodium or potassium chloride salts, simply filter out wax and iron filings from your aqueous waste jar. Flush the filtrate down the drain with plenty of water, and throw away the solids in the trash can.

For your acetone waste jar, filter out wax and metal filings and allow them to dry before throwing the solids in the trash can. Pour the filtrate over paper towels, and allow the towels to dry before throwing away.

THE BIG REVEAL

Students will need to know the identity of their unknowns to finish answering the questions in this activity. Discretely reveal the identities and categories, perhaps by using small folded pieces of paper. Avoid letting the students know whether you have given them all the same unknowns.

Question 12 Answer

Answers will vary. In the first place, all life on Earth as we currently know it would not exist since all life depends on ions operating in solutions. Life in a world free of ionic solutions would therefore be very different. Students may also mention that an absence of dissolved salts would make all surface water fresh water and prevent the use of ionic compounds as preservatives and disinfectants. Other answers are possible.

Analysis

8. Identify the bond type in each of your three unknowns. Check with your teacher to learn the identities of the unknowns and to see whether your hypotheses are correct.

   table salt—ionic bond; iron filings—metallic bond; wax—covalent bond

9. Did the substances that you used in this lab activity follow the "like dissolves like" rule? Explain.

   Yes. Water is polar, and it dissolved the ionic table salt. Acetone is nonpolar, and it dissolved the wax.

10. Did the "solution" that you formed by mixing the metal and water conduct electricity? How do your results compare with those for the solid metal?

    No. Only the solid metal conducted electricity.

11. Why does an ionic solid conduct electricity only in the molten state or in solution?

    Only in a solution or in the molten state are the charged particles free to move.

Going Further

12. Imagine that ionic compounds such as table salt did *not* dissolve in water. How would that affect your life?

    See TE margin for answer.

13. Think of some ways that we can create substances with customized physical properties for special uses, like airplane wings, electric wires, and insulation for homes.

    We can use chemical changes to form bonds that have the properties we need for specific purposes.

14. Considering what you wrote in Question 13, explain how God uses chemists to provide for our needs.

God uses chemists to save lives, for example, in the medical industry.
He also uses them to improve our ease of living, as with cooking,
clothing, and fuel. Materials in the things we buy and use, such as
plastic, metal, ceramic, and even fabric, are important products of
chemistry that God provides for us through the work of chemists.

TABLE 1 *Properties of Unknown Substances*

	Unknown 1	Unknown 2	Unknown 3
Melting	hard to melt	hard to melt	easy to melt
Solubility in Water (H_2O)	soluble	insoluble	insoluble
Solubility in Acetone (CH_3COCH_3)	insoluble	insoluble	soluble
Conductivity of Aqueous Mixtures	conductive	nonconductive	nonconductive
Conductivity of Acetone Mixtures	nonconductive	nonconductive	nonconductive
Conductivity of the Solids	nonconductive	conductive	nonconductive

8A LAB

The Shape of Things

Modeling Molecules

Scale models are helpful to show the important features of an object. We use scale models for both large and small objects. We scale large objects down so that we can look at the entire object at a glance. These scaled-down models include model cars, trains, and ships, models of the solar system, and models of buildings. The specific scale that we use depends on the size of the actual object and the size of model that we want to make. For models of cars it is common to use a scale of 1:25; that is, 1 cm on the model represents 25 cm on the actual car. For ships, you would need a smaller scale, such as 1:200. Of course, if you were making a tabletop model of the solar system, you would need a tiny scale—1:1.5 × 10^{13}.

What determines the shape of a molecule?

When we study objects that are very small, such as molecules, we need scaled-up models. You may have done this before in a biology class, for example, when making a model of a cell. In this lab activity you will make scaled-up models of molecules. This is a very large-scale model because 1 cm on your models will represent $2 × 10^{-9}$ cm in a real-world molecule.

In this lab activity you will make physical three-dimensional models of molecules. These models will give you the opportunity to make connections between the Lewis structures that you have been drawing in class and the three-dimensional molecules that they represent.

EQUIPMENT

- **foam balls, 7.5 cm (9)**
- **foam balls, 2.5 cm (29)**
- **toothpicks (36)**

QUESTIONS

» How do I draw Lewis structures of molecules?

» How do I make 3D models of molecules?

» How does the shape of a molecule affect its polarity?

LAB 8A OBJECTIVES

» Draw Lewis structures of molecules.

» Assemble three-dimensional models of selected molecules.

» Assess the polarity of molecules on the basis of electronegativity and molecule shape.

EQUIPMENT NOTES

Not Modeling Kits

You've probably noticed that this lab activity calls for old-fashioned foam balls and toothpicks rather than the MolyMod set that was used in Chapter 7. The reason for this is that MolyMod parts *already* restrict bonds to the locations within models that will produce the correct bond angles, eliminating an element of discovery from the activity. Foam balls and toothpicks allow the student to experiment with different bond locations and angles.

Other Options

There are many alternatives for the foam balls in this activity. Modeling clay works well. Some teachers like to use marshmallows and gumdrops. If you opt to use food, remind students that they are never to eat in the laboratory.

PRE-LAB CHECK

1. What are the three types of bonds that your textbook discusses? *(metallic, ionic, and covalent)*

2. What type(s) of bond(s) does this lab exercise emphasize? *(covalent polar and nonpolar, or simply, covalent bonds)*

3. What is a covalent bond? *(a chemical bond formed by the sharing of valence electrons in overlapping orbitals)*

4. What two things do you have to consider when determining molecular shape? *(the number of electron regions and the number of bonded nuclei around the central atom)*

5. Describe what is meant by polar bond. *(In a polar bond, the electrons are not shared equally. The result is a semi-ionic condition with partially positive and partially negative charges.)*

6. If a molecule has polar bonds, is it always a polar molecule? Why or why not? *(A molecule containing polar bonds is not polar if the bonds are symmetrically arranged.)*

PHOTOGRAPHING THE MODELS

You may choose to have students take photos or screenshots of the molecules that they build. They could put these in a lab report that they write or simply use them to refer to when answering the questions.

UNDERSTANDING BONDS

To help tie everything together about covalent bonding, you may want to discuss molecular bond theories such as valence bond theory and hybridization. If the overlapping orbitals are both *s* orbitals, then a sigma bond is produced. If they are both *p* orbitals that overlap end to end, then they also produce sigma bonds. If they are both *p* orbitals that are side by side, then a pi bond is produced. But if the original orbitals are of different types such as *s* and *p*, yet they form equivalent energy bonds, then they are said to mix, or hybridize, to form hybrid orbitals of equal energies. See page 176 of the Student Edition for illustrations.

Procedure

Consider formaldehyde (CH_2O). To make a model of this organic compound, we begin by drawing its Lewis structure.

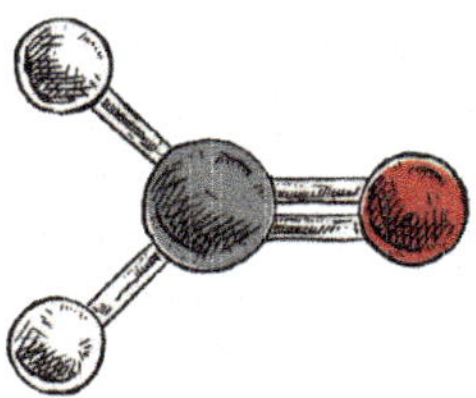

After drawing the Lewis dot structure, we observe that carbon is the central atom. It has three regions of electrons around it, two single bonds and a double bond, the double bond acting as a single region of electrons. Each of the single bonds connects a hydrogen atom with the central carbon atom, and the double bond connects the carbon atom with the oxygen atom.

It's possible to make a model showing three regions of electrons that are spaced as far apart as possible. This molecule is trigonal planar—the bonds lie in the same plane and point to the three corners of a triangle. By comparing electronegativity values, you can determine that the molecule contains polar bonds and that the molecule is polar overall.

Now it's your turn to explore some molecular models. To begin with, let's look at some of the basic possible shapes.

DETERMINING BOND ANGLE

A. Take two of the large foam balls, representing central atoms, and insert two toothpicks, representing electrons, either bonding or unbonded, into each ball so that the toothpicks are as far from each other as possible. What is the approximate angle between the toothpicks? Record this bond angle in Table 1.

B. Repeat Step A, but use three large foam balls and insert three toothpicks in each.

C. Repeat Step A, but use four large foam balls and insert four toothpicks in each.

1. What determines the bond angles in each of your models? Explain how Steps A–C demonstrate this fact.

   The number of regions of electrons around the central atom determines bond angles. Since there are no atoms bonded to the central atom, it is clear that only the regions of electrons factor into bond angle.

DETERMINING BASIC SHAPES

D. To the first foam ball from Step A attach one small foam ball to one of the toothpicks, representing an atom bonded to the central atom. To the second large foam ball, attach a small foam ball to each of the toothpicks. What shape would you call each of these molecules? Record your observations in Table 1.

E. Repeat Step D for the three large foam balls from Step B, adding one, two, and three small foam balls to each of the molecules.

F. Repeat Step D for the four large foam balls from Step C, adding one, two, three, and then four small foam balls to each of the molecules.

G. Disassemble your models so that you can use the parts to model real molecules.

MODELING ACTUAL MOLECULES

H Table 2 includes the molecules that you will now model. Begin by drawing the Lewis structure for each molecule in Table 2.

I Count the regions of electrons around each central atom. A bond counts as one region, regardless of whether it is a single or multiple bond, and each lone pair counts as one region. Record your counts in Table 2.

2. What do you notice about the number of valence electrons for elements that are usually central atoms?

Atoms that function as central atoms in a molecule usually have

between three and five valence electrons.

J Identify how many atoms are bonded to each central atom and record your counts in Table 2.

K Now let's make some models, starting with formaldehyde. Again, use the large foam balls for the central atom, the toothpicks for regions of electrons, and small foam balls for outer atoms. Record the finished molecule's shape in Table 2.

L Identify any polar bonds in your molecule. Remember that when two different atoms are bonded together, the bond is polar. Record your observations in Table 2.

3. In the formaldehyde molecule on page 66, why do carbon and hydrogen form a slightly polar bond?

Students should focus on the difference in electronegativities and

notice that the difference between carbon and hydrogen is only 0.4,

making this a mildly polar bond.

M Study the shape of your molecule and the locations of any polar bonds to determine whether the entire molecule is polar. Record your conclusions in Table 2.

N Repeat Steps K–M for the additional molecules listed in Table 2. Be sure to read ahead—some of the following questions pertain to a particular model.

4. Consider the model for methane. If carbon and hydrogen form a slightly polar bond, why is methane nonpolar?

Methane has a tetrahedral shape, meaning that the effects of the four

slightly polar bonds cancel each other, resulting in a nonpolar

molecule.

Make sure that students understand the difference between *charge* and *polarity*. The ammonium ion is nonpolar because its four bonded hydrogen atoms are evenly distributed about the central nitrogen atom. But the group as a whole contains an excess positive charge, making it a cation. This excess positive charge is not associated with any particular hydrogen atom within the ion but rather is distributed over the entire structure.

Question 6 Answer

Yes. The pyramidal shape has unbalanced polar bonds, making it a polar molecule. The addition of the hydrogen ion would change it to a tetrahedral shape with balanced polar bonds, resulting in a nonpolar ammonium ion.

Question 8 Answer

Lewis structures can be misleading because they are two-dimensional models. The three-dimensional model for dichloromethane can be represented by either Lewis structure. But in three-dimensions, there is no way to get the pulls of either the chlorine or hydrogen atoms to cancel. Therefore, the asymmetric pulling of both the hydrogen and chlorine atoms will result in a polar molecule.

---

5. If you were to add a hydrogen ion (H^+) to the ammonia model that you made, what substance would you have? Would the additional hydrogen ion cause the shape of the new molecule to be different from ammonia? Explain.

   This addition would form the ammonium ion (NH_4^+). Yes, the shape would change since the shape of ammonia is pyramidal and the ammonium ion is tetrahedral.

6. On the basis of your answer to Question 5, do you think that the addition of a hydrogen ion (H^+) would change the polar nature of ammonia? Explain.

   See TE margin for answer.

7. What is the difference between the carbon-oxygen bond in your model of methanol and the one found in carbon dioxide?

   In methanol, the carbon-oxygen bond is a single bond. In carbon dioxide it is a double bond.

   Suppose that two students who are working through this lab activity show you their Lewis structures for dichloromethane. The first student explains that her Lewis structure (below left) demonstrates that the molecule is polar because the polar chlorine-carbon bonds and polar hydrogen-carbon bonds pull electrons unequally. The second student claims that his Lewis structure (below right) demonstrates that the molecule is nonpolar because the pulls of the two chlorine atoms balance, as do the pulls of the two hydrogen atoms, leaving a balanced, nonpolar molecule.

$$\begin{array}{ccc} & Cl & \\ & | & \\ Cl- & C & -H \\ & | & \\ & H & \end{array} \qquad \begin{array}{ccc} & H & \\ & | & \\ Cl- & C & -Cl \\ & | & \\ & H & \end{array}$$

8. Use what you learned from your three-dimensional model to explain the polarity of this molecule.

   See TE margin for answer.

9. On the basis of what you have observed, state which molecular shape(s) will *always* produce polar molecules when the central atom is bonded to an atom (or atoms) of a different element.

   Bent and trigonal pyramidal molecular shapes will always be polar molecules. Linear molecules of two atoms will always be polar if they contain two different atoms.

10. Which molecular shape(s) will produce nonpolar molecules if a central carbon atom is bonded to atoms of a different element?

    Tetrahedral, trigonal planar, and three-atom linear molecules will always be nonpolar if the central atom is bonded to atoms of the same element.

TABLE 1 *Molecular Shapes*

Number of Electron Regions	Bond Angle	Nuclei around Central Atom	Geometry
2	180°	1	linear
		2	linear
3	120°	1	linear
		2	bent
		3	trigonal planar
4	109.5°	1	linear
		2	bent
		3	pyramidal
		4	tetrahedral

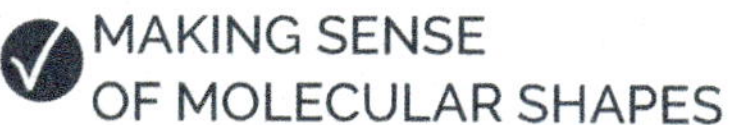

MAKING SENSE OF MOLECULAR SHAPES

Students often have trouble visualizing molecules in space. The goal of this lab activity is to help them do that once they have handled physical models. The last two questions of this activity help them sum up what they have learned so that they can better identify molecular shapes and molecular polarity.

DATA TABLES

The answers given in the data tables are *not* sample data—they are correct answers. Students *should* come up with these answers.

TABLE 2

IUPAC Name and Formula	Formalde-hyde CH_2O	Methane CH_4	Water H_2O	Ammonia NH_3	Carbon Dioxide CO_2	Methanol CH_3OH	Hydridoni-tridocarbon HCN	Dichloro-methane CH_2Cl_2	Carbon Tetrachloride CCl_4
Lewis Structure	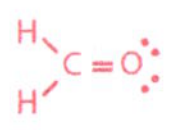	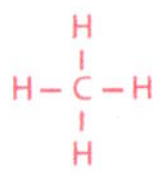	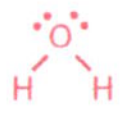	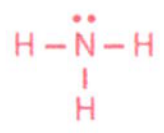	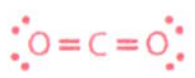	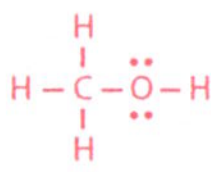	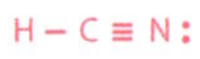	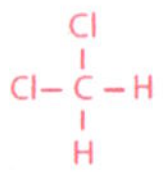	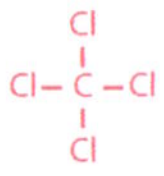
Regions of Electrons around Central Atom	3	4	4	4	2	4	2	4	4
Atoms around Central Atom	3	4	2	3	2	4	2	4	4
Molecular Shape	trigonal planar	tetrahedral	bent	pyramidal	linear	tetrahedral	linear	tetrahedral	tetrahedral
Polar Bonds?	yes	yes	yes	yes	yes	yes	yes	yes	yes
Polar Molecule?	yes	no	yes	yes	no	yes	yes	yes	no

8B LAB

Change of Address

Investigating Molecular Orbitals

Name

Date

One of the great challenges of chemistry is the fact that almost everything we are studying is too small to see, even with magnification. To understand what is going on inside atoms and molecules, scientists observe how matter interacts with its surroundings and then hypothesize about what is happening at the atomic and molecular levels. You can understand why the concept of modeling is so important in chemistry.

QUESTIONS

» How do I model molecular orbitals?

» How does molecular orbital theory explain that so few elements are found unbonded in nature?

EQUIPMENT

- none

How does the molecular orbital theory relate to orbital notation for atoms?

In the last few chapters in your textbook you have been learning about the structure of atoms, with much of your time spent learning about where the electrons are located and what they are doing. The electrons, particularly the valence electrons, are key actors in chemical reactions and chemical bonding. In this lab activity, you will look more closely at the molecular orbital theory, which postulates that the electrons in the atoms can move between atomic orbitals. You might say that they experience a change of address. Let's see what the molecular orbital theory says about the valence electrons in a molecule.

According to the molecular orbital theory, all the valence electrons from bonded atoms become delocalized from their atomic orbitals and relocate to newly formed molecular orbitals. This theory helps us understand why some atoms bond, while other atoms do not. Molecular orbital theory can also explain why some atoms bond with stronger double and triple bonds. One of the most amazing things about this model is that it accurately predicts the magnetic behavior of bonded substances. But let's not get ahead of ourselves!

LAB 8B OBJECTIVES

» Create molecular orbital diagrams of diatomic molecules.

» Evaluate the evidence that few substances are found in nature as individual atoms on the basis of the molecular orbital theory.

✓ MORE MODELING

This lab activity is designed as a follow-up to Lab 8A. In it students will be going beyond the shape of the molecules and modeling how the electrons are arranged in the molecule. The activity is designed to have students think about the way that atomic orbitals change into molecular orbitals and how that affects the structure and behavior of the molecules.

PRE-LAB CHECK

1. What is a model? What is the purpose of using models in science? *(A model is a simplified representation of something in the real world. The purpose of using models in science is to explain or describe the natural world.)*

2. Why do we spend so much time working with models in chemistry? *(Most of what occurs in chemistry is on the molecular, atomic, or subatomic level. These are too small to be seen visually, so we create models to explain or describe what is happening.)*

3. What are sigma bonds? *(bonds in which the orbitals overlap end to end)*

4. What are pi bonds? *(bonds in which the orbitals overlap side to side)*

5. What bond order(s) result(s) in a stable molecule? *(Any bond order of 1 or higher results in a stable molecule.)*

ANTIBONDING ORBITAL

The concept of antibonding orbitals will be new for students. It should be emphasized that these orbitals work against a stable molecule forming due to their position around the molecule.

Procedure

MOLECULAR ORBITALS FOR THE FIRST ENERGY LEVEL

Let's start by looking at a simple example, H_2. We draw the orbital notation for each of the hydrogen atoms that will form H_2.

Hydrogen 1 $\quad \uparrow \atop 1s$ $\qquad\qquad$ $\uparrow \atop 1s$ $\quad$ *Hydrogen 2*

Since each hydrogen atom contributes an orbital, there will be two molecular orbitals in H_2. We draw the two orbitals in between the atomic orbitals of the hydrogen atoms. Just like atomic orbitals, molecular orbitals have different energy levels. When drawing molecular orbitals, we indicate the lower energy orbitals by placing them at the bottom of the diagram. Therefore, starting at the bottom we have a σ1s orbital. The Greek letter *sigma*, σ, indicates that the electrons in this orbital will bond via a sigma bond. The 1s indicates that this orbital formed from the 1s orbitals of the bonding atoms. The σ1s orbital is a bonding orbital—one that works toward successful bonding of the atoms. The second molecular orbital is designated σ1s* and is located above the first orbital because it is a higher energy orbital. The asterisk indicates that this is an antibonding orbital and actually works against the atoms' attempt at bonding. Let's take a look at these orbitals.

Notice that the σ1s orbital is lower than the atomic orbitals, indicating a lower energy state, while the nonbonding orbital is above the atomic orbitals—a higher energy state.

Now we need to put electrons in the orbitals. Again, just like with atomic orbitals, we always fill lower energy orbitals first. Therefore, the electron from each atom will fill the σ1s orbital. Our finalized diagram is shown below.

What does this diagram tell us? From it, we can tell whether the molecule is stable; if so, it will likely form and stay together. Stability is dependent on the number of electrons in bonding orbitals versus the number of electrons in nonbonding orbitals. To determine stability, we calculate the bond order.

$$BO = \frac{\text{electrons in bonding orbitals} - \text{electrons in antibonding orbitals}}{2}$$

For our hydrogen molecule we calculate the bond order to be 1.

$$BO = \frac{2-0}{2} = 1$$

Any bond order greater than 0 indicates a stable, likely-to-form molecule. Also notice that there are no unpaired electrons in the molecular orbitals. This tells chemists that the material is likely to be nonmagnetic, or *diamagnetic*. If there were any unpaired electrons, the material would be attracted to magnets; that is, it would be *paramagnetic*.

A Use Drawing Area 1 to draw the orbital notation for each of two helium atoms that attempt to form He_2. As you become more comfortable with drawing molecular orbital diagrams, you will be able to skip this step.

1. How many total atomic orbitals are there in the two helium atoms? What does this tell us about the number of molecular orbitals in He_2?

There is one orbital in each of the helium atoms for a total of two orbitals. Therefore, there will be two molecular orbitals in He_2.

____

B Draw the molecular orbitals for the He_2.

C Add the electrons to the molecular orbitals.

D Determine the bond order and indicate whether it is stable or unstable and paramagnetic or diamagnetic. Record your observations with your drawings.

E Using Drawing Area 1, repeat Steps A–D for H_2^-. Don't forget the extra electron that gives it a charge of 1−.

MOLECULAR ORBITALS FOR THE SECOND ENERGY LEVEL

The molecular orbital diagrams that we have considered so far have been fairly simple. How do the diagrams change once we are in the second energy level, with its s and p suborbitals? Let's look at the molecular orbital diagram for C_2.

We start by drawing the orbital notation for each of the carbon atoms that will form C_2, but we include only the orbitals from energy level 2 since that is where carbon's valence electrons are.

Each carbon atom has four orbitals; therefore the C_2 molecule will have eight molecular orbitals, four bonding and four nonbonding. Starting at the bottom, we have the $\sigma 2s$, and $\sigma 2s^*$, a pair of $\pi 2p$ orbitals, and a $\sigma 2p$ orbital. We finish with a pair of $\pi 2p^*$ orbitals and a $\sigma 2p^*$ orbital. The p orbitals are often designated with x, y, and z subscripts to signify the three dimensional axes. As in the earlier orbital diagrams, the nonbonding orbitals are at a higher energy state than their associated bonding orbitals. A note about the order of the orbitals is important to state here: The arrangement shown here works for boron, carbon, and nitrogen; when dealing with oxygen, fluorine, or neon, the $\pi 2p$ orbitals and the $\sigma 2p$ orbital are reversed.

✓ MOLECULAR ORBITAL THEORY HELPS EXPLAIN MAGNETIC PROPERTIES

Point out to students the additional workability of this model in that it helps to explain another observation about matter. This model explains why different molecular substances have different magnetic properties. The more observations that a model can explain, the more workable it is.

✓ MOLECULAR ORBITAL ORDER

Students will often have trouble remembering the order of the molecular orbitals. Remind them that the order is dependent on the atoms involved in bonding.

Filling the orbitals is very much like drawing atomic orbital notation. Fill lower energy levels first, adding one electron to each orbital prior to adding a second electron to any of them. Electrons in any particular orbital must have opposite spins.

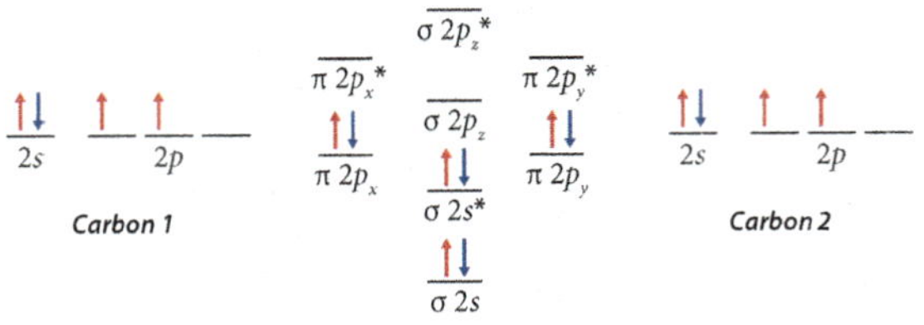

$$\text{Carbon 1} \qquad \qquad \text{Carbon 2}$$

We calculate the bond order for dicarbon to be 2.

$$BO = \frac{6 - 2}{2} = 2$$

The bond order of 2 indicates that the molecule is stable. There is also a relationship between bond order and multiple bonds. While not a hard and fast rule, the relationship does hold true for dicarbon. Dicarbon has a bond order of 2 and indeed does have a double bond in it. We can also see that, since there are no lone electrons in dicarbon, it is diamagnetic.

F Using Drawing Area 1, repeat Steps A–D for for Li_2, Be_2, O_2^-, and N_2.

MOLECULAR ORBITALS FOR MOLECULES WITH DISSIMILAR ATOMS

Up to this point we have looked only at diatomic molecules that contain two of the same type atom. Now let's consider the molecular orbital diagrams for a molecule made of two different atoms. We'll start by drawing the molecular orbital diagram for boron monofluoride (BF), beginning by drawing the orbital notation for each of its atoms.

Notice that the orbitals for fluorine are lower than those for boron. Due to

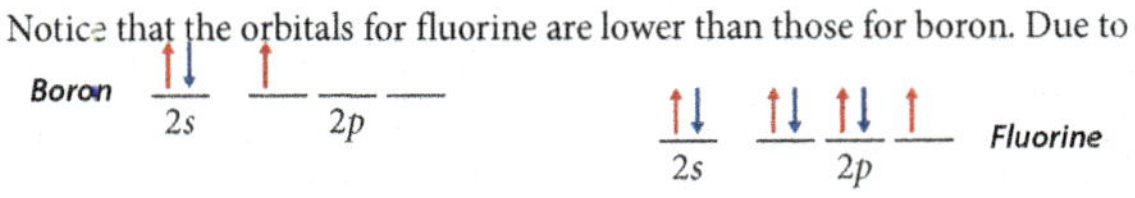

$$\text{Boron} \qquad \qquad \text{Fluorine}$$

fluorine's higher electronegativity, those orbitals are lower in energy than boron's orbitals are.

Because the molecule includes fluorine, we use the molecule orbital order for fluorine. We then fill the orbitals with electrons.

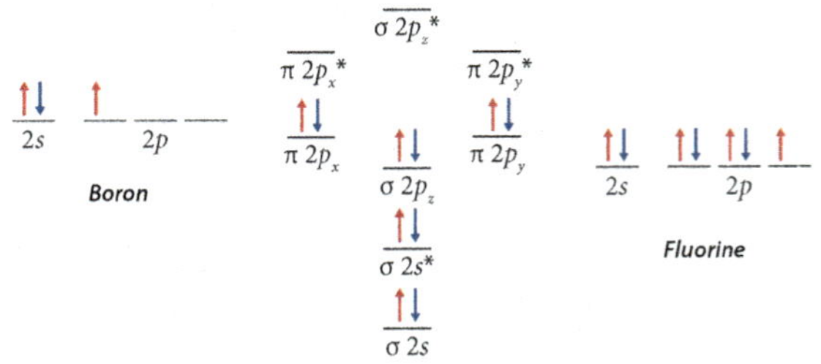

$$\text{Boron} \qquad \qquad \text{Fluorine}$$

The bond order for boron monofluoride is 3.

$$BO = \frac{8-2}{2} = 3$$

This indicates that the molecule is very stable and may have a triple bond. A look at one possible Lewis structure for boron monofluoride confirms that a triple bond is possible.

$$:B \equiv F:$$

Again, there are no lone electrons in boron monofluoride, so it is diamagnetic.

 Using Drawing Area 2, repeat Steps A–D for CO, NO, CN⁻, and OF⁺.

2. Very few substances are found in nature as individual atoms. According to molecular orbital theory, particularly bond order, does it make sense that most substances naturally exist as bonded molecules rather than individual atoms? Explain.

Any bond order greater than 0 indicates that the molecule should be stable. Since molecular orbitals fill the bonding orbitals first, most combinations will have a nonzero bond order, indicating that most bonds are stable. Therefore, we should expect most materials to naturally occur in a bonded form.

 NOW WHICH ORBITAL ORDER?

When working with dissimilar atoms, the issue of orbital order becomes even more complex. As a general rule, the orbital order will follow that of the atom with the higher electronegativity value.

ASSESSING STUDENT WORK

Don't grade students' work strictly on the correctness of the drawing. Consider the student's progress in understanding this challenging topic.

DRAWING AREA 1	

H_2^-

$$\frac{\uparrow}{\sigma 1s^*}$$

$$\frac{\uparrow\downarrow}{\sigma 1s}$$

Bond Order?	½		
Stable?	yes		
Magnetism?	paramagnetic		

He_2

$$\frac{\uparrow\downarrow}{\sigma 1s^*}$$

$$\frac{\uparrow\downarrow}{\sigma 1s}$$

Bond Order?	0		
Stable?	no		
Magnetism?	n/a		

Li_2

$$\frac{\ }{\sigma 2p_z^*}$$
$$\frac{\ }{\pi 2p_x^*} \quad \frac{\ }{\sigma 2p_z} \quad \frac{\ }{\pi 2p_y^*}$$
$$\frac{\ }{\pi 2p_x} \qquad \frac{\ }{\pi 2p_y}$$
$$\frac{\ }{\sigma 2s^*}$$
$$\frac{\uparrow\downarrow}{\sigma 2s}$$

Bond Order?	1		
Stable?	yes		
Magnetism?	diamagnetic		

Be_2

$$\frac{\ }{\sigma 2p_z^*}$$
$$\frac{\ }{\pi 2p_x^*} \quad \frac{\ }{\sigma 2p_z} \quad \frac{\ }{\pi 2p_y^*}$$
$$\frac{\ }{\pi 2p_x} \quad \frac{\uparrow\downarrow}{\sigma 2s^*} \quad \frac{\ }{\pi 2p_y}$$
$$\frac{\uparrow\downarrow}{\sigma 2s}$$

Bond Order?	0		
Stable?	no		
Magnetism?	n/a		

O_2^-

$$\frac{\ }{\sigma 2p_z^*}$$
$$\frac{\uparrow\downarrow}{\pi 2p_x^*} \qquad \frac{\uparrow}{\pi 2p_y^*}$$
$$\frac{\uparrow\downarrow}{\pi 2p_x} \quad \frac{\uparrow\downarrow}{\sigma 2p_z} \quad \frac{\uparrow\downarrow}{\pi 2p_y}$$
$$\frac{\uparrow\downarrow}{\sigma 2s^*}$$
$$\frac{\uparrow\downarrow}{\sigma 2s}$$

Bond Order?	1.5		
Stable?	yes		
Magnetism?	paramagnetic		

N_2

$$\frac{\ }{\sigma 2p_z^*}$$
$$\frac{\ }{\pi 2p_x^*} \quad \frac{\uparrow\downarrow}{\sigma 2p_z} \quad \frac{\ }{\pi 2p_y^*}$$
$$\frac{\uparrow\downarrow}{\pi 2p_x} \qquad \frac{\uparrow\downarrow}{\pi 2p_y}$$
$$\frac{\uparrow\downarrow}{\sigma 2s^*}$$
$$\frac{\uparrow\downarrow}{\sigma 2s}$$

Bond Order?	3		
Stable?	yes		
Magnetism?	diamagnetic		

Name ______________________

DRAWING AREA 2

CO	NO

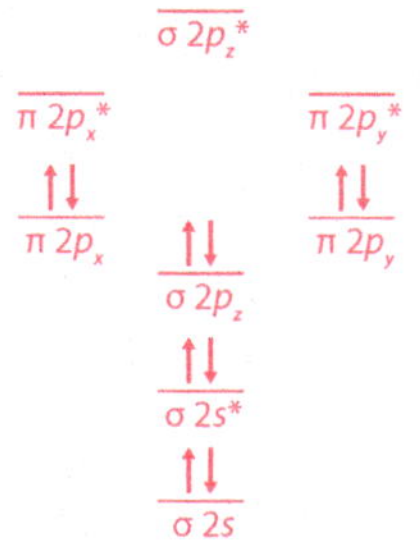

CO

$\overline{\sigma\,2p_z{}^*}$

$\overline{\pi\,2p_x{}^*}$ $\overline{\pi\,2p_y{}^*}$

$\overline{\uparrow\downarrow}\atop{\pi\,2p_x}$ $\overline{\uparrow\downarrow}\atop{\sigma\,2p_z}$ $\overline{\uparrow\downarrow}\atop{\pi\,2p_y}$

$\overline{\uparrow\downarrow}\atop{\sigma\,2s^*}$

$\overline{\uparrow\downarrow}\atop{\sigma\,2s}$

Bond Order?	3
Stable?	yes
Magnetism?	diamagnetic

NO

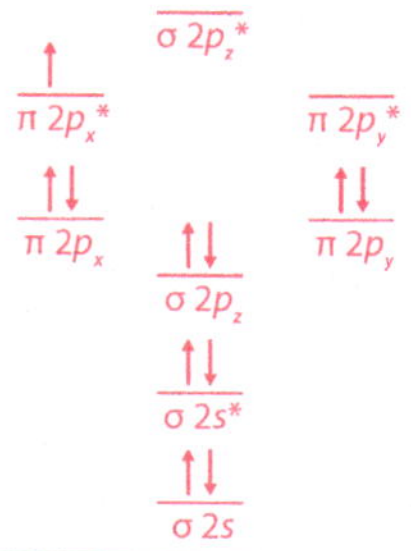

$\uparrow\atop{\overline{\sigma\,2p_z{}^*}}$

$\overline{\uparrow\downarrow}\atop{\pi\,2p_x{}^*}$ $\overline{\pi\,2p_y{}^*}$

$\overline{\uparrow\downarrow}\atop{\pi\,2p_x}$ $\overline{\uparrow\downarrow}\atop{\sigma\,2p_z}$ $\overline{\uparrow\downarrow}\atop{\pi\,2p_y}$

$\overline{\uparrow\downarrow}\atop{\sigma\,2s^*}$

$\overline{\uparrow\downarrow}\atop{\sigma\,2s}$

Bond Order?	2.5
Stable?	yes
Magnetism?	paramagnetic

CN⁻	OF⁺

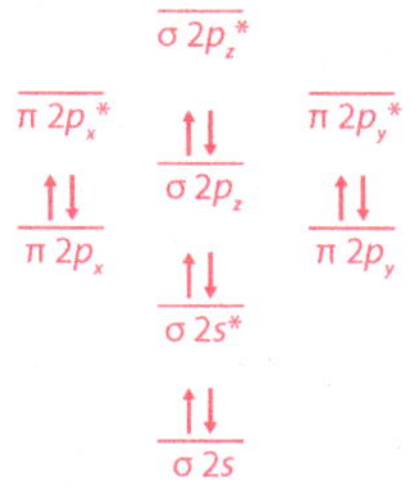

CN⁻

$\overline{\sigma\,2p_z{}^*}$

$\overline{\pi\,2p_x{}^*}$ $\overline{\uparrow\downarrow}\atop{\sigma\,2p_z}$ $\overline{\pi\,2p_y{}^*}$

$\overline{\uparrow\downarrow}\atop{\pi\,2p_x}$ $\overline{\uparrow\downarrow}\atop{\pi\,2p_y}$

$\overline{\uparrow\downarrow}\atop{\sigma\,2s^*}$

$\overline{\uparrow\downarrow}\atop{\sigma\,2s}$

Bond Order?	3
Stable?	yes
Magnetism?	diamagnetic

OF⁺

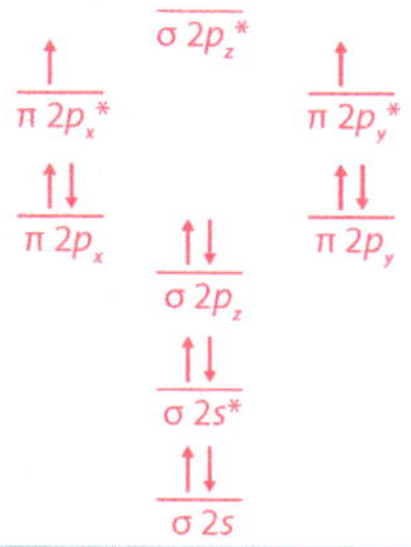

$\overline{\sigma\,2p_z{}^*}$

$\uparrow\atop{\overline{\pi\,2p_x{}^*}}$ $\uparrow\atop{\overline{\pi\,2p_y{}^*}}$

$\overline{\uparrow\downarrow}\atop{\pi\,2p_x}$ $\overline{\uparrow\downarrow}\atop{\sigma\,2p_z}$ $\overline{\uparrow\downarrow}\atop{\pi\,2p_y}$

$\overline{\uparrow\downarrow}\atop{\sigma\,2s^*}$

$\overline{\uparrow\downarrow}\atop{\sigma\,2s}$

Bond Order?	2
Stable?	yes
Magnetism?	paramagnetic

9 LAB

Name ___________________

Date ___________________

Compounds Scavenger Hunt

Naming Chemical Compounds

Do you have a hobby? Most people do. And most hobbies come with a lot of jargon—words that mean something to people inside the hobby but might sound like gibberish to everyone else. For instance, do the terms *frontside air*, *Haakon flip*, or *switchstance* mean anything to you? They probably do if you are into snowboarding. If they don't, you probably are not deeply immersed into snowboarding culture. The names of chemical compounds are similar—the more chemistry you do, the more second nature compound names become. In this activity you are going to be given ample opportunity to practice naming compounds and writing formulas. This activity will pay dividends as you start studying chemical reactions and equations in the next chapter.

How do I name chemical compounds?

1. When looking at a chemical formula, what are some questions that you can ask that will help identify the type of compound as an acid, covalent compound, ionic compound that doesn't use the Stock system, or an ionic compound that *does* use the Stock system?

   Does the formula start with a hydrogen as the cation? *(acid)* Does the formula consist of two nonmetals? *(covalent)* Does the cation have only one possible oxidation number? *(ionic non-Stock system)* Does the cation have more than one possible oxidation number? *(ionic Stock system)*

EQUIPMENT

- compound name and formula cards
- compound names and formulas worksheet

QUESTIONS

» How do I write the chemical formulas for compounds on the basis of their names?

» How do I name chemical compounds on the basis of their formulas?

NAME AND FORMULA CARDS

The card set for this activity, found in Appendix H, can also be used as flash cards for drill-and-practice, though you will need to write the appropriate name or formula on the reverse of each card. As an alternative, you may choose to have students create their own flash card sets.

MEMORIZATION

This activity is intended to give students additional practice at naming chemical compounds and writing their formulas. Emphasize to students that it is to their advantage as they advance in their study of chemistry to have common names and formulas committed to memory.

Students should do this activity in pairs.

CREATING FLOW CHARTS

These questions are a great starting point if you intend to have your students develop flow charts for naming compounds and writing formulas. Students should think of specific questions related to names and formulas that will help them differentiate between the types of compounds. An example of a question that is not helpful is, "Is it an acid?" Better questions relate to the name, such as, "Does it have the word acid in the name?" or to its formula, such as, "Does the formula start with an H?" Then lead the student to the proper classification.

PRE-LAB CHECK

1. What are the four categories into which you will be classifying compound names? *(acid, covalent compound, ionic non-Stock system compound, ionic Stock system compound)*

2. Where are the answers to your worksheet located? *(on the cards around the room)*

3. What steps should you take if you can't locate a card with the desired compound information? *(First double-check your work. If still stuck, consult with your partner, another team, or the teacher.)*

2. When looking at a chemical name, what are some questions that you can ask that will help identify the type of compound as an acid, covalent compound, ionic compound that doesn't use the Stock system, or an ionic compound that *does* use the Stock system?

See TE margin for answer.

Procedure

Though you are working in teams, each team member should do all the problems. Teammates should help each other work through naming compounds and writing formulas. If you get stuck, see what your teammate thinks. If you are both stuck, see what your classmates and teacher think.

A Complete the table on pages 81–85. In each row you are given either the name or the formula for a compound. Identify the type of compound as either an acid (A), a covalent compound (C), an ionic non-Stock system compound (I), or an ionic Stock system compound (S). Then give the missing formula or name for that compound.

B When you think you have the names and formulas correct, check your work in the scavenger hunt. The solutions are printed on cards around the room. The cards are grouped by type of compound. Work your way around the room to find the card with each compound. On the card is a card number; record this in the appropriate column of the table.

C If you can't find a card with the desired compound, double-check the type and formula/name. Consider discussing with other groups or your teacher the ones that you are having difficulty with.

Analysis

3. Which were the most difficult compounds to name? Explain.

Answers will vary. Many students struggle with the Stock system. Other students have difficulty with the prefixes and suffixes on the acids.

4. Which were the most difficult compounds to write formulas for? Explain.

See TE margin for answer.

Compound Names and Formulas

Given Name or Formula	Type	Answer: Name or Formula	Card Number
aluminum oxide	I	Al_2O_3	65
NO	C	nitrogen monoxide	23
iron(II) iodide	S	FeI_2	22
MgS	I	magnesium sulfide	32
H_3BO_3	A	boric acid	17
LiBr	I	lithium bromide	26
nitrogen trifluoride	C	NF_3	30
sodium dihydrogen phosphate	I	NaH_2PO_4	77
chromic acid	A	H_2CrO_4	63
$SrCl_2$	I	strontium chloride	71
tin(IV) sulfide	S	SnS_2	52
sodium iodide	I	NaI	56
$NaHCO_3$	I	sodium hydrogen carbonate	85
phosphoric acid	A	H_3PO_4	36
CBr_4	C	carbon tetrabromide	35
hydroiodic acid	A	HI	29
aluminum phosphide	I	AlP	81
$HClO_4$	A	perchloric acid	18
Ag_2CO_3	I	silver carbonate	7
phosphorus trichloride	C	PCl_3	94
$Al(OH)_3$	I	aluminum hydroxide	45

Given Name or Formula	Type	Answer: Name or Formula	Card Number
calcium oxide	I	CaO	98
$HClO_1$	A	hypochlorous acid	20
K_2HPO_4	I	potassium hydrogen phosphate	83
$Ca(ClO_3)_2$	I	calcium chlorate	84
SnF_2	S	tin(II) fluoride	13
nitrous acid	A	HNO_2	74
$Mn(NO_3)_2$	S	manganese(II) nitrate	66
hydrofluoric acid	A	HF	88
Al_2S_3	I	aluminum sulfide	58
carbonic acid	A	H_2CO_3	33
carbon dioxide	C	CO_2	24
magnesium hydride	I	MgH_2	64
H_3PO_3	A	phosphorous acid	70
nitrogen dioxide	C	NO_2	60
$KMnO_4$	I	potassium permanganate	91
magnesium nitride	I	Mg_3N_2	72
carbon monoxide	C	CO	92
ammonium nitrate	I	NH_4NO_3	8
magnesium hydroxide	I	$Mg(OH)_2$	79
sodium oxalate	I	$Na_2C_2O_4$	78
sodium chromate	I	Na_2CrO_4	21

Given Name or Formula	Type	Answer: Name or Formula	Card Number
sulfur hexafluoride	C	SF_6	50
$HgBr_2$	S	mercury(II) bromide	90
AlN	I	aluminum nitride	46
$Ca(C_2H_3O_2)_2$	I	calcium acetate	43
H_2SO_3	A	sulfurous acid	44
sulfur trioxide	C	SO_3	55
NaClO	I	sodium hypochlorite	57
$RaBr_2$	I	radium bromide	6
calcium carbonate	I	$CaCO_3$	97
strontium fluoride	I	SrF_2	27
lead(II) phosphate	S	$Pb_3(PO_4)_2$	48
barium cyanide	I	$Ba(CN)_2$	4
mercury(II) oxide	S	HgO	99
copper(I) fluoride	S	CuF	75
sulfuric acid	A	H_2SO_4	89
iron(III) oxide	S	Fe_2O_3	37
tungsten(V) bromide	S	WBr_5	62
SnS	S	tin(II) sulfide	59
tin(IV) iodide	S	SnI_4	73
$FeCl_2$	S	iron(II) chloride	67
KI	I	potassium iodide	54

Given Name or Formula	Type	Answer: Name or Formula	Card Number
Co_3P_2	S	cobalt(II) phosphide	96
$HClO_2$	A	chlorous acid	53
copper(II) sulfide	S	CuS	19
xenon hexafluoride	C	XeF_6	68
potassium sulfate	I	K_2SO_4	34
magnesium sulfide	I	MgS	80
potassium nitrite	I	KNO_2	5
CO	C	carbon monoxide	49
HBr	A	hydrobromic acid	93
Cu_2S	S	copper(I) sulfide	15
NI_3	C	nitrogen triiodide	38
aluminum phosphate	I	$AlPO_4$	41
dinitrogen monoxide	C	N_2O	39
lithium sulfide	I	Li_2S	1
iodic acid	A	HIO_3	2
AuI	S	gold(I) iodide	28
XeO_3	C	xenon trioxide	11
tin(II) chloride	S	$SnCl_2$	10
PbI_2	S	lead(II) iodide	42
radium chloride	I	$RaCl_2$	87
PBr_7	C	phosphorus heptabromide	9

Given Name or Formula	Type	Answer: Name or Formula	Card Number
potassium bromate	I	$KBrO_3$	47
Li_2CO_3	I	lithium carbonate	12
Na_2S	I	sodium sulfide	95
$Mg(HCO_3)_2$	I	magnesium hydrogen carbonate	100
HCN	A	hydrocyanic acid	69
nitric acid	A	HNO_3	31
Ca_3N_2	I	calcium nitride	76
Cr_2O_3	S	chromium(III) oxide	40
SF_4	C	sulfur tetrafluoride	51
$HClO_3$	A	chloric acid	3
N_2O_3	C	dinitrogen trioxide	25
MgO	I	magnesium oxide	14
hydrosulfuric acid	A	H_2S	16
$HC_2H_3O_2$	A	acetic acid	86
S_4N_4	C	tetrasulfur tetranitride	82
P_2F_4	C	diphosphorus tetrafluoride	61

Name ___________________

Date ___________________

Expeditions in Chemical Equations

Investigating Chemical Reactions and Equations

Many people love soda; in fact, in 2018 the average American drank almost thirty-nine gallons of it. But as much as they love the sweetness, many would like to avoid the Calories. So soft drink manufacturers have turned to chemistry to find sweeteners without the Calories.

Today you can get drinks sweetened with aspartame, saccharin, cyclamate, or sucralose with all the sweetness but with no Calories. Each of these sweeteners has gone through a chemical process to produce it. In most cases the process involves a number of chemical reactions. For example, to produce saccharin, chemists react toluene with nitrous acid. The product of this reaction is then reacted with sulfur dioxide. A third reaction combines the product of the second reaction with chlorine and ammonia, resulting in saccharine.

You will do something similar in this lab activity as you start with copper and then conduct a series of four reactions. You will see a number of evidences for chemical reactions and observe a number of the different types of reactions that you have learned about in class.

How can I tell whether a chemical reaction has occurred?

QUESTIONS

» How do I know whether a chemical change occurred?

» How can I differentiate types of chemical reactions?

» How do I model chemical reactions?

Procedure

REACTION 1

A Measure about 1 g of copper wool and form it into a loose wad.

B Set up the Bunsen burner. Light the burner and adjust the flame.

C Using the crucible tongs, hold the copper in the Bunsen burner flame for about 5 minutes. A black product results from the reaction of copper with atmospheric oxygen.

D Place the blackened copper into the 150 mL beaker.

1. Describe anything that you observed during the reaction.

Students may observe some copper that is unreacted during the burning. If the copper ignites, they may notice sparks or flashes.

EQUIPMENT

- laboratory balance
- laboratory burner and lighter
- crucible tongs
- beaker, 150 mL
- graduated cylinders, 25 mL (2)
- glass stirring rod
- filtering funnel
- clay triangle
- ring stand and ring
- filter paper
- beaker, 250 mL
- pipette
- wire gauze
- copper wool
- sulfuric acid (H_2SO_4), 1 M
- sodium hydroxide (NaOH), 3 M
- goggles
- laboratory apron
- nitrile gloves

LAB 10A OBJECTIVES

» Identify evidences of chemical reactions.

» Record the changes that take place during chemical reactions.

» Write chemical equations to describe chemical reactions.

⚠ ACID AND BASE SAFETY

In the event of acid or base contact with skin or eyes, rinse with water for 15–20 minutes. Remove contact lenses, if present. If contact is made with clothing, remove clothing prior to the spill reaching the skin.

You should have the following available in the laboratory: a saturated boric acid solution (H_3BO_3) for neutralizing spilled bases and a saturated sodium bicarbonate solution (NaHCO$_3$) for neutralizing spilled acids. Neutralize the spill prior to cleaning with paper towels. These solutions are not intended to treat contact with skin, eyes, or clothing.

To prepare a saturated solution of boric acid, add boric acid powder to warm water until boric acid settles out to the bottom of the solution. You will be able to add up to 27 g of boric acid powder to 100 mL of warm water.

To prepare a saturated solution of sodium bicarbonate, add the solid to warm water until sodium bicarbonate settles out on the bottom of the solution. For a volume of 100 mL of water, you will need about 15 g of sodium bicarbonate.

🧪 PREPARING SOLUTIONS

Sulfuric Acid

To prepare 1 M H_2SO_4, pour approximately 70 mL of distilled water into a volumetric flask. Add 5.6 mL of concentrated (18 M) sulfuric acid. Stopper the flask and swirl gently. Dilute with distilled water up to the 100 mL mark. Use caution because the glassware may become hot while you mix the acid and water. This process can be scaled to make other quantities of acid.

PRE-LAB CHECK

1. List common evidences that a chemical reaction may have occurred. *(change in color, change in energy, formation of a gas, formation of a precipitate)*

2. Write the balanced chemical equation for reacting hydrogen and oxygen to form water.

   [*Answer:* $2H_2(g) + O_2(g) \longrightarrow 2H_2O(l)$]

3. Which type of chemical reaction often produces a precipitate? *(a double-replacement reaction)*

4. What two types of chemical reactions are opposites of each other? *(synthesis and decomposition reactions)*

5. What law is the basis for balancing chemical equations? *(law of conservation of mass [or matter])*

6. How do you know when you have a balanced equation? *(The number of each type of atom is the same on both sides of the equation.)*

Sodium Hydroxide

To prepare 3 M NaOH, pour approximately 70 mL of distilled water into a volumetric flask. Add 12.00 g of NaOH pellets. Stopper the flask and swirl gently until dissolved. Dilute with distilled water up to the 100 mL mark. Use caution because the glassware may become hot while you mix the base and water. This process can be scaled to make other quantities of base.

Question 3 Answer

It is both a synthesis reaction and a combustion reaction. Some students may also recognize it as a redox reaction.

 OBSERVING THE REACTIONS

This lab activity depends on color changes, and sometimes students may get mixed results. If students don't produce enough copper oxide at the beginning, the blue color will be faint in the next step, and then the precipitate is often nonexistent in the next. In the third step, the acid-base neutralization at the end can get so hot that it cooks the reaction, turning the product black before students even heat it on the burner. This means that the synthesis reaction in the first part of this experiment is the most significant. If students are careful, this activity will be fun to do and watch and will give great results.

Even if students fail to observe dramatic reactions in all four steps, they can actually fill in all the equations without getting perfect results in the activity. So if the procedure does not go perfectly, it is not a total loss.

 QUESTION 5: COLLISION THEORY

The answer to this question is not covered until the Chapter 16 discussion of kinetics. The purpose now is to get students thinking about what happens on the particle level during a chemical reaction. The main idea here is the collision theory, the concept that particles must collide in the right orientation and with enough energy for a reaction to occur. Therefore, heating the substance will increase the energy, increasing the likelihood of effective collisions, which in turn means a faster reaction.

---

2. Write the balanced chemical equation for Reaction 1 in which copper and atmospheric oxygen combine to form a single product. (*Hint*: Copper in this compound has a +2 oxidation number.)

$$2Cu(s) + O_2(g) \longrightarrow 2CuO(s)$$

3. What type(s) of reaction is this first reaction?

See TE margin for answer.

4. What compound is formed in this reaction?

copper(II) oxide

5. Copper reacts spontaneously with oxygen in the air. Why do you think the instructions include heating the copper?

Students should recognize that reactions happen when particles collide. If particles move faster due to a gain in thermal energy, they are more likely to react, and the reaction will progress faster.

REACTION 2

E. Pour 20 mL of 1 M H_2SO_4 into a 25 mL graduated cylinder.

F. Add the sulfuric acid to the beaker with the black product from Reaction 1 and carefully stir the mixture with the glass stirring rod. You may need to give this solution a minute or two to react.

6. Describe anything that you observed during the reaction.

Students may notice that the sulfuric acid reacts with the black copper(II) oxide to produce a blue liquid.

G. Place the filtering funnel in the clay triangle and set it on the ring. Fold a piece of filter paper and place it in the funnel. Consult Appendix C for procedural guidelines if needed.

H. Adjust the height of the filtering apparatus so that the tip of the funnel is touching the inside of the beaker. Filter the solution into the 250 mL beaker and save the filtrate in the beaker for the next step. The filter paper containing any unreacted material from Step F can be thrown out.

7. What was the color of the solution after the sulfuric acid was added to the black compound and the resulting solution was filtered?

blue

8. Write the balanced equation for Reaction 2 resulting from the addition of sulfuric acid to the black compound.

$$CuO(s) + H_2SO_4(aq) \longrightarrow CuSO_4(aq) + H_2O(l)$$

9. What type(s) of reaction is the reaction in Question 8?

a double-replacement reaction; Some students may also recognize it
as an acid-base neutralization reaction.

10. What ion do you think caused the color change in the solution of sulfuric acid and the black compound? What evidence do you have for your answer?

The Cu^{2+} ion caused the color change. The ions of the sulfuric acid
were already in solution, and none of them had any color. When the
copper(II) oxide (the black compound) was added to the solution of
sulfuric acid, the Cu^{2+} ions changed places with the H^+ ions of the
sulfuric acid and caused the blue color.

REACTION 3

I. Pour approximately 20 mL of 3 M NaOH into the second 25 mL graduated cylinder.

J. Add 10 mL of the sodium hydroxide from the graduated cylinder to the filtrate in the beaker gradually as you stir the mixture with the glass stirring rod. A precipitate will form.

K. Using a pipette, gradually add sodium hydroxide, stirring continuously until no more precipitate forms.

11. Describe anything that you observed during the reaction.

Students should observe the formation of a blue precipitate in a clear
liquid.

12. Write the balanced equation for Reaction 3 resulting from the addition of sodium hydroxide to the colored filtrate.

$CuSO_4(aq) + 2NaOH(aq) \longrightarrow Cu(OH)_2(s) + Na_2SO_4(aq)$

13. What type(s) of reaction is the reaction in Question 12?

a double-replacement reaction

REACTION 4

L. Place the wire gauze on the iron ring. Place the beaker containing the mixture from Reaction 3 on the gauze and *cautiously* heat the mixture until it boils. Stir constantly until a reaction takes place, keeping in mind that ***alkaline solutions tend to splatter!***

14. When you heated the mixture, what compound from Reaction 3 changed into copper(II) oxide (CuO) and water?

copper(II) hydroxide, $Cu(OH)_2$

**WATCH OUT
FOR SPLATTERS!**

Make sure that you are wearing gloves, goggles, and an apron when you handle this caustic solution. If it splatters and lands on your skin, it could cause a chemical burn. If this happens, go to your teacher, who can rinse the area with a boric acid solution that will neutralize the base so that it will stop stinging and burning your skin.

FILTERING PROCEDURES (STEP G)

Remind students that Appendix C contains information about laboratory techniques, including gravity filtration.

 AMOUNT OF NaOH

Sometimes students may need more than 20 mL of NaOH depending on the amount of copper wool that they began the procedure with.

WASTE DISPOSAL

Have readily available a labeled container for collecting chemical wastes. To treat these wastes, filter the waste to remove the CuO, which may be discarded in the trash. Neutralize the filtrate with 6 M hydrochloric acid, using phenolphthalein or universal indicator until a neutral pH is indicated. Flush the treated filtrate down the drain with plenty of water.

15. Describe anything that you observed during the reaction.

Answers will vary.

16. Write the balanced equation for Reaction 4 that resulted in the products copper(II) oxide and water.

$Cu(OH)_2(s) \longrightarrow CuO(s) + H_2O(l)$

17. What type(s) of reaction is the reaction in Question 16?

a decomposition reaction

M Pour the remaining mixture into the waste container provided.

18. What substance(s) was (were) present in your mixture from Reaction 3 that did not react in Reaction 4?

Sodium and sulfate ions were present in solution but did not react as the mixture of sodium sulfate and copper(II) hydroxide was heated.

19. If you had filtered out the precipitate produced in Reaction 3, would you have gotten the same results when heating the filtrate? Explain.

After writing the reactions in Questions 12 and 16, students should recognize that if they had filtered out the precipitate, it would have been impossible to produce copper oxide since there would have been no copper hydroxide to decompose.

20. Compare the result from Reaction 4 to the result from Reaction 1. Did you observe any similarities in these two substances?

Students should notice that both reactions produced the same product: black, solid copper(II) oxide.

21. How can chemical reactions like the four that you have just done be useful to people?

Students should recognize that we need certain chemicals for a variety of purposes and that we can use chemical reactions to produce the substances we need. Students may mention that copper oxide is used to produce ceramics and rayon.

Name

Date

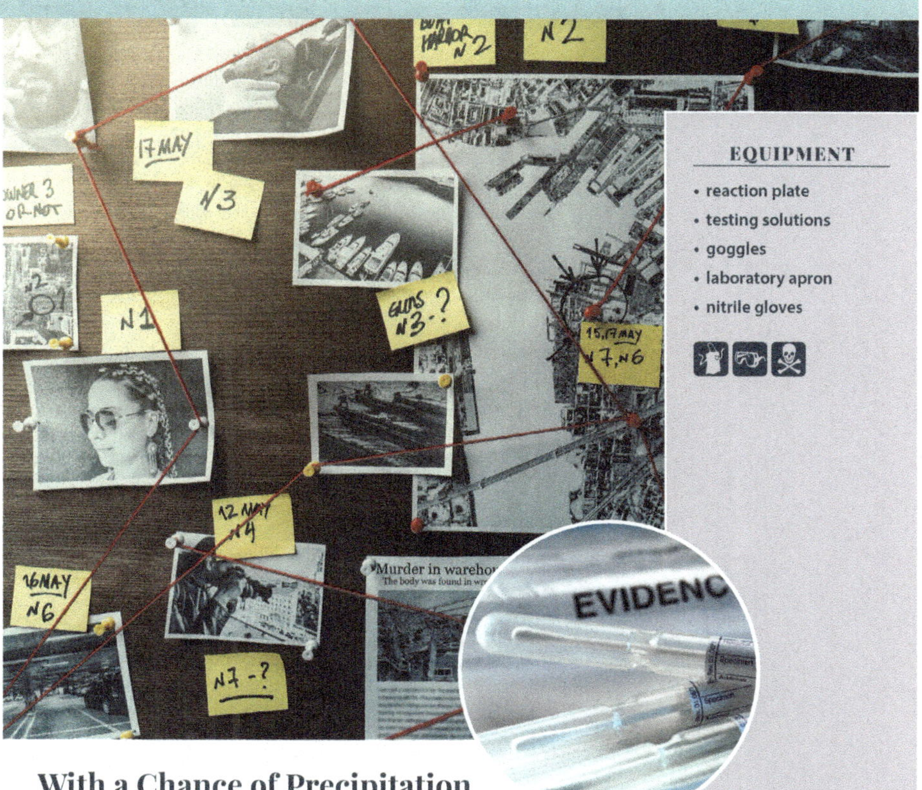

With a Chance of Precipitation

Inquiring into Solubility

For years, the lack of any eyewitnesses resulted in many crimes going unsolved. As science has advanced, it has helped police departments solve many such challenging cases through DNA evidence. One method for obtaining sufficient DNA is called *ethanol precipitation*. When a scientist adds ethanol to a DNA-containing solution, the DNA comes out of solution and gives the scientist more DNA with which to make an identification.

How can we generate solubility rules for ionic compounds?

Precipitation is one of the indicators that a chemical reaction has occurred. In this inquiry lab activity, you will develop microscale (small-scale) chemistry procedures to develop solubility rules for some common anions and cations in certain given solutions.

EQUIPMENT

- reaction plate
- testing solutions
- goggles
- laboratory apron
- nitrile gloves

QUESTIONS

» How do I know whether a reaction occurred between two solutions?

» How were solubility rules developed?

» How can I model reactions between ions in double-replacement reactions?

» Develop procedures for microscale chemical reactions to test for precipitates in double-replacement reactions.

» Develop solubility rules on the basis of empirical data.

» Write ionic equations for double-replacement reactions.

SOLUTIONS

All the solutions are 0.1 M solutions. Most of these chemicals can be purchased as solutions.

To make your own solutions, the following instructions will make 100 mL of each solution. Pour approximately 70 mL of distilled water into a volumetric flask. Add the appropriate mass of salt. Stopper the flask and swirl gently until dissolved. Dilute with distilled water up to the 100 mL mark. This process can be scaled to make other quantities of the solutions.

aluminum nitrate, 3.75 g

ammonium nitrate, 0.80 g

barium nitrate, 2.61 g

calcium nitrate, 2.36 g

copper(II) nitrate, 2.42 g

iron(III) nitrate, 4.04 g

silver nitrate, 1.70 g

zinc nitrate, 2.97 g

sodium carbonate, 1.24 g

sodium chloride, 0.58 g

sodium hydroxide, 0.40 g

sodium iodide, 1.50 g

sodium phosphate, 1.42 g

sodium sulfate, 1.42 g

CHEMICAL SAFETY

Ammonium nitrate, copper(II) nitrate, and silver nitrate solutions are slightly toxic when ingested. Silver nitrate, sodium carbonate, and sodium hydroxide solutions are skin and eye irritants; silver nitrate will stain skin and clothing. Avoid contact of all chemicals with eyes and skin. Wear chemical splash goggles. Wash hands thoroughly before leaving the laboratory.

Question 1 Answer

List 1 includes the cations that we will be testing, and List 2, the anions. List 1 includes all nitrate anions, which means that the cation is what is changing—what we are testing. Similarly, List 2 includes compounds with sodium as the cation. Therefore, we must use the List 2 compounds to test the anions. Some students may recognize that sodium nitrate, which will form in all the reactions, is always soluble.

Procedure

PLANNING/WRITING SCIENTIFIC QUESTIONS

You are to test combinations of anions and cations from the following two lists so that you can determine the solubility for different ions.

Solution List 1	Solution List 2
aluminum nitrate, $Al(NO_3)_3$	sodium carbonate, Na_2CO_3
ammonium nitrate, NH_4NO_3	sodium chloride, $NaCl$
barium nitrate, $Ba(NO_3)_2$	sodium hydroxide, $NaOH$
calcium nitrate, $Ca(NO_3)_2$	sodium iodide, NaI
copper(II) nitrate, $Cu(NO_3)_2$	sodium phosphate, Na_3PO_4
iron(III) nitrate, $Fe(NO_3)_3$	sodium sulfate, Na_2SO_4
silver nitrate, $AgNO_3$	
zinc nitrate, $Zn(NO_3)_2$	

1. Which list represents the cations that you are testing? Which represents the anions? Explain how you know this.

   See TE margin for answer.

A Research the types of reactions in Chapter 10 of your textbook. Determine which one or ones relate to solubility and how you will know whether a reaction took place for this type (or these types) of reactions.

B Consider the reaction plate and the two lists of substances to be tested. Brainstorm how you can test all the possible combinations of cations and anions.

C With your lab group, brainstorm ways to organize your data to determine the solubility rules for the cations and anions tested.

D Write specific questions related to precipitation reactions, solubility rules, and microscale chemistry that you could answer by collecting data.

DESIGNING SCIENTIFIC INVESTIGATIONS

E Write procedures to conduct microscale reactions with all the possible combinations of cation-containing compounds with every anion-containing compound to answer the questions that you formulated in Step D above.

F Have your teacher approve your procedures.

CONDUCTING SCIENTIFIC INVESTIGATIONS

G Following the procedures that you have written, collect the data to answer the questions that you wrote.

DEVELOPING MODELS

H From the data that you collected, make note of trends regarding which combinations of anions and cations reacted and which didn't. Note any exceptions to these general trends. Also note similarities in these trends within chemical groups.

I Develop solubility rules for the anions and cations that you tested, including appropriate exceptions to the rules.

SCIENTIFIC ARGUMENTATION

J Compare your solubility rules with those in your textbook.

K State a claim about the effectiveness of your procedures and support your claim with evidence from your data.

2. Pick three of the combinations that reacted and write the chemical equation, the complete ionic equation, and the net ionic equation for each.

Answers will vary.

Aluminum nitrate and sodium carbonate:

$$2Al(NO_3)_3(aq) + 3Na_2CO_3(aq) \longrightarrow Al_2(CO_3)_3(s) + 6NaNO_3(aq)$$

$$2Al^{3+}(aq) + \cancel{6NO_3^-(aq)} + \cancel{6Na^+(aq)} + 3CO_3^{2-}(aq) \longrightarrow Al_2(CO_3)_3(s) + \cancel{6Na^+(aq)} + \cancel{6NO_3^-(aq)}$$

$$2Al^{3+}(aq) + 3CO_3^{2-}(aq) \longrightarrow Al_2(CO_3)_3(s)$$

Silver nitrate and sodium chloride:

$$AgNO_3(aq) + NaCl(aq) \longrightarrow AgCl(s) + NaNO_3(aq)$$

$$2Ag^+(aq) + \cancel{NO_3^-(aq)} + \cancel{Na^+(aq)} + Cl^-(aq) \longrightarrow AgCl(s) + \cancel{Na^+(aq)} + \cancel{NO_3^-(aq)}$$

$$Ag^+(aq) + Cl^-(aq) \longrightarrow AgCl(s)$$

Copper(II) nitrate and sodium hydroxide:

$$Cu(NO_3)_2(aq) + 2NaOH(aq) \longrightarrow Cu(OH)_2(s) + 2NaNO_3(aq)$$

$$Cu^{2+}(aq) + \cancel{2NO_3^-(aq)} + \cancel{2Na^+(aq)} + 2OH^-(aq) \longrightarrow Cu(OH)_2(s) + \cancel{2Na^+(aq)} + \cancel{2NO_3^-(aq)}$$

$$Cu^{2+}(aq) + 2OH^-(aq) \longrightarrow Cu(OH)_2(s)$$

INQUIRING INTO SOLUBILITY

MANAGING INQUIRY LAB ACTIVITIES

An inquiry lab activity allows students a degree of freedom, but the teacher has to guide the process to achieve the desired educational outcomes. As teachers start using inquiry lab activities, they often struggle with this process of guiding students. The best method is to think through the goals that you want students to achieve and then use questions to guide them to reach those goals. You want to balance students' freedom to investigate the question in the manner they choose with the need to meet your educational goals.

This activity is all about double-replacement reactions and solubility rules. The students will combine pairs of compounds in the two lists of solutions: cations in List 1 and anions in List 2. They will test each combination of solutions from the lists to see whether they react (i.e., form a precipitate) or not (i.e., no reaction occurs). From this data they should be able to formulate some solubility rules for the anions and cations that they tested.

Procedure

PLANNING/WRITING SCIENTIFIC QUESTIONS

Lab groups should research the types of reactions in Chapter 10 in their textbook to determine which ones relate to solubility and how they will know whether a double-replacement reaction takes place.

Groups will then brainstorm how they can use the reaction plate to conduct microscale reactions to test the anions and cations. They also have to think through making sure that they test all the possible combinations of cations and anions. The groups will also have to think through the organization of the large amount of data that they will accumulate. Some groups will want to make lists, while others may want to make a data table. Some groups may struggle with the fact that they are not dealing with numerical data.

The groups will then write specific questions related to double-replacement reactions, solubility rules, and microscale chemistry.

DESIGNING SCIENTIFIC INVESTIGATIONS

Once the groups have written their questions, they need to write specific procedures to test all possible combinations of the anions and cations. Make sure that you review their procedures prior to their starting to collect data.

Help students think through how they will organize their procedures and data table to ensure that they test all possible combinations of testing solutions. If they make a table with the testing solutions as the column and row headers, they can then make a grid that correlates to the reaction plate. This will show them into which well they should put each solution. It will also help them make sure that they test all combinations of solutions.

CONDUCTING SCIENTIFIC INVESTIGATIONS

Once the groups have good procedures, allow them to collect data.

Teacher Guide

DEVELOPING MODELS

Students will analyze their data to determine whether trends are suggested on the basis of which cations and anions reacted (products are insoluble and precipitate out of solution) and which didn't (ions remain in solution). From these trends, they should be able to develop solubility rules for the anions and cations tested.

SCIENTIFIC ARGUMENTATION

Students will compare their solubility rules with those given in the Student Edition. They will then state a claim about their procedures and analysis. As always, claims should be supported with specific evidence from their data.

Sample Procedure

A Using a pipette, put 4–5 drops of aluminum nitrate solution into the well (Row 1, Column 1) according to Table 1. Continue by adding 4–5 drops of aluminum nitrate solution to the next five wells in Row 1.

B Repeat Step A for each List 1 compound (Rows 2–8). Use a new pipette for each row.

C Using a new pipette, add 4–5 drops of the first anion testing solution, sodium carbonate, into the first well (R1, C1). Continue by adding sodium carbonate solution to the next seven wells in Column 1.

D When all eight mixtures in Column 1 have been made, go back and stir each one with a clean toothpick. For each mixture, write a "P" (precipitate/reaction occurred) if a precipitate forms or the mixture appears cloudy, or "N" (no reaction) if the liquid remains clear.

E Repeat Steps C–D with the remaining List 2 compounds (Columns 2–6), using Table 1 to guide which compounds are put into which wells. Be sure to use a new pipette for each column.

DISPOSAL

F Place a few folded pieces of paper towel over the entire reaction plate. Then invert the plate to catch the contents in the paper towels. Discard the paper towels in the trash.

G Use distilled water to refill the reaction plate wells. Place a few folded pieces of paper towel over the entire reaction plate, then invert the plate to catch the contents in the paper towels. Discard the paper towels in the trash.

H Repeat Steps F–G until the wells are clean. You may need to use toothpicks or cotton swabs to clean the wells completely.

EQUIPMENT

- **reaction plate**
- **pipettes (14)**
- **toothpicks**
- **paper towels**
- **distilled water**
- **testing solution**
- **goggles**
- **laboratory apron**
- **nitrile gloves**

Sample Data

The data given in Table 1 is representative of the data students should obtain. Student results may vary some.

TABLE 1

	Sodium Carbonate	Sodium Chloride	Sodium Hydroxide	Sodium Iodide	Sodium Phosphate	Sodium Sulfate
Aluminum Nitrate	P	N	P	N	P	N
Ammonium Nitrate	N	N	N	N	N	N
Barium Nitrate	P	N	N	N	P	P
Calcium Nitrate	P	N	N	N	P	P
Copper(II) Nitrate	P	N	P	N	P	N
Iron(III) Nitrate	P	N	P	N	P	N
Silver Nitrate	P	P	P	P	P	N
Zinc Nitrate	P	N	P	N	P	N

Name ___________________

Date ___________________

» Write chemical equations after observing chemical reactions.

» Relate the law of definite composition to atomic masses.

» Calculate percent composition.

» Determine an empirical formula.

EQUIPMENT

- laboratory burner and lighter
- laboratory balance
- crucible and cover
- ring stand and ring
- clay triangle
- crucible tongs
- wire gauze
- transfer pipette
- sandpaper
- magnesium ribbon (30 cm)
- goggles
- laboratory apron
- nitrile gloves

SANDPAPER

Sandpaper will be used to sand the surface of the magnesium ribbon before burning. Sandpaper that is 150 grit or finer works best.

Torching Metals

Determining Empirical Formulas

The law of definite composition states that the ratio of elements in a compound is constant for every particle of that compound. These ratios can be expressed by formulas. For example, the compound potassium chlorate is always made up of a ratio of one potassium atom to one chlorine atom to three oxygen atoms, or 1:1:3. We express this in the formula $KClO_3$. Since this formula expresses the simplest whole-number ratio for potassium chlorate, it is potassium chlorate's *empirical formula*. Sometimes a compound's empirical formula is the same as its molecular formula, as with potassium chlorate and water. But in other instances they are different, as in glucose (molecular formula $C_6H_{12}O_6$, empirical formula CH_2O).

In this lab activity, you will synthesize magnesium oxide and experimentally determine its empirical formula.

How can burning a substance make it heavier?

QUESTIONS

» What does the law of definite composition tell me about chemical reactions?

» How can I analyze the composition of a compound?

» How are empirical formulas determined?

PRE-LAB CHECK

1. What is the ratio of particles in potassium chlorate? *(Since the formula is $KClO_3$, the ratio is 1:1:3.)*

2. Define *empirical formula*. *(An empirical formula is the formula that expresses the simplest whole-number ratio of components in a compound.)*

3. How will we synthesize magnesium oxide? *(by burning magnesium ribbon in oxygen, that is, in the air)*

4. Once the masses of magnesium and oxygen in magnesium oxide are determined, what must be done next prior to determining magnesium oxide's empirical formula? *(The masses of each element must be converted to moles.)*

5. What is the empirical formula of $C_6H_{12}O_6$? *(CH_2O)*

6. How will we calculate the actual percent composition of magnesium oxide? *(by assuming a 1 mol sample of MgO)*

Procedure

A Check your crucible for cracks or chips. Clean your crucible and crucible cover with a damp paper towel. Support them on a ring with a clay triangle. The crucible cover should be tilted on the top of the crucible, leaving a small opening (see below).

From this point forward, handle the crucible and its cover with the crucible tongs!

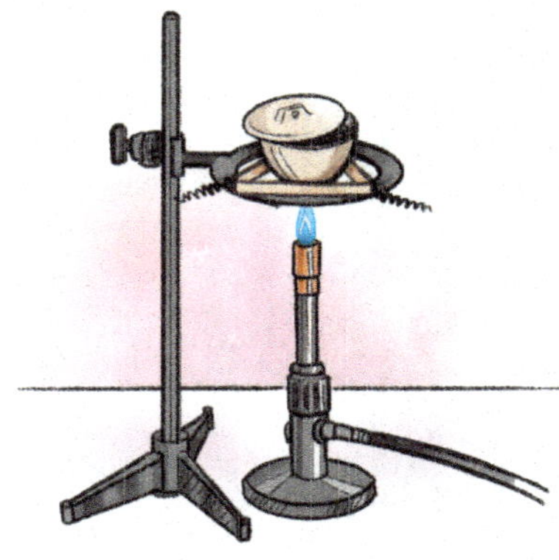

B Use the laboratory burner to heat the crucible and its cover until they are dry. Allow the crucible to cool until it is comfortable to the touch.

C Using the crucible tongs, move the crucible and cover to the laboratory balance. Measure the mass of the crucible and cover, and record your data in Table 1.

D Fold a piece of sandpaper in half. Then, holding the sandpaper so that it opens away from you, clean a strip of magnesium ribbon approximately 30 cm long by gently pulling it through the sandpaper to remove any oxide coating. Wipe it off with a dry paper towel.

1. Why is it necessary to remove any existing oxide coating from the magnesium ribbon before performing the experiment?

Leaving the oxide coating on the ribbon would reduce the accuracy of the experiment since we assume that we're starting out with pure magnesium, not a mixture of magnesium and magnesium oxide.

E While wearing gloves, roll up the magnesium ribbon into a tight spiral and place it flat on the bottom of the crucible. It's important that the magnesium be heated uniformly so that it reacts completely.

F Replace the cover on the crucible and find the mass of the crucible, cover, and magnesium. Record your data in Table 1.

G Calculate the mass of magnesium and record your answer in Table 1.

H Place the crucible and its contents on the clay triangle and begin heating them. Heat the magnesium uncovered, but hold the crucible cover nearby with your tongs. Use it only to stop flare-ups and to keep the magnesium burning in a controlled way. *The moment the magnesium starts to burn, place the cover on the crucible.*

I Continue heating the magnesium until the magnesium fails to glow. At this point, heat the covered crucible as hot as possible for several additional minutes.

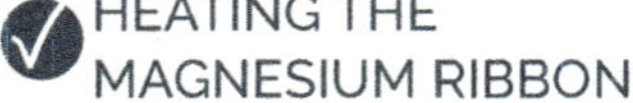

✓ **HEATING THE MAGNESIUM RIBBON**

If students apply the small inner cone of a well-adjusted flame to the bottom of the crucible, the ribbon will glow bright red within three minutes. Avoid a yellow flame because it deposits soot on the bottom of the crucible, affecting the mass.

2. Why do you want to keep the cover off to the maximum extent? Why do we cover the magnesium when it starts to burn?

   We keep the crucible uncovered to the maximum extent to allow as much oxygen to react as possible. Covering the magnesium when it begins to burn is done to control the combustion. We don't want to lose any material in the form of smoke.

J Using the crucible tongs, carefully move the crucible to the wire gauze. Allow the covered crucible to cool for about 10 minutes.

3. Write the balanced equation for the reaction of magnesium with oxygen gas to form magnesium oxide. What do you predict will be the empirical formula of magnesium oxide?

   $2Mg(s) + O_2(g) \longrightarrow 2MgO(s)$; Students should predict the empirical formula of magnesium oxide to be MgO.

Air is a mixture of mostly nitrogen and oxygen gases. When magnesium burns in air, most of the magnesium combines with oxygen to form magnesium oxide, but some of the magnesium combines with nitrogen to form magnesium nitride (Mg_3N_2). So what you have in your crucible right now is a mixture of magnesium oxide and magnesium nitride. We need it all to be magnesium oxide.

4. Suggest a general way to deal with the magnesium nitride.

   Answers will vary. Students should suggest reacting it with something to convert it to magnesium oxide.

To deal with the magnesium nitride, you will add water. This step will convert the magnesium nitride to magnesium hydroxide and ammonia gas (NH_3).

5. Write the balanced equation for the reaction between magnesium nitride and water.

   $Mg_3N_2(s) + 6H_2O(l) \longrightarrow 3Mg(OH)_2(s) + 2NH_3(g)$

There is still a problem though! Instead of a mixture of magnesium oxide and magnesium nitride, we now have a mixture of magnesium hydroxide and magnesium oxide. But there's a very easy way to fix this. If we heat magnesium hydroxide, it decomposes into solid magnesium oxide and water vapor.

6. Write the balanced equation for the decomposition of magnesium hydroxide.

   $Mg(OH)_2(s) \longrightarrow MgO(s) + H_2O(g)$

K Now let's do what you just wrote about. ***Make sure that your crucible is cool enough to touch.*** Using the transfer pipette, add 10 drops of water uniformly over the crucible's contents.

Do not add water to a hot crucible!

MAGNESIUM FIRES

Magnesium is extremely flammable! And magnesium fires can't be put out with water or even with carbon dioxide fire extinguishers—both of these substances make magnesium fires worse. They are usually put out with dry sand or other special fire extinguishers designed to extinguish metal fires.

 DIATOMIC ELEMENTS

Students may need to be reminded about the Hydrogen 7 diatomic elements. Since atmospheric oxygen is a diatomic element, they need to use O_2 for writing the reaction.

 USING DISTILLED WATER

If your local tap water is very hard, have students use distilled water for this procedure since impurities in hard water can hinder the effectiveness of this step in the chemical reaction.

✓ SAMPLE CALCULATIONS

The calculations provided are based on the sample data provided in Table 1. Student answers may vary.

7. Do you detect any recognizable odor? If so, describe the odor. Can you identify the substance?

   See TE margin for answer.

L   Carefully heat the crucible without its cover until the water evaporates; then heat strongly for several minutes.

M   Allow the crucible, cover, and contents to cool to room temperature. Measure their combined mass and record your data in Table 1.

N   Calculate the mass of the magnesium oxide and record your answer in Table 1.

O   Finally, calculate the mass of the oxygen that chemically combined with the magnesium to form magnesium oxide. Record your answer in Table 1.

Analysis

8. Use your starting mass of magnesium and your mass of oxygen to calculate the moles of each.

   Answers will vary. Students should use the molar mass of each element to convert the mass of each to moles. Sample calculations are shown here and throughout the remainder of the activity.

   What we know: $m_{Mg} = 0.27$ g, $m_O = 0.18$ g

   Unknown: n_{Mg}, n_O

   Convert grams to moles.

   $$0.27 \text{ g Mg}\left(\frac{1 \text{ mol Mg}}{24.31 \text{ g Mg}}\right) = 0.011 \text{ mol Mg}$$

   $$0.18 \text{ g O}\left(\frac{1 \text{ mol O}}{16.00 \text{ g O}}\right) = 0.011 \text{ mol O}$$

9. Divide both mole values by the smaller value to get the ratio in whole numbers.

   What we know: $n_{Mg} = 0.011$ mol, $n_O = 0.011$ mol (both from Question 8)

   Unknown: mole ratio

   Solve for whole-number ratio.

   $$\frac{0.011 \text{ mol Mg}}{0.011} = 1 \text{ mol Mg}$$

   $$\frac{0.011 \text{ mol O}}{0.011} = 1 \text{ mol O}$$

   1 mol Mg: 1 mol O

10. Determine the empirical formula for magnesium oxide. Was it what you predicted in Question 3?

See TE margin for answer.

11. Using the masses of magnesium and magnesium oxide obtained during the experiment, calculate the observed (experimental) percent composition of magnesium in your magnesium oxide.

What we know: $m_{Mg} = 0.27$ g, $m_{MgO} = 0.45$ g

Unknown: $\%_{Mg}$

Write the formula and solve for the unknown.

$$\%_{Mg} = \left(\frac{m_{Mg}}{m_{MgO}}\right)100\%$$

Evaluate.

$$\%_{Mg} = \left(\frac{0.27 \text{ g}}{0.45 \text{ g}}\right)100\%$$

$$= 60\%$$

12. Calculate the actual percent composition of magnesium in magnesium oxide from the formula MgO by assuming a 1 mol sample (see Example 11-6 on page 247 of your textbook).

What we know: 1 mol MgO, Mg: 24.31 g/mol, MgO: 40.31 g/mol

Unknown: $\%_{Mg}$

Write the formula and solve for the unknown.

$$\%_{Mg} = \left(\frac{m_{Mg}}{m_{MgO}}\right)100\%$$

Evaluate.

$$\%_{Mg} = \left(\frac{24.31 \text{ g}}{40.31 \text{ g}}\right)100\%$$

$$= 60.31\%$$

Question 10 Answer

MgO; Hopefully students will have predicted this formula on the basis of their experience in writing chemical formulas and equations.

13. Calculate your percent error.

What we know: $\%_{Mg\ measured} = 60\%$, $\%_{Mg\ expected} = 60.31\%$

Unknown: $\%_{error}$

Write the formula and solve for the unknown.

$$\%_{error} = \left(\frac{value_{measured} - value_{expected}}{value_{expected}}\right)100\%$$

Evaluate.

$$\%_{error} = \left(\frac{60\% - 60.31\%}{60.31\%}\right)100\%$$

$$= \left(\frac{-0.31\%}{60.31\%}\right)100\%$$

$$= -0.51\%$$

14. How does this experiment demonstrate the law of definite composition?

This experiment demonstrated that the composition of magnesium oxide was definite and could be related by ratios.

TABLE 1

	Mass (g)
Crucible and Cover	113.52
Crucible, Cover, and Magnesium	113.78
Magnesium	0.26
Crucible, Cover, and Magnesium Oxide—First Mass	113.98
Magnesium Oxide Produced	0.45
Oxygen	0.18

✓ SAMPLE DATA

The data shown in Table 1 is sample data. Students' actual data values may vary.

Name ___________________

Date ___________________

Chymestry

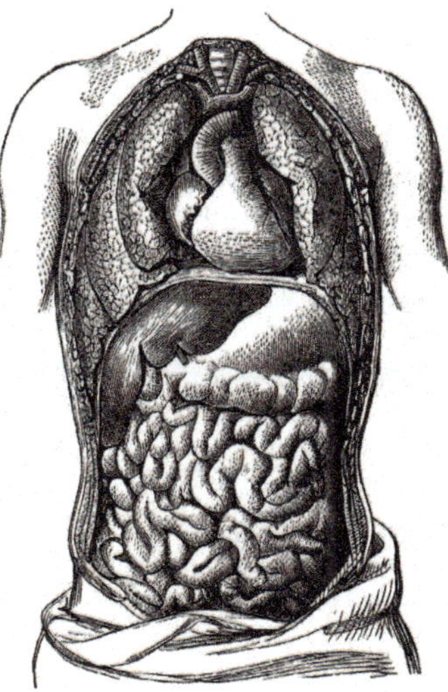

What chemistry takes place in my small intestine?

Using Stoichiometric Relationships

When food particles enter your stomach, they encounter gastric acids and enzymes that turn them into a slurry called *chyme*. Gastric acids are fairly strong, with a pH of 1–2, and they are mostly hydrochloric acid (HCl). At the beginning of your small intestine, your pancreas injects some sodium hydrogen carbonate (NaHCO₃) into the mixture. This neutralizes the acids so that they don't irritate the rest of your digestive system.

In this lab activity, you are going to re-create the contents of your small intestine! You will be reacting HCl with sodium hydrogen carbonate to see what happens. The goal is for you to measure the mole ratios of the reactants and the products so that you can figure out the stoichiometric relationships in the balanced chemical reaction of these two substances. You will measure the masses of both products and reactants to help you determine the mole ratios of each.

Procedure

1. Write the chemical reaction of NaHCO₃ with HCl and predict what you think the products will be.

   See TE margin for answer.

A Clean an evaporating dish and rinse it with water from a wash bottle.

B Using the crucible tongs, hold the evaporating dish in a well-adjusted burner flame for several minutes to remove all the moisture.

QUESTION

» How are the moles of reactants and products related in a chemical reaction?

EQUIPMENT

- laboratory burner and lighter
- laboratory balance
- evaporating dish
- wash bottle
- crucible tongs
- spatula
- small watch glass
- test tube
- test tube rack
- transfer pipette
- wire gauze
- ring stand and ring
- sodium hydrogen carbonate (NaHCO₃)
- hydrochloric acid (HCl), 6 M
- goggles
- laboratory apron
- nitrile gloves

» Collect data on the masses of reactants and products in a chemical reaction.

» Relate the moles of reactants and the moles of products in a chemical reaction.

PREPARING ACID

You can purchase a prepared 6 M HCl solution. To make your own solution, add approximately 40 mL of distilled water to a 100 mL volumetric flask. Add 50.0 mL of 12 M HCl and swirl gently. Add distilled water to the graduation mark. If you need more solution, you can scale these instructions.

Have sodium carbonate (soda ash) available to neutralize any acid spills.

Question 1 Answer

Answers will vary. Students may suggest either of the following two reactions:

$$NaHCO_3(s) + HCl(aq) \longrightarrow NaCl(aq) + H_2CO_3(aq)$$

$$NaHCO_3(s) + HCl(aq) \longrightarrow NaCl(aq) + CO_2(g) + H_2O(l)$$

PRE-LAB CHECK

1. What is the basic goal of this exercise? *(to determine the mole ratios of the reactants and products in a reaction of sodium hydrogen carbonate and hydrochloric acid)*

2. What experimental procedure, if carefully done, will cause this experiment to yield accurate results? *(determining the masses of the reactant NaHCO₃ and the product NaCl)*

3. Why will you rinse the watch glass into the evaporating dish? *(to make sure that no sodium chloride is lost)*

4. How is the reaction done in this laboratory exercise similar to what happens in the human digestive system? *(Hydrochloric acid is neutralized with sodium hydrogen carbonate in the small intestine.)*

C After the dish is cool, measure its mass. Record your data in Table 1.

2. Why is it important for the evaporating dish to be cool and dry?

The evaporating dish needs to be cool because hot dishes can damage a balance and affect weight readings. It needs to be dry so that no water affects its weight or contents.

D Using the spatula, add about 3 g of the $NaHCO_3$ to the evaporating dish while it is still on the balance. Record the combined mass of the dish and $NaHCO_3$ in Table 1.

E Calculate the mass of sodium hydrogen carbonate and record it in Table 1.

REACTING NaHCO₃ WITH HCl

F Cover the evaporating dish with a small watch glass to keep chemicals from splattering during the reaction.

G Pour about 6 mL of 6 M HCl into a clean test tube. Gradually add the acid to the $NaHCO_3$ with a transfer pipette or dropper. Allow the drops to enter the lip of the evaporating dish so that they flow down the side *gradually and slowly* (see left).

H Continue adding the acid drop by drop until the reaction stops and there is no more fizzing. Do not add more acid than is necessary. Tilt the dish from side to side to make sure that the acid has reached all the solid.

3. What substance do you think is producing the fizzing that you are observing?

Students may conclude that the fizz is carbon dioxide gas.

There are actually two reactions going on here. In the first reaction, the hydrochloric acid reacts with the sodium hydrogen carbonate to produce sodium chloride and carbonic acid. In the second reaction, the carbonic acid decomposes to form carbon dioxide gas and water.

4. Does this description of the reactions support your conclusion in Question 3? Explain.

Students should recognize that carbon dioxide gas is the most likely product to readily come out of solution.

What you have left in your evaporating dish is a mix of salt and water.

I Remove the watch glass and, using a wash bottle, rinse any splattered material from the underside of the watch glass with a small amount of distilled water. Be careful to wash all the material into the evaporating dish so that no NaCl is lost (see left).

This is how you *slowly* add acid with a pipette.

⚠ WASTE DISPOSAL OF HCL

To properly dispose of the unused HCl, collect it in a container and dilute it to 1 M by adding a volume of water equal to five times your acid volume. Then neutralize the acid by cautiously adding 1 M sodium carbonate. Check the pH. Once neutralized, flush the neutralized mixture down the drain with plenty of running water.

Ⓐ WHAT THE FIZZ IS

If students need help guessing what the fizz is, have them think of a carbonated beverage and the fizz that it can produce. This may be enough to get them thinking of carbon dioxide.

WRITING CHEMICAL EQUATIONS

Name ___________________

5. Write the balanced equation for the reaction between sodium bicarbonate and hydrochloric acid to produce sodium chloride and carbonic acid.

$NaHCO_3(s) + HCl(aq) \longrightarrow NaCl(aq) + H_2CO_3(aq)$

6. Write the balanced equation for the decomposition of carbonic acid.

$H_2CO_3(aq) \longrightarrow CO_2(g) + H_2O(l)$

7. Now add these two equations together, canceling out any products from the first reaction that were consumed in the second reaction.

$NaHCO_3(s) + HCl(aq) \longrightarrow NaCl(aq) + \cancel{H_2CO_3(aq)}$

$\cancel{H_2CO_3(aq)} \longrightarrow CO_2(g) + H_2O(l)$

$NaHCO_3(s) + HCl(aq) \longrightarrow NaCl(aq) + CO_2(g) + H_2O(l)$

8. What do you predict will be the stoichiometric ratio of sodium hydrogen carbonate to sodium chloride? How do you know? Record your prediction in Table 2.

Students should predict a 1:1 ratio, which they can derive from the stoichiometric coefficients.

FINDING THE MASS OF NaCl

Let's see whether your answer to Question 8 is right.

J Place the evaporating dish, supported by the wire gauze, on the ring stand. Heat the water in the evaporating dish until it boils gently. Do not let the water boil over or you will lose some of the NaCl and spoil the experiment.

K Continue to heat the dish until most of the water has evaporated. Continue heating gently until the NaCl is dry.

L Turn off the burner and allow the dish to cool; then weigh it and record its mass in Table 1.

M Calculate the mass of NaCl produced and record your answer in Table 1.

WRITING CHEMICAL EQUATIONS

Students may need some help and review in writing chemical equations. If you choose to show them the ionic equations, you may want to do that here or during your pre-lab discussion.

SAMPLE CALCULATIONS

The calculations provided are based on the sample data provided in Table 1. Student answers may vary.

Analysis

9. Calculate the number of moles of $NaHCO_3$ used and the number of moles of NaCl produced in the reaction. Show your work below and record your answers in Table 2.

   Answers will vary slightly. Sample calculations are shown.

   What we know: $m_{NaHCO_3} = 3.00$ g, $m_{NaCl} = 2.09$ g

   Unknown: n_{NaHCO_3}, n_{NaCl}

   Convert mass to moles.

   $$3.00 \text{ g NaHCO}_3 \left(\frac{1 \text{ mol NaHCO}_3}{84.01 \text{ g NaHCO}_3} \right) = 0.0357 \text{ mol NaHCO}_3$$

   $$2.09 \text{ g NaCl} \left(\frac{1 \text{ mol NaCl}}{58.44 \text{ g NaCl}} \right) = 0.0358 \text{ mol NaCl}$$

10. Divide both mole amounts of $NaHCO_3$ and NaCl by the smaller of the two to give your experimental ratio between $NaHCO_3$ and NaCl. Record the mole ratio of $NaHCO_3$ to NaCl in Table 2.

    What we know: $n_{NaHCO_3} = 0.0357$ mol, $n_{NaCl} = 0.0358$ mol

    Unknown: mole ratio

    Divide both mole values by 0.0357.

    $$\frac{0.0357 \text{ mol NaHCO}_3}{0.0357} = 1.00 \text{ mol NaHCO}_3$$

    $$\frac{0.0358 \text{ mol NaCl}}{0.0357} = 1.00 \text{ mol NaCl}$$

    $NaHCO_3$:NaCl = 1:1

11. How did this ratio compare to what you predicted in Question 8?

    Answers will vary.

12. So what did you use to determine the stoichiometric ratios between $NaHCO_3$ and NaCl?

    the masses measured before and after the reaction progression

Going Further

13. Now that you've seen how sodium hydrogen carbonate reacts with hydrochloric acid, explain how eating too much acidic foodstuffs can cause heartburn.

    An excess of acidic foods in the digestive tract can overwhelm the

    body's ability to produce sufficient amounts of sodium hydrogen

    carbonate to react with the acid. The excess acid irritates the digestive

    tract, causing the burning sensation of heartburn.

14. What is the usual treatment for heartburn? What makes that treatment effective?

    The usual treatment for heartburn is to take an over-the-counter

    antacid. Antacids include substances that can react with excess acid in

    the digestive tract.

TABLE 1 *Data*

	Mass (g)
Evaporating Dish	22.87
Evaporating Dish and NaHCO$_3$	25.87
NaHCO$_3$	3.00
Evaporating Dish and NaCl	24.96
NaCl	2.09

TABLE 2 *Results*

Predicted Ratio $n_{NaHCO_3} : n_{NaCl}$	1:1
Moles NaHCO$_3$	0.0357
Moles NaCl	0.0358
Experimental Ratio $n_{NaHCO_3} : n_{NaCl}$	1:1

✓ SAMPLE DATA

The data shown in Tables 1 and 2 is sample data. Students' actual data values may vary.

Name ______________________
Date ______________________

Cold and Calculating

Finding Absolute Zero

Some people live in locations with extremely low temperatures. For example, in Dudinka, Russia, the average low temperature in January is −33 °C (240 K). How close is this to the coldest possible temperature?

In 1665 Robert Boyle suggested that there was a limit to how cold temperatures could go—a *primum frigidum*. Today we know this temperature as *absolute zero*. For centuries people theorized about the existence of absolute zero and made guesses about its value on various temperature scales. In 1848 Lord Kelvin devised his temperature scale, placing absolute zero at the lowest end. Since then scientists have tried to get measured temperatures as close as possible to absolute zero. They theorize that there could be temperatures in space that are smaller than a quadrillionth of a kelvin.

This lab activity recreates a historical experiment in an attempt to find absolute zero.

How can we determine an impossibly cold temperature?

QUESTIONS

» How does changing temperature affect the volume of a sample of gas?

» How can I model the relationship between the volume and temperature of a gas?

» How do scientists know what temperature is absolute zero?

EQUIPMENT

- laboratory burner and lighter
- watch glass
- crucible tongs
- laboratory thermometer
- beaker, 1000 mL
- glass stirring rod
- metric ruler
- melting point capillary tubes (2)
- masking tape
- vegetable oil
- ice cubes
- goggles
- laboratory apron
- nitrile gloves

» Measure the volume of a sample of air at varying temperatures.

» Create a model of gas volume as a function of absolute temperature.

» Extrapolate from the empirical data to determine absolute zero.

 CAPILLARY TUBES

For this lab activity you will need to use glass capillary tubes. The tubes should be 90–100 mm long with a width of 1.3–1.4 mm. If yours are open on both ends, you will need to heat one end enough to fuse the glass shut for this experiment. Some melting point capillary tubes are already sealed for you.

 USING CHARLES'S LAW WITH HEIGHT

In this lab activity students will be using Charles's law to determine absolute zero. They may not realize the connection until they get to the questions at the end of the activity. The radius of the two tubes containing air does not change, so students will be looking at the change in height to determine the change in volume as shown in the derivation below, where V_1 is the initial volume, V_2 is the final volume, h_1 is the initial height of the air column, and h_2 is the final height of the air column.

$$\frac{V_1}{T_1} = \frac{V_2}{T_2}$$

$$\frac{\pi r^2 h_1}{T_1} = \frac{\pi r^2 h_2}{T_2}$$

$$\frac{h_1}{T_1} = \frac{h_2}{T_2}$$

PRE-LAB CHECK

1. If the volume of a gas changed from 255 mL to 282 mL while the amount of gas and the pressure on it remained constant, what must have happened to the temperature of the gas? *(It must have increased.)*

2. What type of relationship exists between volume and absolute temperature for ideal gases? *(a direct relationship)*

3. Why must the entire length of the air column be submerged in the water bath? *(All of the air column must be submerged so that the entire volume of air is at the same temperature.)*

4. Why do you *not* need to calculate V_2 in this experiment to test Charles's law? *(The volume of trapped air is proportional to the height of the air column because the diameter of the tube is assumed to be constant over the entire length of the tube.)*

5. Why do you need to take temperature and height measurements? *(to determine absolute zero)*

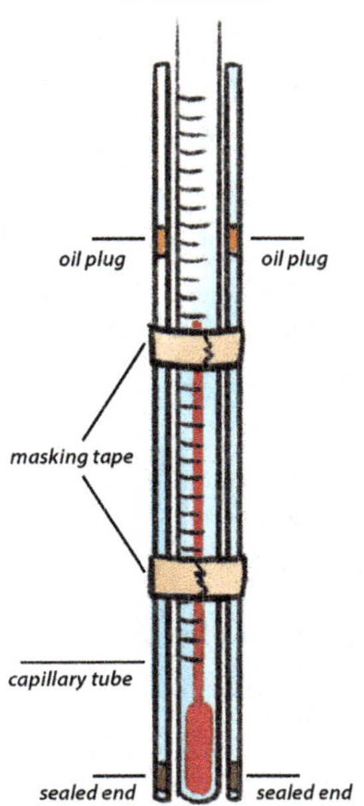

This is the setup that you will use to determine the value of absolute zero.

Procedure

SETTING UP

A Pour about 10 drops of vegetable oil onto the watch glass. Light the laboratory burner.

B Using tongs, carefully hold one of the capillary tubes with the open end slanted upward. Pass the entire length of the tube through the burner flame several times. Immediately dip the open end of the heated tube into the oil on the watch glass and allow it to cool and draw up a "plug" of oil about 1 cm long. You may need to hold the tube in the oil for a second or two to draw up the oil. When cooled to room temperature, the length of the column of air trapped in the tube should be about 5–7 cm.

C Repeat Step B with the second capillary tube.

1. What is the purpose of the oil plug?

   Students should recognize that they just formed a pocket or column of air in the capillary tube sealed by the oil plug.

2. Describe the temperature, pressure, and volume conditions inside the trapped air in the capillary tube.

   The temperature of the air will take on the temperature of its surroundings. The oil plug will move to vary the volume of the air so that the pressure continues to match atmospheric pressure.

D Attach the capillary tubes containing trapped air—with their open ends upward—to the thermometer using small pieces of tape, as shown at left.

3. Which gas law applies to these conditions? Explain.

   See TE margin for answer.

E Fill the beaker about two-thirds full of water and ice. Hold the thermometer in the water close to the beaker wall so that you can read the temperature scale and see the location of the oil plugs in the capillary tubes. The entire air columns of both tubes must be submerged, but their open ends must stay above the water line.

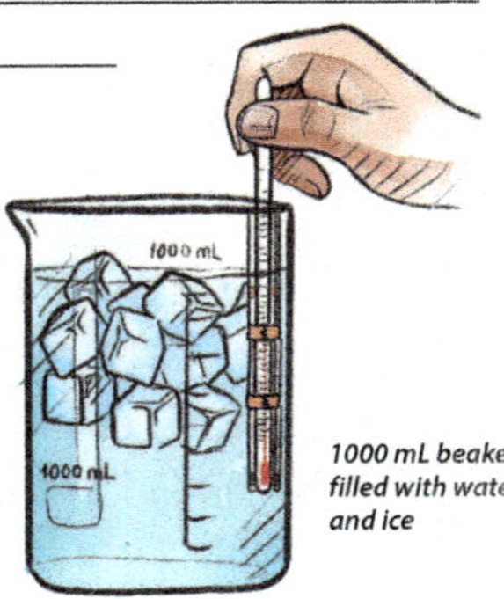

1000 mL beaker filled with water and ice

GETTING NUMBERS

F Allow the temperature to stabilize. Stirring the bath periodically with a stirring rod will speed up this process.

4. Why do you think that stirring the ice water promotes temperature stabilization?

Stirring the bath speeds up the equilibrium between the thermometer
assembly and the ice-water bath because it dissipates the heat that is
transferred from the thermometer assembly to the bath.

5. Predict the temperature for the air column after it stabilizes in the ice-water bath.

See TE margin for answer.

G Once the temperature has stabilized, measure and record the temperature in Table 1.

H Wait an additional minute. Carefully place the ruler in the water next to one of the capillary tubes. Measure the height of the air column, from the top of the melted glass to the bottom of the oil plug. Do *not* include either the melted glass itself or the oil plug in the measurement. Record your data in Table 1. Repeat for the other capillary tube.

6. What are some possible sources of error in your measurements of the air columns?

Both water and glass can distort images and refract light, making the
measurement of the height of the air columns less certain.

I Repeat Steps E–H with cold tap water (approximately 20 °C), warm tap water (approximately 35 °C), and hot tap water (approximately 50 °C). If you can't get sufficiently hot water from the tap for the fourth data point, use a hot plate or laboratory burner to heat the water. Do not exceed 50 °C.

Analysis

DETERMINING ABSOLUTE ZERO

J Using a graphing calculator or spreadsheet program, create scatterplots of the height values (*y*-axis) and temperatures (*x*-axis) for each tube.

7. What relationship do you think would fit the graphs?

Both graphs should be linear.

K Draw a curve of best fit for each set of data points.

8. Determine the equation for the best-fit curves that you drew for Step K.

Answers will vary. See TE margin for worked-out solution.

Cold and Calculating | **109**

MEASURING THE AIR COLUMN

In this cold-water bath, you will probably need to wipe off the condensation from the beaker to get a clear view of the capillary tubes. Be careful that you do not knock the beaker off the wire gauze and ring when making your measurements! Put the ruler into the water with the capillary tubes to measure the height of your column. When you measure the height of your air column, be sure not to include the oil plug.

Question 5 Answer

Students should predict that the temperature is near the freezing point of water.

MEASURING THE AIR COLUMN

The air column is underwater in the beaker, so it is important that the temperature apparatus is close enough to the side of the beaker so that students can see the column of air well enough to measure it. There may be some distortion due to the water and the glass. This and the condensation on the beaker may make it difficult for students to get reliable results for absolute zero.

Question 8 Answer

Tube 1: $h = 0.023T + 6.20$

Tube 2: $h = 0.024T + 6.35$

Example for Tube 1:

What we know: (0, 6.2), (30, 6.9)

Unknown: m, b

Write the formula and solve for the unknown.

$$m = \frac{\Delta y}{\Delta x} = \frac{(y_2 - y_1)}{(x_2 - x_1)}$$

$$= \frac{(6.9 - 6.2)}{(30 - 0)} = \frac{0.7}{30}$$

$$= 0.023$$

$$b = 6.2$$

Evaluate.

$$y = mx + b$$

$$h = .023T + 6.2$$

You may want to have a discussion regarding interpolation and extrapolation. Focus the discussion on the effect of a small variation in the measured data on values derived from these mathematical tools. Extrapolation is valuable in that it gives a useful estimate of the value of absolute zero while also illustrating the gas laws at work.

Question 9 Answer

Tube 1: −270 °C

Tube 2: −265 °C

Example for Tube 1:

What we know: $h = 0.023T + 6.20$, $h = 0$

Unknown: T,

Write the formula and solve for the unknown.

$$h = 0.023T + 6.20$$

$$h - 6.20 = 0.023T + \cancel{6.20} - \cancel{6.20}$$

$$\frac{h - 6.20}{0.023} = \frac{\cancel{0.023}T}{\cancel{0.023}}$$

$$T = \frac{h - 6.20}{0.023}$$

Evaluate.

$$T = \frac{0 - 6.20}{0.023}$$

$$= \frac{-6.20}{0.023}$$

$$= -270$$

Question 10 Answer

Example:

What we know: $T_{measured} = -267\ °C$, $T_{accepted} = -273\ °C$

Unknown: $\%_{error}$

Write the formula and solve for the unknown.

$$\%\ error = \left(\frac{value_{measured} - value_{accepted}}{value_{accepted}} \right) 100\%$$

Evaluate.

$$\%error = \left(\frac{(-267°C) - (-273°C)}{-273°C} \right) 100\%$$

$$= 2\%$$

9. Use the equations that you determined in Question 8 to calculate the temperature (absolute zero) at which the height would be zero. Average the two values for your final answer.

Answers will vary. See TE margin for worked-out solution.

10. Calculate the percent error for your value, using −273 °C as the theoretical, or actual value, of absolute zero.

Answers will vary. See TE margin for worked-out solution.

Going Further

11. To calculate absolute zero, you assume that the height of the air column is zero. Is this condition realistic? What could this imply?

No. For the air column height to drop to zero, the matter contained in that space would have no volume at all. This could imply that absolute zero can never be attained or that our model of matter does not apply under those conditions.

TABLE 1

Temperature (*t*) in °C	Length (*h*) in cm	
	Tube 1	Tube 2
0.5	6.2	6.3
19.4	6.7	6.9
34.4	7.0	7.2
42.6	7.2	7.3

 REACHING ABSOLUTE ZERO

As an enrichment activity or perhaps as a homework assignment during this chapter, have students do some research on whether anyone has reached absolute zero in the laboratory. If they find that it has not been reached, have them find out how close scientists have gotten. Lasers have been able to cool things down to a billionth of a kelvin, so scientists have been able to get very close to absolute zero.

 SAMPLE DATA

Table 1 shows sample data for this activity. The data that students generate may vary with equipment and how well they perform the procedure.

12B LAB

An Aquanaut's World

Predicting the Production of Oxygen

About sixty feet down on the sandy floor of the Florida Keys by a coral reef rests a yellow metal structure. It isn't a sunken ship or a submarine —it's an underwater laboratory.

Aquarius is one of the few underwater laboratories on Earth dedicated to scientific pursuits. The scientists who work there—called *aquanauts*—live at higher pressures, spending long amounts of time exploring the ocean floor. They don't decompress until their mission is done.

Air is supplied to the laboratory by a surface buoy that houses a compressor. The divers use and recharge their air tanks during missions to explore for up to eight hours. A room in Aquarius maintains pressure to match that of the ocean so that a moon pool can be used to easily enter the marine environment. Life and work at Aquarius is all about two things that you'll explore in this lab activity—pressure and oxygen.

In this experiment, you will predict the mass of oxygen gas generated in a reaction. Then you will compare this experimental value to a theoretical value to determine your percent error.

How can we predict the volume of gas produced in a reaction?

EQUIPMENT

- barometer
- laboratory balance
- graduated cylinder, 25 mL
- large test tube
- test tube rack
- spatula
- Erlenmeyer flask, 500 mL
- pinchcock clamp
- glass and rubber tubing
- rubber stoppers, 1-hole and 2-hole
- beaker, 600 mL
- graduated cylinder, 100 mL
- laboratory thermometer
- weighing paper
- 3.00% hydrogen peroxide (H_2O_2), 15.0 mL
- manganese(IV) oxide (MnO_2), 1 g
- goggles
- laboratory apron
- nitrile gloves

QUESTIONS

» How do I collect a sample of gas over water?

» How can I determine the amount of a dry gas that was produced over water?

LAB 12B OBJECTIVES

» Collect a sample of gas over water.

» Determine the amount of dry oxygen produced empirically.

» Evaluate the experimental procedure by assessing the accuracy of the measured volume of oxygen gas.

✔ OPTION FOR LAB 12B

You may want to perform this experiment as a demonstration because it is so involved. If so, explain what you are doing in each step. If you choose to have students do this lab activity, make sure that they carefully read the procedure ahead of time. Be sure to have a good pre-lab discussion with your students to prepare them well.

CHEMICAL SOURCES

A solution of 3% hydrogen peroxide (H_2O_2) is available at pharmacies and grocery stores. Manganese dioxide can be obtained from a science supplier.

PRE-LAB CHECK

1. Define and give the value of the molar volume of a gas. *(The molar volume is the space 1 mole of gas would occupy at STP. In this case, it would be 22.4 L.)*

2. Define *molar mass*. *(Molar mass is the mass of 1 mole of a substance.)*

3. Under the same conditions, will the molar volumes of all gases be the same? Will the molar masses be the same? *(Molar volumes will be the same, but molar masses will not.)*

4. Why does raising and lowering the beaker equalize the air pressure in the flask with that of the atmosphere? *(When the levels of the liquids in the flask and the beaker are at the same height or elevation, the same amount of atmosphere is above both; thus, the same pressure is above both.)*

5. Why do you siphon water back and forth between the beaker and the flask? *(to check for leaks)*

ATMOSPHERIC PRESSURE

You may want to take the atmospheric pressure reading and write it on the board for the entire class to see. If you have no barometer, find your local barometric pressure online.

Procedure

A. Determine today's atmospheric pressure and record your data in Table 1.

B. In the 25 mL graduated cylinder, measure out exactly 15.0 mL of hydrogen peroxide (H_2O_2). Record this volume in Table 1. Pour the H_2O_2 into the test tube, which we will refer to as the *reaction tube*.

C. Given that the density of 3.00% hydrogen peroxide solution is 1.01 g/mL at room temperature, determine the mass of the H_2O_2 solution. Show your work below and record your answer in Table 1.

What we know: $\rho_{H_2O_2} = 1.01$ g/mL, $V_{H_2O_2} = 15.0$ mL

Unknown: $m_{H_2O_2\,solution}$

Write the formula and solve for the unknown.

$$\rho = \frac{m_{solution}}{V_{solution}}$$

$$\rho V_{solution} = \left(\frac{m_{solution}}{V_{solution}}\right) V_{solution}$$

$$m_{solution} = \rho V_{solution}$$

Evaluate.

$$m_{solution} = (1.01 \text{ g/mL})15.0 \text{ mL}$$

$$= 15.15 \text{ g} = 15.2 \text{ g}$$

D. Determine the mass of the H_2O_2 (solute) in the solution. Show your work below and record your answer in Table 1.

What we know: $m_{solution} = 15.15$ g, $w_{solute} = 3.00\%$

Unknown: m_{solute}

Write the formula and solve for the unknown.

$$w_{solute} = \left(\frac{m_{solute}}{m_{solution}}\right)100\%$$

$$\left(\frac{w_{solute}}{100\%}\right) m_{solution} = \left(\frac{m_{solute}}{m_{solution}}\right)\left(\frac{100\%}{100\%}\right) m_{solution}$$

$$m_{solute} = \left(\frac{w_{solute}}{100\%}\right) m_{solution}$$

Evaluate.

$$m_{solute} = \left(\frac{3.0\%}{100\%}\right)15.15 \text{ g}$$

$$= 0.4545 \text{ g} = 0.455 \text{ g}$$

E On a piece of weighing paper, use a spatula to obtain and measure out about 1 g of manganese(IV) oxide. The exact amount is not critical. *Do not put it into the reaction tube yet.*

1. The manganese(IV) oxide isn't actually going to react with the hydrogen peroxide. It is speeding up the reaction in which H_2O_2 releases oxygen. What kind of substance is MnO_2 in this reaction?

   The MnO_2 is acting as a catalyst in this reaction since it doesn't

   participate in the reaction.

2. In this reaction, you will be using hydrogen peroxide in the presence of manganese(IV) oxide to produce oxygen gas and water. Write the balanced chemical equation for this reaction, showing where all these chemicals and conditions fit in.

   See TE margin for answer.

3. What kind of chemical reaction is this?

   See TE margin for answer.

4. Now predict the volume of oxygen gas that will be produced at STP. Record your answer in Table 1.

F Assemble the rest of the setup as shown below. Fill the 500 mL Erlenmeyer flask with water to just below the neck. Make sure that the bottom end of the reaction side glass tube is about 0.5 cm below the rubber stopper. Check to see that the water exit side glass tube is 0.5 cm above the bottom of the flask. Keep the water exit tubing clamped with the pinchcock to keep water from running out. Wet the rubber stopper before you insert it in the flask's mouth to improve the seal.

G Ask your teacher to check your setup.

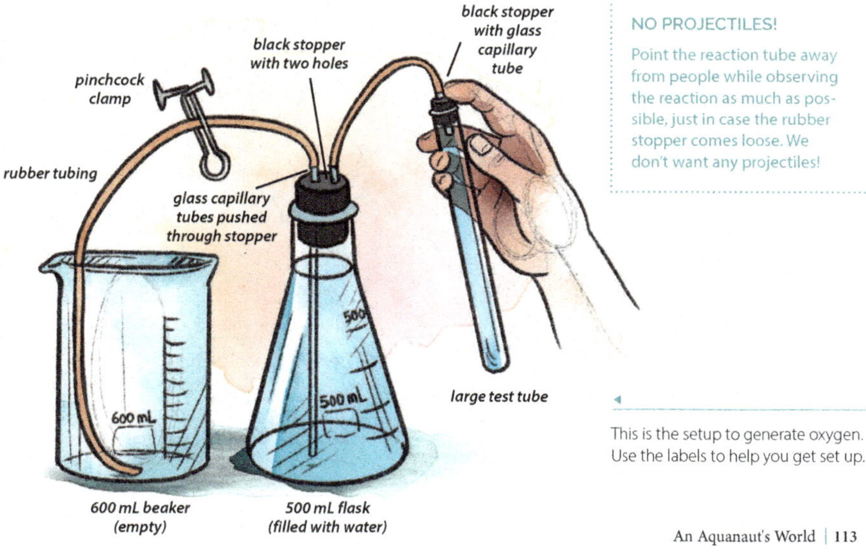

 REVIEWING REACTIONS

You may need to remind students of catalysts and the four types of chemical reactions from Chapter 10—synthesis, decomposition, single-replacement, and double-replacement.

Question 2 Answer

$$2H_2O_2(aq) \xrightarrow{MnO_2} 2H_2O(l) + O_2(g)$$

Question 3 Answer

It is a decomposition reaction. Some students will also recognize it as a redox reaction.

QUESTION 4

A sample worked-out calculation is shown in the bottom margin.

QUESTION 4 ANSWER

$$0.4545 \text{ g } H_2O_2 \left(\frac{1 \text{ mol } H_2O_2}{34.02 \text{ g } H_2O_2} \right) \left(\frac{1 \text{ mol } O_2}{2 \text{ mol } H_2O_2} \right) \left(\frac{22.4 \text{ L } O_2}{1 \text{ mol } O_2} \right) = 0.1496 \text{ L}$$

Since this lab activity includes a fairly involved procedure, you might want to demonstrate the setup and operation of the apparatus. Point out the following:

1. The mouth of the reaction tube should be higher than the closed end. It should be pointing away from people as much as possible.

2. The gas inlet tube should be just above the level of the water.

3. The water exit tube should be about 6 mm from the bottom of the Erlenmeyer flask. This can be adjusted by sliding the glass rod up or down as needed.

4. The water exit tube should be clamped, or water will spill out.

Also demonstrate how to make the apparatus leakproof and bubble-proof (by siphoning water from the flask to the beaker and from the beaker to the flask) and how to equalize the pressure.

H Place about 100 mL of water in the beaker.

I Disconnect the reaction tube by removing the stopper with the tube attached. Remove the pinchcock clamp from the water exit tube.

J Tip the flask about 90° to the right so that water can flow through the water exit tube to produce a siphon (see below). The water will not siphon if there is a leak.

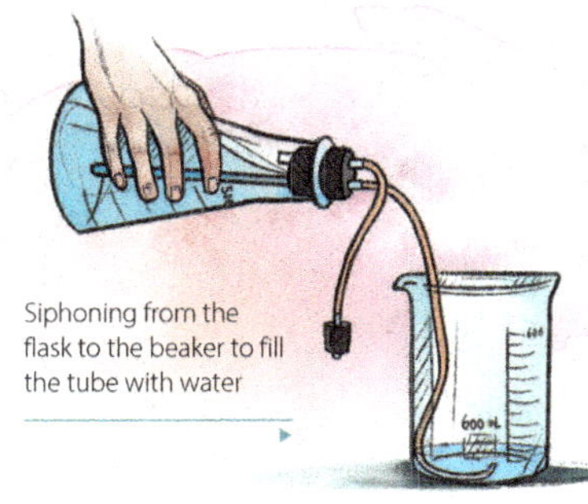

Siphoning from the flask to the beaker to fill the tube with water

K When the water exit tube has filled with water, set the flask down. At this point, water should continue siphoning from the flask into the beaker. Stop the siphoning by clamping the water exit tube with the clamp.

L Now siphon the water back into the flask so that the water level in the flask is just below the reaction side inlet tube. To do this, raise the beaker above the water level in the flask and remove the clamp (see below).

Siphon water from the beaker to the flask until the water level is just below the gas inlet tube.

M As soon as the flask fills to the desired level, replace the clamp and put the beaker back on the desk.

N Wet the reaction tube stopper and insert it into the mouth of the test tube. Confirm that both stoppers are snug.

O Remove the clamp. The water level in the flask will fall slightly. After this change, the water level should remain at the new position.

P Equalize the air pressure in the flask with that of the atmosphere by lifting the beaker so that its water level is even with the water level in the flask. Replace the clamp while the water levels are matched (see below). Once you've clamped the tube, you can put the beaker back on the table.

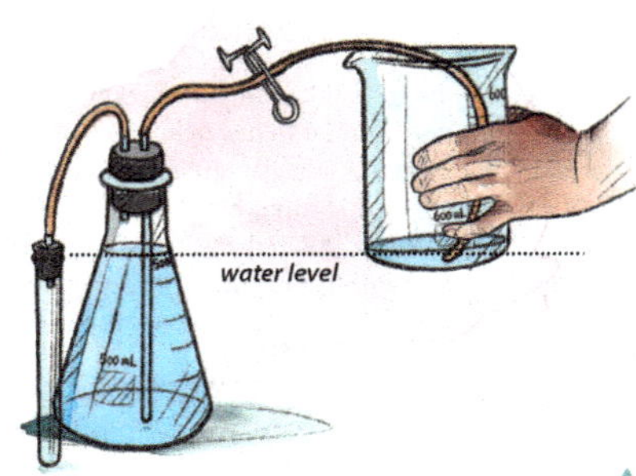

Equalize the pressure in your tubes by lining up the water levels in the Erlenmeyer flask and the beaker. Put on the pinchcock clamp when water levels are equal.

5. Why is it important that there are no leaks or bubbles in your setup and that the air pressure in the flask be equal to atmospheric pressure?

   Bubbles, leaks, or unequal pressure will change the amount of displaced water, making the volume of gas generated by the reaction unequal to the volume of displaced water.

Q Empty and dry the beaker and return it to the assembly.

MAKING OXYGEN

R Remove the stopper from the reaction tube and pour the manganese(IV) oxide into the hydrogen peroxide solution. Quickly replace the stopper and remove the clamp. Hold the reaction tube vertically near the top. With one finger, hold the stopper firmly in place. Swirl the tube to speed up the reaction. Every few seconds, give the tube another swirl.

6. Record anything that you observe during the reaction, including changes in the reaction mixture, the space of air over the water in the flask, and the amount of water in the beaker.

   See TE margin for answer.

   ____

   ____

7. Are the reaction conditions for your production of oxygen gas STP conditions? Explain.

   See TE margin for answer.

   ____

   ____

S After about 10 minutes, very little water should be coming out of the water exit tube. Consider the reaction finished at this point even though the mixture in the reaction tube may still be bubbling slightly.

8. The volume of water in the beaker represents the volume of displaced water. What else does this volume equal?

   the volume of the oxygen produced

T Lift the beaker so that its water level matches the water level in the flask. Clamp the water exit tube with the clamp. Immediately remove the stopper from the reaction tube so that it doesn't pop out under pressure!

U Using the 100 mL graduated cylinder, determine the volume of the water displaced by the oxygen produced and record the result in Table 1.

V Measure the temperature of the water in the graduated cylinder. Convert the result to kelvins and record this temperature in Table 1.

W Determine the water vapor pressure at the temperature measured in Step V. Record this value in Table 1.

9. Why do you need to take the temperature of the water?

   Since the temperature in the laboratory fluctuates considerably, the

   most representative temperature for the oxygen from the reaction is

   that of the water in the beaker.

Question 6 Answer

Students should observe bubbles form in the hydrogen peroxide as the manganese(IV) oxide is added. The test tube may become slightly warm. The space over the water in the Erlenmeyer flask should increase as oxygen gas is produced, pushing down the water level. This makes water flow through the water exit tube into the beaker.

Question 7 Answer

Students should recognize that both temperature and pressure conditions are not standard since STP is defined as 101.325 kPa at 273.15 K. If your current atmospheric pressure happens to be 101.325 kPa, students should note that only the temperature is nonstandard.

Students may have to interpolate between values on Table 12-2 on page 281 of the Student Edition. For example, if the temperature is 62.0 °C, what is the vapor pressure of water?

Temperature (°C)	Vapor Pressure (kPa)
60.0	19.9320
62.0	x
65.0	25.0220

$$\frac{62.0°C - 60.0°C}{65.0°C - 60.0°C} = \frac{x}{25.0220 \text{ kPa} - 19.9320 \text{ kPa}}$$

$$\left(\frac{2.0 \,\cancel{°C}}{5.0 \,\cancel{°C}}\right) 5.0900 \text{ kPa} = \left(\frac{x}{\cancel{5.0900 \text{ kPa}}}\right) \cancel{5.0900 \text{ kPa}}$$

$$x = 2.0360 \text{ kPa}$$

The vapor pressure of water at 62.0 °C is 19.9320 kPa + 2.0360 kPa, or 21.968 kPa.

10. Look back at Question 2. Where did the oxygen, water, and manganese(IV) oxide go after the hydrogen peroxide began to react?

    The oxygen traveled out of the reaction tube and through the gas inlet tube until it was collected over the water in the Erlenmeyer flask. The water and manganese(IV) oxide both remained in the reaction tube.

11. Use Dalton's law to determine the partial pressure of oxygen. Show your work below and record your answer in Table 1.

    Answers will vary.

    What we know: P_{total} = 100.89 kPa, P_{H_2O} = 2.196 kPa

    Unknown: P_{O_2}

    Write the formula and solve for the unknown.

    $$P_{total} = \Sigma P_i$$

    $$P_{total} - P_{H_2O} = \cancel{P_{H_2O}} + P_{O_2} - \cancel{P_{H_2O}}$$

    $$P_{O_2} = P_{total} - P_{H_2O}$$

    Evaluate.

    $$P_{O_2} = 100.89 \text{ kPa} - 2.196 \text{ kPa}$$

    $$= 98.694 \text{ kPa} = 98.69 \text{ kPa}$$

12. Now use gas laws to find the volume of the oxygen if the conditions were changed to STP. Show your work below and record your answer in Table 1.

    What we know: V_1 = 0.171 L, P_1 = 98.694 kPa, T_1 = 292.2 K, P_2 = 101.325 kPa, T_2 = 273.15 K

    Unknown: V_2

    Write the formula and solve for the unknown.

    $$\frac{P_1 V_1}{T_1} = \frac{P_2 V_2}{T_2}$$

    $$\left(\frac{P_1 V_1}{T_1}\right)\frac{T_2}{P_2} = \left(\frac{\cancel{P_2} V_2}{\cancel{T_2}}\right)\frac{\cancel{T_2}}{\cancel{P_2}}$$

    $$V_2 = \frac{P_1 V_1 T_2}{T_1 P_2}$$

    Evaluate.

    $$V_2 = \frac{(98.694 \,\cancel{\text{kPa}})(0.171 \text{ L O}_2)(273.15 \,\cancel{\text{K}})}{(292.2 \,\cancel{\text{K}})(101.325 \,\cancel{\text{kPa}})}$$

    $$= \frac{4609}{29\,607} \text{ L O}_2$$

    $$= 0.156 \text{ L O}_2$$

13. Look back at your predicted value for the volume of oxygen gas generated at STP from the decomposition of H_2O_2. Do a yield calculation for this experiment. Show your work below and record your answer in Table 1.

    Answers will vary.

    What we know: $yield_{theoretical}$ = 0.150 L, $yield_{actual}$ = 0.156 L

    Unknown: $\%_{yield}$

    Write the formula and solve for the unknown.

    $$\%_{yield} = \left(\frac{yield_{actual}}{yield_{theoretical}} \right) 100\%$$

    Evaluate.

    $$\%_{yield} = \left(\frac{0.156 \, \text{L}}{0.150 \, \text{L}} \right) 100\%$$

    $$= 104\%$$

14. Explain possible causes for quantities of oxygen smaller or larger than predicted.

    See TE margin for answer.

    ____

    ____

    ____

    ____

    ____

    ____

    ____

    ____

    ____

Going Further

15. Not only was your experiment today *not* at STP, the aquanaut's world isn't at STP either! In fact, if these research scientists don't consider the pressure and temperature conditions, it could cost them their lives. How do you think aquanauts need to adjust for these conditions?

    Students should recognize that temperature and pressure could affect

    some key functions, such as the oxygen that aquanauts carry during

    dives, the amount of time that they stay submerged, the way that

    they decompress at the end of a mission, and the amount of oxygen

    dissolved in their bloodstream.

Question 14 Answer

There are many places for error to enter this procedure. For a quantity of gas that is smaller than predicted, possible causes include leaks, the reaction not running to completion, measurement errors, or a hydrogen peroxide solution that contains less than 3% H_2O_2. For a quantity of gas that is larger than predicted, possible causes include measurement errors or a hydrogen peroxide solution that contains greater than 3% H_2O_2. Inaccurate temperature or barometric pressure measurements or improperly equalizing pressures can result in errors in either direction. Unobserved gas in the apparatus prior to starting the reaction would also contribute to a higher-than-expected value.

16. Can you think of another career in which changes to pressures and temperatures of gases must be considered?

TABLE 1

Atmospheric Pressure (kPa)	100.89
Volume H_2O_2 Solution (mL)	15.0
Mass of H_2O_2 Solution (g)	15.2
Mass of H_2O_2 (g)	0.455
Predicted Volume Oxygen (L)	0.150
Measured Volume Oxygen (L)	0.171
Temperature (K)	292.2
Water Vapor Pressure (kPa)	2.196
Pressure of Oxygen (kPa)	98.69
Volume Oxygen at STP (L)	0.156
Percent Yield	104%

 SAMPLE DATA

Table 1 shows sample data for this activity. The data that students generate may vary with equipment and how well they perform the procedure.

13A LAB

Name _______________

Date _______________

EQUIPMENT

- computer with internet access
- magnifying glass
- stereomicroscope
- minerals

Cracking the Crystal

Relating Geology to Chemistry

Along the shores of an acid lake in Indonesia, miners eke out a living by gathering sulfur from naturally occurring deposits, despite the risk of breathing air laced with toxic sulfuric acid. The miners sell the sulfur, some of it later carved into statues and figures.

The sulfur that these miners gather grows in characteristic crystals. You've spent some time in your textbook learning how crystal shapes affect the formation of minerals. Now you get an opportunity to explore minerals on your own, looking at some real minerals and thinking about what is happening at the atomic level to produce the crystals that you hold in your hand.

What determines the shape of a crystal?

Your goal for this activity is to explain the crystal shapes of different minerals using their chemical crystal structures.

Procedure

A Make sure that you are thinking about minerals and not rocks. Minerals are more often chemically pure substances.

B Think about the kind of chemical that you are dealing with: an atomic crystal, a covalent molecular crystal, a covalent network crystal, an ionic crystal, or a metallic crystal.

QUESTIONS

» What distinguishes one type of crystal from another?

» How is the shape of a crystal related to its molecular or ionic structure?

Cracking the Crystal | 119

» Observe the crystal structures of certain minerals.

» Identify types of crystals.

» Relate the molecular and ionic structure of compounds to their crystal shape.

SCHEDULING

This lab activity asks students to consider structures within solids that are not covered in the Student Edition until Section 13.2. Lab 13A should be scheduled after they have worked through the material in that section.

 MINERALS UNDER STUDY

Table 1 includes some sample minerals for students to examine: calcite, pyrite, galena, quartz, graphite, and silver. These minerals were picked for their distinct appearance, ease of use, and ready availability. You can consider other minerals, especially if your students are using the internet to examine minerals.

 NOTES FOR DOING INQUIRY

The goal of this lab activity is to have students explore the area where chemistry meets geology. This activity is more open-ended. You may choose to have students build models of their formula units if it will help them visualize the structure.

If possible, have samples for students to examine. Students should also have access to the internet to do research and look at different pictures of the same mineral.

You may choose to review the types of crystals with students, showing them some examples of each.

You may need to clarify the difference between rocks and minerals with students. Native minerals are pure, naturally occurring elements, such as diamond, silver, gold, and iron. Rocks are mixtures of minerals.

If students find it difficult to think about the three-dimensional shapes of crystals, it may be helpful to have plastic building bricks or molecular models available for them to use in visualizing the crystal shape that results from a particular formula unit.

PRE-LAB CHECK

1. What is your goal for this activity? *(to describe the crystal shapes of minerals)*

2. What is the difference between a rock and a mineral? *(Minerals are normally pure substances, while rocks are mixtures of minerals.)*

3. What are allotropes? *(variations of a substance with different physical forms)*

Sulfur crystals (yellow) in aragonite

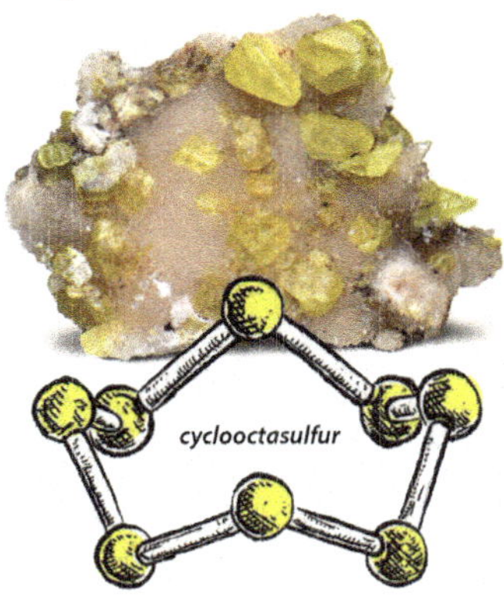

C. Try to determine the formula for the crystal that you are examining. Sketch it out, identifying the atoms in your formula unit somehow.

D. Some minerals may be allotropes. Watch out for these!

E. Try to get an actual sample. If you can't, harness the power of the internet to look at lots of different pictures of your mineral.

F. If you have a sample, examine it with a magnifying glass or a stereomicroscope. If you are working with digital pictures of minerals, zoom in to see lots of detail.

G. In Table 1, record your observations or notes on the minerals that you examine.

THINK ABOUT IT

1. Sulfur is an example of an atomic crystal that has many allotropes. In its solid state, it can appear in granular deposits or as an orthorhombic crystal. The number of sulfur atoms that form rings in the formula unit of these crystals can range from six to as many as twenty. Why do you think sulfur forms such a wide range of crystals?

   See TE margin for answer.

2. What do you think determines the form that sulfur will take?

   See TE margin for answer.

3. One of the most common forms of sulfur is cyclooctasulfur, S_8. In nature, it forms crystals with a rhombohedral shape. Using the photograph shown at left, write a paragraph that relates the shape of the formula unit to its crystal shape.

   The shape of the formula unit is a sort of crown shape. If "crowns" are linked together and stacked on top of each other, the crystal will take on a bumpy surface similar to that of a sample of amethyst. The crystal would be more likely to break between crowns.

4. The opener to this lab activity shows how people are mining sulfur from a lake in Indonesia. Why is it important to investigate crystals like sulfur and their properties?

   It is important to identify different kinds of minerals and their properties so that we can use them for specific purposes and find ways to make and purify them.

TABLE 1

Mineral	Observations	Sketches
calcite	Calcite forms a rhombohedral crystal. Its formula unit is $CaCO_3$. It is an ionic crystal.	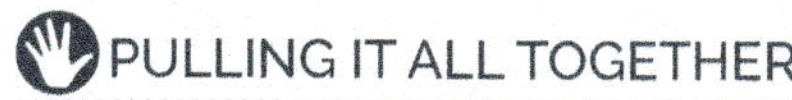 calcium (Ca), oxygen (O), carbon (C)
pyrite	Pyrite forms a simple cubic crystal. Its formula unit is FeS_2. It is also an ionic crystal.	iron (Fe), sulfur (S)
galena	Galena forms a simple cubic crystal. Its formula unit is PbS. It is another ionic crystal.	lead (Pb), sulfur (S)
quartz	Quartz forms a trigonal crystal. Its formula unit is SiO_2, one of the largest families of minerals. Quartz has bonds that are somewhere in between covalent and ionic, as would be expected for the metalloid silicon.	silicon (Si), oxygen (O)
graphite	Graphite forms a hexagonal crystal that breaks in sheets. Its formula unit is C. Graphite is a network covalent crystal.	carbon (C)
silver	Silver forms a face-centered cubic crystal. Its formula unit is Ag. It is a metallic crystal.	silver (Ag)

SAMPLE RESULTS

Table 1 includes examples of minerals that students can examine and the kinds of observations that they should make about each one. Sketches of the formula units are included for your reference.

Consider using the following simple rubric for evaluating student sketches and assign point values accordingly.

0: no sketch

1: very basic or inaccurate sketch

2: sketch with average detail and accuracy

3: very detailed and accurate sketch

✋ PULLING IT ALL TOGETHER

Consider having a discussion time in which students present their findings to each other or in which lab groups compare their results. A reflection period on learning can be valuable as an instructional strategy. You may include the answers to the "Think about It" questions (p. 120). Have students reflect on the procedures they chose to use and evaluate their effectiveness, including ways to make them better. Have students talk about what they learned and about their thought processes in customizing their procedure to solve a problem. Consider extending this discussion to the general art of problem-solving to engage students in metacognition.

13B LAB

Name

Date

Forces of Nature

Exploring Intermolecular Forces in Liquids

Many ponds across North America freeze over in winter. Yet in many of those ponds, safe within their icebound lodges, beavers carry on with daily activities. When not lounging in their lodges, they can exit through underwater doors to swim beneath the ice.

Why does ice form on the surface of ponds? If water behaved like most liquids, ice would be denser than liquid water, making it sink to the bottom of the pond. The intermolecular forces in water—London dispersion forces, dipole-dipole forces, and, most of all, hydrogen bonds—make ice less dense than water, causing it to float. Because of this, beavers can survive under the ice.

What determines the physical properties of liquids?

In this lab activity, you will investigate the intermolecular forces in four liquids—water, acetone (CH_3COCH_3), ethanol (CH_3CH_2OH), and mineral oil. The four liquids have different properties because they are made of molecules containing different arrangements of atoms. Let's see how intermolecular forces affect a liquid's density, viscosity, surface tension, solubility, and ease of evaporation.

QUESTIONS

» How do the viscosity, surface tension, solubility, and volatility of different liquids compare?

» How are the physical properties of liquids related to the polarity of their molecules?

» How are the physical properties of liquids related to the molecular masses of their molecules?

» Can we explain how intermolecular forces affect the characteristics of liquids on the basis of empirical data?

EQUIPMENT

- probeware with temperature probe
- prepared test tubes with marbles (4)
- stopwatch
- beakers, 100 mL (5)
- marker or labeling tape
- watch glass
- pipettes (4)
- small test tubes (3)
- food coloring
- water (20 mL)
- acetone (20 mL)
- ethanol (20 mL)
- mineral oil (20 mL)
- pennies (4)
- food coloring
- filter paper
- goggles
- laboratory apron
- nitrile gloves

LAB 13B OBJECTIVES

» Compare physical properties of four liquids.

» Relate physical properties to the polarity of liquid molecules.

» Relate physical properties to the molecular mass of molecules.

» Explain how intermolecular forces affect the physical properties of liquids.

SCHEDULING

Because of its length, you may want to do Lab 13B over the course of two class periods. Do the viscosity and surface tension tests on the first day, followed by the solubility and volatility tests on the second. Alternatively, you may consider setting up stations for each test; if time is limited, have students do two stations during each class period.

PROBEWARE

Ideally, the volatility portion of this lab activity is done with probeware. If you have access to probeware, follow the manufacturer's instructions for monitoring temperature. The volatility testing can be done with a glass thermometer, but in two of the tests (water and mineral oil), the temperature change is small and students may have difficulty correctly noting the amount of change.

VISCOSITY TEST TUBES

For the viscosity test, set up four stoppered, large test tubes, each with one marble inside and filled almost to the bottom of the stopper with one of the four liquids. The marble should be slightly smaller than the inside diameter of the test tube; loose-fitting marbles will not demonstrate the viscosity differences as clearly. Put all four tubes in a test tube rack. If you use the station approach, you will need only one or two of these setups for the viscosity test.

PRE-LAB CHECK

1. How will a marble show the differences in viscosity between the four liquids? *(The slower the marble sinks, the more viscous the liquid is.)*

2. Why do the surface tension tests need to be done fairly quickly? *(Some of the liquids evaporate quickly.)*

3. What word describes a liquid's ability to dissolve into another liquid? *(miscibility)*

4. If a liquid evaporates faster than a second liquid, is the first liquid more or less volatile than the second? *(more)*

PHYSICAL PROPERTIES OF THE FOUR LIQUIDS

See the table of data below to help you guide students in this activity. This table does not include information given in the Student Edition.

	Molecular Weight (g/mol)	Boiling Point (°C)	Solubility in Water (g/L)
Water	18.01	100	NA
Acetone	58.09	56	miscible
Ethanol	46.08	79	miscible
Penta-decane	212.47	270	2.866×10^{-6}

 ORDER OF PROCEDURE

You can do the parts of this lab activity in any order, so if one group is using something that another group needs, have that group skip to another part of the procedure.

Procedure

MAKING PREDICTIONS

Before you get started, let's think about the substances that you will be working with. Notice their Lewis structures below. Mineral oil is represented by pentadecane, a 15-carbon molecule.

water   *acetone*   *ethanol*

pentadecane

1. Which of the four substances are polar?

water, acetone, and ethanol

An *electric dipole moment*—a measure of polarity—can be measured for a molecule and assigned a number. While the SI unit for electric dipole moment is the *coulomb meter*, a unit called the *debye (D)* is more commonly used. The debye is the measure of charge and the distance that separates areas of charge. The dipole moments for pure samples of the four liquids are provided in the box shown at left.

> **Dipole Moment Data**
>
> water: 1.85 D
> acetone: 2.91 D
> ethanol: 1.69 D
> pentadecane: 0 D

2. Using the dipole moment measurements in the box, which substance do you think is the most polar?

acetone

3. Which molecule has the highest molecular mass? Which one has the lowest?

pentadecane; water

4. How do you think the mass of the molecule will affect the way that it responds to intermolecular forces?

More massive molecules will be less influenced by intermolecular forces, and less massive molecules will be more influenced by intermolecular forces.

Think for a moment about the polarity and molecular weight of acetone. Throughout this lab activity, you will make some predictions about how acetone's physical properties will compare with those of the other three liquids. Then you'll explore how intermolecular forces affect the manner in which these liquids behave.

VISCOSITY

Viscosity, a liquid's resistance to flow, is measured in Pa·s, or pascal-seconds. The pascal is a unit of pressure, and pascal-seconds are related to the rate of flow of a liquid. The higher its viscosity value, the thicker and more resistant to flow a liquid will be. Viscosity is a form of cohesion.

5. With 4 being the highest and 1 being the lowest, predict acetone's viscosity compared with the other three liquids.

   See TE margin for answer.

A. Notice that there is a marble at the bottom of each of the four large, labeled, stoppered test tubes. Swirl the contents of each test tube.

6. Which liquid do you think is the most viscous?

   Students should predict that the mineral oil will be the most viscous.

B. Now let's find out whether you're right. Give the stopwatch to one person designated from your group to be the timer.

C. One at a time, invert each tube so that the marble settles to the stoppered end of the test tube.

D. Invert each tube again, starting the stopwatch as soon as the test tube is in the upright position. Stop the stopwatch as soon as the marble reaches the bottom of the test tube. Record the time in Table 1.

E. Repeat Steps D and E to complete three trials for each labeled test tube.

F. Average your trials to find one value for the time of the marble's fall for each liquid.

7. Why should you do more than one trial for each liquid?

   Having more than one trial makes the average result more representative and more accurate, minimizing the significance of error.

8. In which liquid did the marble take the longest to fall? How does this relate to viscosity? Did this follow your prediction?

   The marble fell the slowest through the mineral oil. This means that the mineral oil is more viscous. Students should comment on how their prediction related to their observations.

9. Explain why you think this liquid was the most viscous.

   Students may notice from the Lewis structures that the mineral oil has the longest molecule. It does not have the strongest intermolecular forces, but the molecular weight is the highest of the four liquids.

10. How do intermolecular forces affect viscosity?

   The greater the intermolecular forces in a liquid, the slower it will flow compared with other liquids of similar mass.

PASCAL-SECONDS

Viscosity is a measure of shear stress, so the units are similar to that of a force. Viscosity is caused by internal friction in a liquid, caused by the intermolecular forces that resist the flow of the liquid. The SI unit for viscosity is the pascal-second (Pa·s).

Question 5 Answer

Answers will vary. The correct answer is 1, but do not penalize students for an incorrect answer.

Consider the viscosities of pure samples of the four liquids at room temperature provided in the box shown at left.

Viscosity Data

water: 1.002 mPa·s

acetone: 0.3311 mPa·s

ethanol: 1.114 mPa·s

pentadecane: 2.863 mPa·s

11. How do these values of viscosity relate to what you observed?

Liquids with a higher number for viscosity show a longer time for the marble to fall to the bottom of the tube.

12. Evaluate your prediction of acetone's viscosity.

Answers will vary.

SURFACE TENSION

If you need to remind your students about what surface tension is, have them review page 314 in the Student Edition. Remind students that surface tension is a demonstration of cohesion, just as viscosity is.

Question 13 Answer

Answers will vary. The correct answer is 1, but do not penalize students for an incorrect answer.

13. With 4 being the highest and 1 being the lowest, predict acetone's surface tension compared with the other three liquids.

See TE margin for answer.

G Fill four 100 mL beakers with fresh samples of each of the four liquids; label each one.

H Wash four pennies with detergent, then dry. Put a large watch glass on top of your laboratory bench and put a paper towel on it. Place the four pennies on the paper towel. Label the paper towel to show which liquid will be placed atop each penny.

14. How is the surface tension of a liquid related to the number of drops of that liquid that you could fit on a penny?

Liquids with higher surface tension will hold together better, allowing more drops to fit on a penny.

I Using a clean pipette, add the first liquid to its penny one drop at a time until the liquid overflows onto the paper towel. Keep a careful count of the drops and record them in Table 2. Do this step as quickly as possible since some of the liquids evaporate fairly rapidly. Repeat for the other three liquids.

J Clean all four pennies well, then dry. Repeat Step I a second time for all four liquids, and record your results in Table 2.

K Average your results and record this data in Table 2.

15. Which liquid has the highest surface tension? Why do you think this is so?

Water has the highest surface tension. It has a high dipole moment and is capable of hydrogen bonding. Its low molecular mass also makes it more responsive to intermolecular forces than ethanol and acetone.

16. How do intermolecular forces affect surface tension?

The higher the intermolecular forces in a liquid, the stronger the surface tension will be.

17. Evaluate your prediction of acetone's surface tension.

Answers will vary.

SOLUBILITY

18. Do you think acetone will mix with water?

See TE margin for answer.

L Fill three small test tubes halfway, one with acetone, another with ethanol, and the last with mineral oil. Label your test tubes.

M Fill a 100 mL beaker three-quarters full with water. Add a few drops of food coloring.

N Add the colored water to the three test tubes until they are almost full.

19. What do you observe about the three liquids?

The water mixes with the acetone and ethanol, but it forms a layer with the mineral oil.

Question 18 Answer

Answers will vary. Acetone is miscible, but do not penalize students for an incorrect answer.

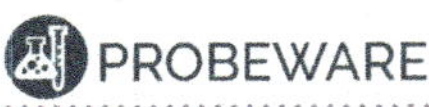 PROBEWARE

Some probeware includes onboard data analysis software; other programs allow you to download data to a computer or tablet to be analyzed in a spreadsheet. If your probeware or software allows you to do so, consider displaying all four temperature trends on a single graph, which will make clear the differences between the four liquids.

Question 23 Answer

Answers will vary. The correct answer is 4, but do not penalize students for an incorrect answer.

---

20. Explain why your liquids behaved the way that they did on the basis of intermolecular forces. (*Hint*: See page 315 of your textbook.)

    The mineral oil formed a layer with the water rather than mixing with the water because the oil is a nonpolar liquid. The water is on the bottom because it is denser than the oil. The acetone and ethanol each mixed with the water because they are polar liquids.

    When two liquids mix together or one dissolves in the other, they are said to be *miscible*.

21. How do intermolecular forces relate to miscibility?

    If two liquids are polar, they can attract each other, making them miscible. If one is polar and one is not, each will attract only itself, making the two liquids immiscible.

22. Evaluate your prediction of acetone's ability to dissolve in water.

    Answers will vary.

Setup for testing the volatility of your liquids

 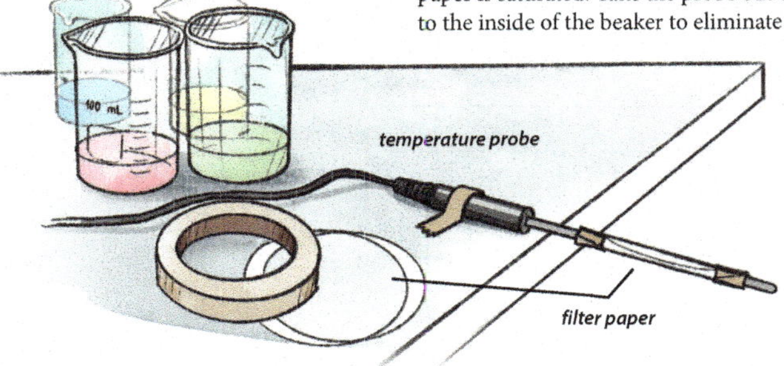

VOLATILITY

The ease at which a liquid evaporates into a gas is called its *volatility*.

23. With 4 being the highest and 1 being the lowest, predict acetone's volatility compared with the other four liquids.

    See TE margin for answer.

- Plug in the temperature probe for your probeware.

- Wrap a single layer of filter paper around the end of the temperature probe and secure it with tape. Trim or tear off any excess.

- Dip the tip covered in filter paper in one of the four liquids until the paper is saturated. Take the probe out of the beaker, touching the probe to the inside of the beaker to eliminate any drips.

- Tape the probe onto the edge of the laboratory bench as shown at left. Press the probeware's **Collect** button or equivalent. If your probeware displays real-time data, watch how the temperature changes over time.

S Collect data for three minutes. Follow the instructions for your probeware to analyze the data and calculate statistical trends.

T Locate the maximum and minimum temperature values and subtract the maximum value from the minimum. The result represents the temperature change caused by the evaporating liquid. Record this value in the appropriate row of Table 3.

U Remove the filter paper and wipe the probe with a paper towel.

Repeat steps P through U for the other three liquids and record the temperature changes in Table 3. Make sure that you change the filter paper each time!

24. Why did the temperature drop noticeably for some samples over time?

The liquid used thermal energy from the environment to change into a gaseous form.

25. According to your data, which liquid did you observe to evaporate the most quickly? How do you know?

Acetone evaporated the most quickly, as seen by the greatest temperature drop.

26. Check the boiling point data of the four liquids provided in the box shown at right. How does what you observed relate to their boiling points?

Students should notice that the more easily evaporated liquids have lower boiling points.

> **Boiling Point Data**
>
> water: 100 °C
> acetone: 56–57 °C
> ethanol: 78.5 °C
> pentadecane: 212.41 °C

27. How do intermolecular forces affect volatility?

The greater the intermolecular forces in a liquid, the less volatile it will be.

28. Evaluate your prediction of acetone's volatility.

Answers will vary.

29. Look back at the Lewis structures of the four liquids. What properties of these molecules affected their viscosity, surface tension, solubility, and volatility?

The polarity of the molecule, along with its potential for hydrogen bonding, and the molecular weight of the molecule determine how viscous, soluble, and volatile it will be, as well as the strength of its surface tension.

On the basis of the data that you have been given in this lab activity and the experimentation that you have done, you will now rank the properties of the four liquids. Record your results in Table 4.

V Rank the polarity of all four liquids, with 4 being the highest and 1 being the lowest.

W Rank the molecular weight of all four liquids, with 1 being the highest and 4 being the lowest. This scale has been reversed because molecular weight and the influence of intermolecular forces are inversely related.

X Rank the viscosity of all four liquids, with 1 representing the most resistance to flow and 4 the lowest resistance to flow.

Y Rank the surface tension of all four liquids, with 4 being the highest and 1 being the lowest.

Z Assign water a miscibility score of 4. For the other liquids, if the liquid is miscible in water, assign a score of 4. If it is not, assign a score of 0.

AA Rank the volatility of all four liquids, with 1 being the least volatile (having the highest boiling point) and 4 being the most volatile (having the lowest boiling point).

BB Add up your scores.

30. What kind of scores do the polar molecules have? How does this compare with nonpolar molecules?

    Polar molecules have high scores. Nonpolar molecules have low scores.

31. How do these scores relate to what you observed about water and acetone?

    These scores show that they have very similar properties, though acetone is more volatile and less viscous than water.

32. How do these scores relate to how ethanol compares with water and acetone?

    Ethanol has very similar properties to water and acetone, though slightly less dramatic.

33. How do these scores relate to what you observed about mineral oil?

    The scores show just how different mineral oil is from water, acetone, and ethanol. Mineral oil shows properties that are much different from the three other liquids.

34. Relate all these comparisons to the types of intermolecular forces that these liquids experience.

Water, acetone, and ethanol can all experience dipole-dipole forces and hydrogen bonding, along with London dispersion forces. Mineral oil demonstrates London dispersion forces only.

TABLE 1 *Time of the Marble's Fall (s)*

	Trial 1	Trial 2	Trial 3	Average
In Water	varies	varies	varies	varies
In Acetone	varies	varies	varies	varies
In Ethanol	varies	varies	varies	varies
In Mineral Oil	varies	varies	varies	varies

TABLE 2 *Number of Drops*

	# of Drops to Cover a Penny		Average # of Drops to Cover a Penny
	Trial 1	Trial 2	
Water	varies	varies	varies
Acetone	varies	varies	varies
Ethanol	varies	varies	varies
Mineral Oil	varies	varies	varies

TABLE 3 *Temperature Change during Evaporation*

	Change (°C)
Water	−3.0
Acetone	−11.7
Ethanol	−5.5
Mineral Oil	−1.9

TABLE 4 *Ranking the Four Liquids on the Basis of Experimentation*

	Water	Acetone	Ethanol	Mineral Oil
Polarity	3	4	2	1
Molecular Weight	4	2	3	1
Viscosity	3	4	2	1
Surface Tension	4	3	2	1
Miscibility	4	4	4	0
Volatility	2	4	3	1
Sum of Scores	20	21	16	5

14A LAB

Name _______________

Date _______________

EQUIPMENT

- hot plate
- laboratory balance
- beaker, 250 mL
- medium-sized test tubes (4)
- test tube rack
- spatula
- weighing paper
- graduated cylinder, 10 mL
- glass stirring rod
- test tube clamp
- laboratory thermometer
- masking tape or grease pencil
- ammonium chloride (NH_4Cl), 21.00 g
- distilled water
- goggles
- laboratory apron
- nitrile gloves

One Giant Solution

Making a Solubility Curve

You've learned that solubility is the amount of solute that will dissolve in a given amount of solvent to make a saturated solution. Think of the ocean as one giant solution! And in that solution, the fish and the plankton and algae they feed on need dissolved oxygen. Colder waters are able to hold more oxygen since the solubility of gases increases with decreasing temperature. So the solubility of ocean water varies with different conditions such as temperature. You will get to see this for yourself in this lab activity as you vary the temperature of a solution and observe how solids dissolved in the solution respond to those changes.

QUESTIONS
» How does changing the temperature of a solvent alter the solubility of a salt?
» How do scientists make a solubility curve?

One Giant Solution | 133

» Demonstrate how the solubility of a salt varies with temperature.
» Plot the solubility curve of a salt on the basis of observed data.

MEASURING SOLUBILITY

Scientists must specify the exact conditions when measuring solubility because solubility varies so much with differing conditions.

PRE-LAB CHECK

1. Define solubility. *(Solubility is the amount of solute that will dissolve in a given amount of solvent at a given temperature to produce a saturated solution.)*

2. What factors can affect the solubility of solid or gaseous solutes? *(Temperature affects the solubility of both solids and liquids; pressure affects the solubility of gases only.)*

3. What change will you observe that signals you to record the temperatures in this experiment? *(Crystals will begin to form.)*

4. Why will stirring help dissolve the solute? *(A major factor that affects the rate at which a solute dissolves is the number of collisions that occur between the solute and solvent particles. Stirring increases the number of collisions by bringing more solvent in contact with the solute.)*

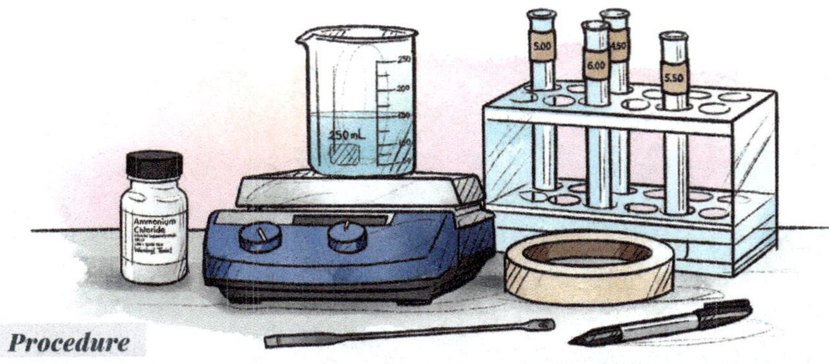

Procedure

A Add about 150 mL of water to the beaker. Check the water level by immersing the four test tubes in the water. Water level should be near the top of the beaker. Move the test tubes back to the test tube rack.

B Place the beaker on the hot plate and set the hot plate on medium-high. Adjust the temperature of the hot plate to keep the hot water bath near boiling.

C Label the test tubes as follows: "6.00," "5.50," "5.00," and "4.50."

D Using the spatula, weighing paper, and laboratory balance, measure approximately 6.00 g of ammonium chloride and record the exact mass in Table 1. Add this to Tube 6.00.

E Repeat Step D with 5.50 g, 5.00 g, and 4.50 g of ammonium chloride in the appropriate test tubes.

1. Which sample will dissolve the quickest? Explain.

   The 4.50 g sample will dissolve the quickest since it is the smallest amount to dissolve.

F Using the graduated cylinder, add exactly 10.0 mL of distilled water to each test tube. You may want to use a pipette to control the addition of water.

2. Which test tube do you think will form crystals the quickest when cooled? Why?

   Students should predict that Tube 6.00 will crystallize the quickest since it is the most concentrated.

G Place the four test tubes into the hot-water bath. Do not allow any water from the beaker to get into the test tubes!

3. Why is it important that no water from the hot-water bath get into the test tubes?

   Increasing the water in the test tubes would decrease the concentration of ammonium chloride in the test tubes, affecting measurements of solubility.

H Stir the solutions with the stirring rod to help dissolve the ammonium chloride. Be sure to rinse and dry the stirring rod before putting it into a different solution. When the solute in all four tubes has completely dissolved, turn off the hot plate.

4. Use the kinetic-molecular theory to describe why the hot-water bath speeds up the solvation process for solids.

   Temperature speeds up particle motion, allowing the solute particles to dissociate and dissolve much more quickly.

I. Using the test tube clamp, remove Tube 6.00 from the water bath and place it in your test tube rack. Place a thermometer into the test tube and allow the solution to cool.

J. When crystals start forming in the liquid, note the temperature and record it in Table 1. Double-check this temperature by reheating the test tube just enough to dissolve the solute again. If the temperatures for the first and second crystallizations differ by more than a few degrees, carefully repeat the reheating and recooling process.

Name ___________________________

K. Repeat Steps I and J for Tubes 5.50, 5.00, and 4.50.

5. Which tube formed crystals at the highest temperature? How did this compare with your prediction in Question 2?

See TE margin for answer.

____

____

____

____

CREATING A SOLUBILITY CURVE

L. Create a scatterplot of the data from Table 1, including a curve of best fit that extends down to 0 °C and up to 100 °C.

6. Using your solubility curve, determine the solubility of ammonium chloride (in g/10.0 mL H_2O) at both 90 °C and 10 °C. Record your answers in Table 2.

Answers will vary. *Example* (for 90 °C):

What we know: = 90 °C

Unknown: S

Write the formula and solve for the unknown.

$$S = 0.0468 \frac{g}{10\ mL \cdot °C} T + 2.503 \frac{g}{10\ mL}$$

Evaluate.

$$S = \left(0.0468 \frac{g}{10\ mL \cdot °C}\right)(90.0\ °C) + 2.503 \frac{g}{10\ mL}$$

$$= 4.212\ g/10\ mL + 2.503\ g/10\ mL$$

$$= 6.72\ g/10\ mL$$

7. Obtain the accepted values for the solubility of ammonium chloride at both 90 °C and 10 °C. Calculate your percent error. Record your answers in Table 2.

Answers will vary. *Example* (for 90 °C):

What we know: $S_{measured} = 6.72$ g/10 mL, $S_{accepted} = 7.12$ g/10 mL

Unknown: $\%_{error}$

Write the formula and solve for the unknown.

$$\%_{error} = \left(\frac{value_{measured} - value_{accepted}}{value_{accepted}}\right)100\%$$

Evaluate.

$$\%_{error} = \left(\frac{6.72\ g/10\ mL - 7.12\ g/10\ mL}{7.12\ g/10\ mL}\right)100\%$$

$$= \left(\frac{-0.40\ g/10\ mL}{7.12\ g/10\ mL}\right)100\%$$

$$= -5.6\%$$

Question 5 Answer

Students should see that Tube 6.00 most quickly formed crystals. In most cases, this will match their prediction from Question 2.

⚠ LASER POINTER

Shining a laser pointer into a solution may make it easier to see the early stages of precipitation. Of course, laser pointers can be a hazard if the laser light hits someone in the eye. Consider class size and the maturity of your students if you are considering having students use laser pointers in the laboratory. It may be best for the teacher to operate the laser pointer.

QUESTION 7: ACCEPTED VALUES

The accepted values for the solubility of ammonium chloride are 7.12 g/10 mL at 90 °C and 3.32 g/10 mL at 10 °C.

8. Describe how the shape of the solubility curve would change, if at all, if the actual temperatures were all 7 °C lower than what you measured.

The shape would not change, but the entire curve would be shifted to the left of the actual curve, that is, to larger-than-actual values for solubility.

9. You have learned that the solubilities of gases decrease with increasing temperature. Predict how a solubility curve of a gas would look compared with one for a solid.

A solubility curve for a gas has a negative slope instead of a positive one.

Going Further

10. In the introduction we noted that the ocean is one giant solution. The other solution that animals such as fish rely on is their blood. How are the solutions of ocean water and blood similar? How are they different?

Blood is a solution that contains oxygen, just like the ocean does. It also contains dissolved nutrients that a physical body acquires from food. Unlike in the ocean, living things have processes that work in blood to maintain appropriate levels of substances.

SAMPLE DATA

Student data may vary from the sample data in this table but should be close to these values.

TABLE 1 *Data*

Mass of NH_4Cl (g)	Temperature for Crystallization (°C)	
	Trial 1	Trial 2
6.03	75	73
5.47	65	64
4.98	54	55
4.51	43	41

TABLE 2 *Results*

Temperature (°C)	Solubility$_{measured}$ (g/10 mL)	Solubility$_{accepted}$ (g/10 mL)	Percent Error
90.0	6.72	7.12	-5.6
10.0	2.97	3.32	-11.0

14B LAB

Sugar, Sugar

Determining the Sugar Content in Beverages

Sports drinks, sweet tea, fruit juice, and soda are popular beverages. While these drinks come in many flavors, they all have one thing in common: they are all solutions, and their main ingredients are water and sugar.

How much sugar is really in my favorite beverage?

Have you ever read the nutrition label on these beverages? They all report the grams of sugar per serving. Do you think that the numbers reported are accurate? In this lab activity, you will measure how the density of a solution changes as more sugar is added to it. From this information, we will estimate the sugar in some common beverages. We will then compare this with the sugar content on the label.

QUESTIONS

» What is a calibration curve?

» How can I use density to determine the sugar content of a solution?

EQUIPMENT

- laboratory balance
- beakers, 50 mL (7)
- weighing boat
- stirring rod
- beaker, 100 mL
- volumetric pipette, 10 mL
- pipette bulb
- wash bottle with distilled water
- marker
- distilled water
- food coloring
- sugar
- beverages
- goggles
- laboratory apron

MAKING REFERENCE SOLUTIONS

A Label five of the 50 mL beakers as follows: "0," "5," "10," "15," and "20."

B Fill Beaker 0 with approximately 50 mL of distilled water.

C Using the laboratory balance, add 47.5 g of distilled water to Beaker 5. Add two drops of blue food coloring.

D Using the weighing boat and laboratory balance, measure out 2.50 g of table sugar. Add the sugar to Beaker 5 and stir until the sugar is dissolved.

E Repeat Steps C and D using Beaker 10, 45.0 g of distilled water, two drops of green food coloring, and 5.00 g of sugar.

F Repeat Steps C and D using Beaker 15, 42.5 g of water, two drops of yellow food coloring, and 7.50 g of sugar.

G Repeat Steps C and D using Beaker 20, 40.0 g of water, two drops of red food coloring, and 10.00 g of sugar.

LAB 14B OBJECTIVES

» Create a calibration curve of density and mass fraction.

» Determine the mass fraction of common beverages empirically.

» Evaluate sugar content data provided on nutrition labels.

SUGAR AND NUTRITION

If students are interested in this topic of added sugar in their diet, consider having them do some research on the internet. The Centers for Disease Control, National Institutes of Health, and health.gov all have interesting information on this topic.

 STANDARD SOLUTIONS

If time is a concern, you could have the standard solutions premade for the students. To make the solutions, scale up the instructions in Steps A–G for the amount of solution that you need for the class. Then have students begin the procedures at Step H.

PRE-LAB CHECK

1. How many grams of sugar do you need to add to 50 g of water to make a 5% solution? *(2.63 g)*

2. What is the density of the 5% solution? Assume that $V_{solution} = V_{water}$. *(1.05 g/mL)*

3. According to the nutrition label, orange soda contains 49 g of sugar in a 355 mL serving. The density of the soda is 1.043 g/mL. What is the mass fraction of sugar in the soda? *(13.2%)*

4. What is the molarity of the soda in Question 3 if the sugar in the soda is sucrose ($C_{12}H_{22}O_{11}$)? *(0.40 M)*

VOLUMETRIC PIPETTES

You may need to demonstrate the use of a volumetric pipette to students. If this is a new skill for them, you may want to have them practice with the pipette before doing their data collection. They can pipette the clear solution a number of times until they get proficient at the process.

Question 2 Answer

If you used a clean, dry pipette for each measurement, your results would be even better. Because there is some distilled water left in the pipette, our density measurements are going to be lower than the actual values.

Reference Solutions
0%: Clear
5%: Blue
10%: Green
15%: Yellow
20%: Red

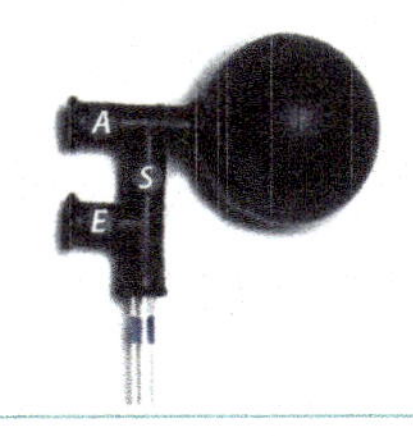

Hold the "A" valve and then squeeze the bulb to release air and create a vacuum. Press the "S" valve with the tip of the pipette in the solution to draw solution into the pipette. Release "S" when the liquid reaches the volumetric level. Use the "E" valve to release the liquid from the pipette.

H Measure the mass of the empty 100 mL beaker and record your data in Table 1.

I Using the volumetric pipette and bulb, transfer 10.00 mL of the clear reference solution (Beaker 0) into the 100 mL beaker.

J Measure the mass of the beaker and solution and record your data in Table 1.

K Repeat Steps H–J for the other four reference solutions. *Be sure to rinse the pipette with distilled water between each sample.*

1. Why is it important to rinse the pipette between each sample?

   By rinsing the pipette each time, we eliminate any residual solution. This will make our measurements more accurate.

2. On the basis of your answer to Question 1, how could you modify the procedures to improve your results even more?

   See TE margin for answer.

3. How can we find the masses and densities of the individual reference solutions when we don't empty the beaker between taking each mass?

   Each time we add 10.00 mL of a reference solution, the mass will increase only by the mass of the new solution. By subtracting each successive mass we can know the mass added with each new solution.

4. Calculate the mass and density for each of the reference solutions. Record your answers in Table 1.

Answers will vary. *Example* (for yellow solution, Beaker 15):

What we know: $V = 10.00$ mL, $m_{initial} = 49.25$ g, $m_{final} = 60.02$ g

Unknown: $m_{solution}$, ρ

Write the formula and solve for the unknown.

$$m_{solution} = m_{final} - m_{initial}$$

Evaluate.

$$m_{solution} = 60.02 \text{ g} - 49.25 \text{ g} = 10.77 \text{ g}$$

Write the formula and solve for the unknown.

$$\rho = \frac{m}{V}$$

Evaluate.

$$\rho = \frac{10.77 \text{ g}}{10.00 \text{ mL}}$$
$$= 1.077 \text{ g/mL}$$

CREATING A CALIBRATION CURVE

L Create a scatterplot of the mass fraction (*x*-axis) and the density (*y*-axis) for the five reference solutions.

5. When plotting data such as that obtained in this experiment, why is it not appropriate to connect the dots?

   As with all scatterplots, we are looking for the relationship between the two variables. By looking at the general trend, we can factor out any small scale variations in our data.

6. Determine the equation of the line of best fit.

   Answers will vary. *Example*:

   What we know: Point 1 (5%, 1.024 g/mL), Point 2 (15%, 1.072 g/mL)

   Unknown: *m*, *b*, line of best fit

   Write the formula and solve for the unknown.

   $$m = \frac{\Delta y}{\Delta x} = \frac{(y_2 - y_1)}{(x_2 - x_1)}$$

   Evaluate.

   $$m = \frac{(1.072 \text{ g/mL} - 1.024 \text{ g/mL})}{(0.15 - 0.05)}$$

   $$= \frac{0.048 \text{ g/mL}}{0.10}$$

   $$= 0.48 \text{ g/mL}$$

   Write the formula and solve for the unknown.

   $$y_2 = mx_2 + b$$

   $$b = y_2 - mx_2$$

   Evaluate.

   $$b = 1.072 \text{ g/mL} - (0.48)(0.15)$$

   $$= 1.00 \text{ g/mL}$$

   Write the formula and solve for the unknown.

   $$y = mx + b$$

   Evaluate.

   $$\rho = 0.48 \text{ g/mL } w_{sugar} + 1.00 \text{ g/mL}$$

7. Does the *y*-intercept make sense? Explain.

   Yes. The *y*-intercept is the density when the mass fraction is 0%, in other words, pure water. The density of pure water is 1.00 g/mL, which is the *y*-intercept of our graph.

MEASURING THE BEVERAGES

M Using your last two 50 mL beakers, obtain approximately 20 mL samples of two beverages. Record which beverages you chose in Table 2.

N Refer to the nutritional label on the beverage and record the volume and grams of sugar. Record this data in Table 2.

O Using the volumetric pipette and bulb, transfer 10.00 mL of the first beverage into the 100 mL beaker that is on the balance.

P Measure the mass of the beaker and solution and record your data in Table 2.

Q Rinse the pipette with distilled water.

R Repeat Steps O–Q for the other beverage.

S Calculate the mass and density for each of the two beverages. Record your answers in Table 2.

8. Using the line of best fit from your scatterplot, estimate the mass fraction of the two beverages. Record your answers in Table 2.

Answers will vary. *Example* (for cola):

What we know: $\rho = 1.048$ g/mL

Unknown: w_{sugar}

Write the formula and solve for the unknown.

$$\rho = 0.48 \text{ g/mL } w_{sugar} + 1.00 \text{ g/mL}$$

$$\rho - 1.00 \text{ g/mL} = 0.48 \text{ g/mL } w_{sugar} + \cancel{1.00 \text{ g/mL}} - \cancel{1.00 \text{ g/mL}}$$

$$\frac{\rho - 1.00 \text{ g/mL}}{0.48 \text{ g/mL}} = \frac{\cancel{0.48 \text{ g/mL}} \, w_{sugar}}{\cancel{0.48 \text{ g/mL}}}$$

$$w_{sugar} = \frac{\rho - 1.00 \text{ g/mL}}{0.48 \text{ g/mL}}$$

Evaluate.

$$w_{sugar} = \frac{1.048 \text{ g/mL} - 1.00 \text{ g/mL}}{0.48 \text{ g/mL}}$$

$$= \frac{0.048 \, \cancel{\text{g/mL}}}{0.48 \, \cancel{\text{g/mL}}}$$

$$= 0.100$$

9. Using your experimental density, experimental mass fraction, and the volume data from the nutrition label, determine the mass of sugar in each beverage.

Answers will vary. *Example* (for cola):

What we know: $\rho = 1.048$ g/mL, $w_{sugar} = 10.00\%$, $V = 591$ mL

Unknown: $m_{beverage}, m_{sugar}$

Convert volume beverage to mass.

$$591 \, \cancel{\text{mL}} \left(\frac{1.048 \text{ g}}{1 \, \cancel{\text{mL}}} \right) = 619.368 \text{ g}$$

Write the formula and solve for the unknown.

$$w_{solute} = \left(\frac{m_{solute}}{m_{solution}} \right) 100\%$$

$$\frac{w_{solute}}{100\%} = \left(\frac{m_{solute}}{m_{solution}} \right) \frac{\cancel{100\%}}{\cancel{100\%}}$$

$$\left(\frac{w_{solute}}{100\%} \right) m_{solution} = \left(\frac{m_{solute}}{\cancel{m_{solution}}} \right) \cancel{m_{solution}}$$

$$m_{solute} = \left(\frac{w_{solute}}{100\%} \right) m_{solution}$$

Evaluate.

$$m_{sugar} = \left(\frac{10.00\%}{100\%} \right) 619.3 \text{ g}$$

$$= 61.9 \text{ g}$$

10. Calculate the percent error for the grams of sugar for each beverage.

Answers will vary. *Example* (for cola):

What we know: $m_{measured} = 61.9$ g, $m_{accepted} = 69.0$ g

Unknown: $\%_{error}$

Write the formula and solve for the unknown.

$$\%_{error} = \left(\frac{value_{measured} - value_{accepted}}{value_{accepted}}\right)100\%$$

Evaluate.

$$\%_{error} = \left(\frac{61.9\text{ g} - 69.0\text{ g}}{69.0\text{ g}}\right)100\%$$

$$= \left(\frac{-7.1\text{ g}}{69.0\text{ g}}\right)100\%$$

$$= -10\%$$

11. State a claim about the accuracy of the grams of sugar reported on the nutritional labels.

See TE margin for answer.

____

____

____

____

____

Going Further

12. This lab activity looks at the relationship between the density of a beverage and its sugar content. How can you apply what you have learned in this lab activity to make appropriate choices in your sugar consumption? Explain.

Point students to the ethics triad. They should be able to apply biblical principles regarding how they should take care of their bodies (as God's temple) by limiting the amount of sugar that they consume. They may also note that the way that they view their bodies is a reflection of their gratitude and love for His creation. These biblical principles and the research collected from this lab activity should lead to a framework that can be used to make decisions about sugar consumption.

Question 11 Answer

Answers will vary. Example: The reported values were reasonable. The apple juice was only slightly off, and this is the one that we would expect to be most accurate. The sports drink reported sugar was 11% less than what was experimentally determined, but we know that sports drinks have other solutes (salts) dissolved in them. The sugar in the soda was reported as 10% higher, but the experimental value could have been off due to any carbon dioxide gas remaining in the sample.

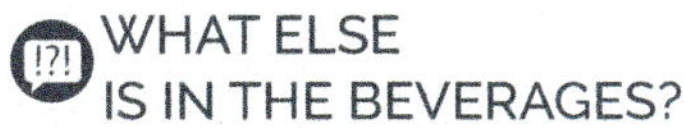

WHAT ELSE IS IN THE BEVERAGES?

Some students may have difficulty thinking about what else may be in the beverages. In the case of soda, there is dissolved gas, which would indicate a lower-than-actual density. This could result in students concluding that there was less than the actual amount of sugar in the soda. In the case of sports drinks, the empirically determined amount of sugar will be higher than actual. This occurs because sports drinks contain dissolved salts. The salts will increase the density. Our assumption that added density is due to sugar results in the faulty values.

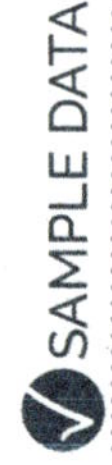

SAMPLE DATA

Student data may vary from the sample data in this table but should be close to these values.

TABLE 1 *Reference Solutions*

Solution	Mass Fraction	V (mL)	Mass (g)	Solution Mass (g)	Density (g/mL)
Empty Beaker			18.58		
Clear	0.00%	10.00	28.63	10.05	1.005
Blue	5.00%	10.00	38.84	10.21	1.021
Green	10.0%	10.00	49.25	10.40	1.040
Yellow	15.0%	10.00	60.02	10.77	1.077
Red	20.0%	10.00	70.98	10.97	1.097

TABLE 2 *Beverages*

Solution	Volume (mL)	Labeled Volume (mL)	Labeled Sugar Mass (g)	Beaker and Solution Mass (g)	Solution Mass (g)	Density (g/mL)	Experimental Mass Fraction	Experimental Sugar Mass (g)	Percent Error
Yellow Sports Drink	10.00	450	26.0	81.28	10.30	1.030	6.25%	29.0	11.0
Cola	10.00	591	69.0	91.76	10.48	1.048	10.00%	61.9	−10.0
Apple Juice	10.00	296	31.0	112.38	10.50	1.050	10.42%	32.4	4.5

Name _______________

Date _______________

Hot Shot

Finding the Specific Heat of a Metal

Have you ever absent-mindedly grabbed the handle of a cast iron skillet that's been sitting on a lit burner? Then you may have noticed how easily iron heats up! This is because of its low specific heat. In very simple terms, it doesn't take a lot of thermal energy to raise iron's temperature by 1 °C.

How can I measure how easily a material heats up?

In this lab activity, you will add some heated metal shot to water inside a simple calorimeter. You'll observe the thermal energy transfer to determine the specific heat of your metal.

QUESTIONS

» How can we observe energy transfer?

» How do I use the law of conservation of energy to account for thermal energy transfer?

» How can I determine the specific heat of a metal?

Procedure

SETTING UP

A Assemble a ring stand and ring, and place the wire gauze on the ring (see below).

B Fill a 250 mL beaker halfway with hot water and place it on the wire gauze, positioning your laboratory burner under the gauze. Light the laboratory burner and begin heating the water.

C While the water is heating, use the weighing dish and laboratory balance to obtain 50–70 g of metal shot. Measure the actual mass of the metal shot and transfer the shot to the test tube. Record the mass in Table 1.

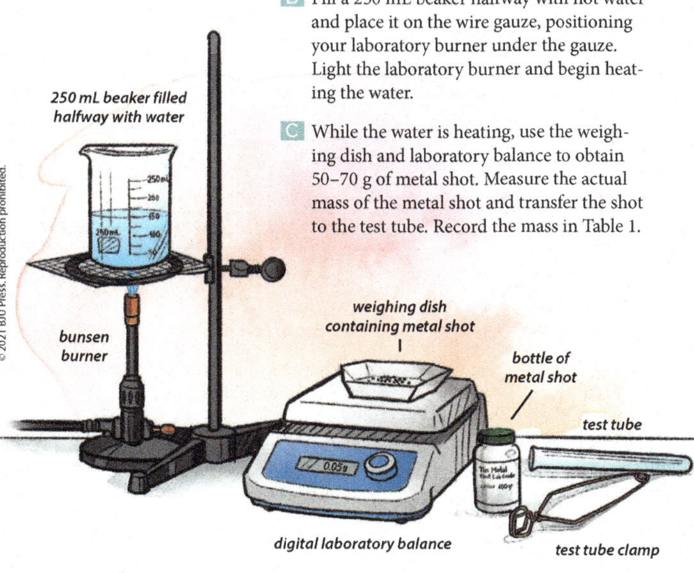

Setup for heating metal shot in a hot-water bath

EQUIPMENT

- laboratory burner and lighter
- laboratory balance
- ring stand and ring
- wire gauze
- beaker, 250 mL
- weighing dish
- medium test tube
- foam cups, 6–8 oz (2)
- graduated cylinder, 100 mL
- room-temperature distilled water
- thermometer
- beaker, 400 mL
- split rubber stopper, 1-hole, #4
- cardboard, 10 × 10 cm
- test tube clamp
- metal shot, 50–70 g
- goggles
- laboratory apron
- puncture-proof gloves

» Measure thermal energy transfer in a calorimeter.

» Analyze thermal energy transfer using the law of conservation of energy.

» Determine the specific heat of a metal empirically.

A NOTE ON TIMING

Lab 15A should be done after students have worked through the material in Section 15.1 of the Student Edition.

EQUIPMENT NOTES

Calorimeters

Foam or electric calorimeters are available from science suppliers. Some 8 oz cups are sufficiently tapered at the bottom to fit into a 250 mL beaker, making the 400 mL beaker unnecessary. Lids with holes for the thermometer or probe dramatically decrease the energy loss, producing more accurate results.

Bunsen Burners

A hot plate works well for heating the metal without the issue of open flame. Hot plates are typically easier to operate.

Probeware

Probeware can be substituted for thermometers.

Metals

Zinc, copper, or lead shot may be used for this activity as well as any of a number of other metals. If you choose to use other metals, look up the literature values for their specific heats ahead of time. Bulk metals in the form of cubes or cylinders can be substituted for metal shot.

PRE-LAB CHECK

1. On what law is this lab activity based? *(the law of conservation of energy)*

2. Why can we assume that a negligible amount of heat energy will be transferred between the water and the surroundings in this experiment? *(An insulated cup will be used to minimize heat transfer.)*

3. Where does the heat energy lost by the hot metal end up? How will you observe the transfer of energy? *(It enters the water and raises the water's temperature.)*

4. What is meant by the symbol ΔT? *(change in temperature)*

MATH IN LAB 15A

The symbol ΔT represents the change in temperature of the water, $T_1 - T_2$, and c_{water} is 4.18 J/g·°C. The calculations assume that no thermal energy is lost or gained by the calorimeter during the experiment.

Some students may have difficulty with Questions 4–6. Answer any questions that students may have as they arise. As an alternative, guide students through the concepts during your pre-lab discussion. The math here is most difficult in Question 6, in which they rearrange the equation to solve for the specific heat of the metal.

Students may need to be reminded that the value for $(c_{sp}m\Delta T)_{metal}$ should be negative; that is, the metal is *losing* heat. But because the ΔT value for the metal will *also* be negative, the overall value for that side of the equation will be positive, as it is for the left side.

Heating Up

D Place the test tube into the beaker of water and bring the water to a boil.

1. How do you think the temperature of the metal shot will be affected during heating?

   The temperature of the metal shot will reach the same temperature as the water.

E While the metal sample is heating, measure the combined mass of the two foam cups and record the value in Table 1. These cups will be used to construct your calorimeter.

F Using the graduated cylinder, measure about 50 mL of distilled water into the inner cup and weigh the cups and the water. Record this mass in Table 1.

G Calculate the mass of water that is in your calorimeter and record your answer in Table 1.

H After the water in the beaker has boiled for at least 10 minutes, measure the temperature of the water. Be sure to hold the thermometer in the center of the beaker. Record this temperature as the initial temperature of the metal in Table 1.

2. Will the metal shot gain or lose thermal energy in the calorimeter? Explain.

   The metal will lose thermal energy because it will be hotter than the water in which it will be placed.

3. Will the water gain or lose thermal energy in the calorimeter? Explain.

   The water will gain thermal energy because it will be cooler than the metal shot that will contact it.

4. Write a word equation that relates the heat of the water and the heat of the metal.

   heat gained by the water = −heat lost by the metal

5. Substitute the formula for thermal energy into your equation, remembering to keep the water on one side and the metal on the other.

$$Q_{water} = -Q_{metal}$$

$$(c_{sp}m\Delta T)_{water} = -(c_{sp}m\Delta T)_{metal}$$

$$[c_{sp}m(T_f - T_i)]_{water} = -[c_{sp}m(T_f - T_i)]_{metal}$$

6. Rearrange this equation to solve for the specific heat of the metal. This is what you are looking for! Show your work in the margin if needed.

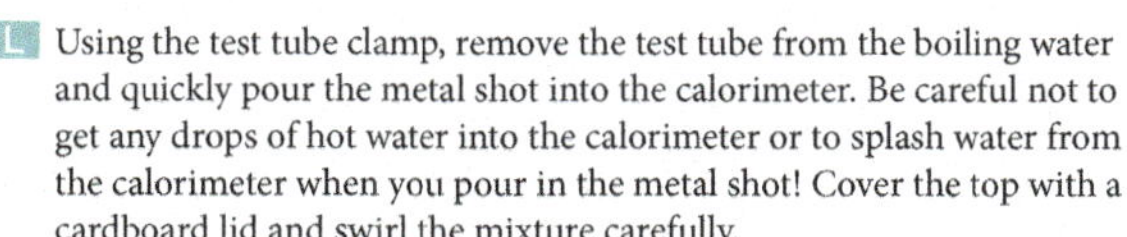

$$c_{sp\,water}\,m_{water}\left(T_{f\,water}-T_{i\,water}\right)=c_{sp\,metal}-m_{metal}\left(T_{f\,metal}-T_{i\,metal}\right)$$

$$\frac{c_{sp\,water}\,m_{water}\left(T_{f\,water}-T_{i\,water}\right)}{-m_{metal}\left(T_{f\,metal}-T_{i\,metal}\right)}=\frac{c_{sp\,metal}-m_{metal}\left(T_{f\,metal}-T_{i\,metal}\right)}{-m_{metal}\left(T_{f\,metal}-T_{i\,metal}\right)}$$

$$c_{sp\,metal}=\frac{c_{sp\,water}\,m_{water}\left(T_{f\,water}-T_{i\,water}\right)}{-m_{metal}\left(T_{f\,metal}-T_{i\,metal}\right)}$$

USING THE CALORIMETER

I Nest the two foam cups together, and place them inside the 400 mL beaker for added stability as shown at right.

J While wearing the puncture-proof gloves, insert the thermometer into the split rubber stopper, using a drop or two of liquid soap to lubricate it. Adjust the position of the stopper on the thermometer so that the bulb of the thermometer does not touch the bottom of the inner cup when it is inserted through the hole in the cardboard lid (used in Step L).

K Measure the initial temperature of the water in the calorimeter and record it in Table 1.

7. Do you need to convert between the Celsius scale and the Kelvin scale for this activity? Explain.

See TE margin for answer.

____

____

L Using the test tube clamp, remove the test tube from the boiling water and quickly pour the metal shot into the calorimeter. Be careful not to get any drops of hot water into the calorimeter or to splash water from the calorimeter when you pour in the metal shot! Cover the top with a cardboard lid and swirl the mixture carefully.

M Note the temperature of the water about every 30 seconds and measure the highest temperature reached. Record this final temperature of both the metal and the water in Table 1.

8. At this point, what has happened to the flow of thermal energy?

The flow of thermal energy has stopped because the water and metal shot are at the same temperature.

9. Consider the change in temperature. Did the water and metal do what you anticipated in Questions 2 and 3?

Students should observe that the metal cooled down and the water heated up as a result of the flow of thermal energy.

N Carefully decant into the sink as much water as possible without losing any metal shot. Pour the wet metal shot into the designated container so that it can dry.

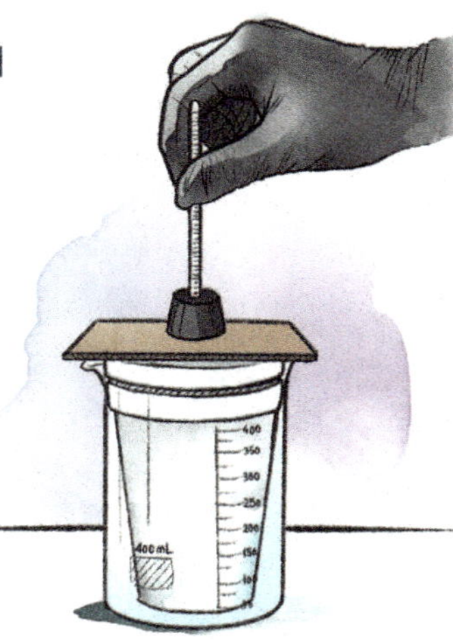

RUBBER STOPPERS AND BROKEN THERMOMETERS

Be sure that you wear gloves and exercise caution when inserting the thermometer into the split rubber stopper. Use soap as a lubricant, and don't put a force on the thermometer that could cause it to break.

Question 7 Answer

There is no need to convert between degrees Celsius and kelvins. Specific heat values are typically reported in Celsius temperatures. Students may also mention that since degrees Celsius and kelvins are of equal size, converting would have no impact on the calculations in this lab activity.

RECYCLING METAL SHOT

There is no reason that you can't recycle metal shot to use for next year. Just make sure that the metal shot dries properly. After it is completely dried, put it in a container. You may want to have separate containers for used and new shot to avoid the possibility of inadvertently contaminating new metal shot.

10. Use your data from Table 1 and the rearranged equation to solve for the specific heat of the metal. Show your work.

Answers will vary. *Example* (for lead shot):

What we know: $m_{water} = 49.85$ g, $c_{sp\,water} = 4.18$ J/g·°C, $T_{i\,water} = 25.8$ °C, $T_{f\,water} = 28.5$ °C, $m_{metal} = 55.20$ g, $T_{i\,metal} = 99.6$ °C, $T_{f\,metal} = 28.5$ °C

Unknown: $c_{sp\,metal}$

Write the formula and solve for the unknown.

$$c_{sp\,metal} = \frac{c_{sp\,water}\,m_{water}(T_{f\,water} - T_{i\,water})}{-m_{metal}(T_{f\,metal} - T_{i\,metal})}$$

Evaluate.

$$c_{sp\,metal} = \frac{\left(4.18\,\frac{J}{g\cdot°C}\right)(49.85\,g)(28.5\,°C - 25.8\,°C)}{-(55.20\,g)(28.5\,°C - 99.6\,°C)}$$

$$= \frac{\left(4.18\,\frac{J}{g\cdot°C}\right)(49.85\,g)(2.7\,°C)}{-(55.20\,g)(-71.1\,°C)}$$

$$= \frac{562}{3924}\,\frac{J}{g\cdot°C}$$

$$= 0.14\,J/g\cdot°C$$

11. Using the accepted value for the specific heat of the metal supplied by your teacher, calculate the percent error for your experimental value.

What we know: $c_{sp\,experimental} = 0.14$ J/g·°C, $c_{sp\,accepted} = 0.13$ J/g·°C (for lead shot)

Unknown: $\%_{error}$

Write the formula and solve for the unknown.

$$\%_{error} = \left(\frac{c_{sp\,experimental} - c_{sp\,accepted}}{c_{sp\,accepted}}\right)100\%$$

Evaluate.

$$\%_{error} = \left(\frac{0.14\,\frac{J}{g\cdot°C} - 0.13\,\frac{J}{g\cdot°C}}{0.13\,\frac{J}{g\cdot°C}}\right)100\%$$

$$= \left(\frac{0.01\,\frac{J}{g\cdot°C}}{0.13\,\frac{J}{g\cdot°C}}\right)100\%$$

$$= 8\%$$

SPECIFIC HEAT VALUES

The literature values for the specific heat of copper, zinc, and lead are provided below.

copper: 0.384 J/g·°C

zinc: 0.388 J/g·°C

lead: 0.127 J/g·°C

12. List three possible sources of error in your experiment.

See TE margin for answer.

____

____

____

13. Determine whether the errors you mentioned in Question 12 would result in an experimental value for the specific heat value that was larger or smaller than expected. Explain why it would have this effect.

See TE margin for answer.

____

____

____

____

Going Further

Maybe you are not used to thinking about heat as something that can be productive! Heat transfer can be very useful because it can be used to generate electricity. This usually happens when heat is used to generate steam, which can turn a turbine.

14. Suggest some ways that heat transfer can be used to generate steam for electricity generation.

Answers will vary. Students may mention a variety of energy generation methods, including the steam engine, solar power, nuclear power, and coal-fired power.

Heat can also be a waste product. When factories put out heated air or heated water, it can cause problems in the environment.

15. Suggest some problems that waste heat can cause in the environment.

Answers will vary. Students may suggest algal blooms in water or harmful effects on animal, plant, and human populations.

16. Suggest a way to deal with waste heat.

This is a major issue. Students may suggest using the heat for something else or finding ways to dissipate it that do not harm people, plants, and animals.

Question 12 Answer

Answers will vary. Accept any three. Students may indicate (1) the loss of heat energy in the transfer of metal to calorimeter, (2) the loss of heat energy to the surroundings, (3) insufficient heating of the metal (the temperature of the metal was less than that of the boiling water), (4) the loss of water from the calorimeter, or (5) the transfer of hot water into the calorimeter. Other answers are possible.

Question 13 Answer

Answers will vary. *Larger*: (1) Any losses during the transfer of metal to the calorimeter would reduce the denominator of the specific heat calculation. (2) If water were lost from the calorimeter, the temperature change of the water would be larger. (3) If hot water from the heating beaker were transferred into the calorimeter, the heating of the water would be greater and it would be assumed that the energy came from the metal. *Smaller*: (1) Loss of heat energy to the surroundings would decrease both the numerator and denominator, but would have a greater impact on the numerator. (2) If the actual initial temperature of the metal were lower than assumed, the water (the numerator) would be changed less and the metal temperature change (the denominator) would be larger.

SAMPLE DATA

The data shown in Table 1 is sample data. Student data values may differ.

17. Most factories produce large quantities of heat, and many facilities operate around the clock. Explain how an engineer would use a calculation similar to that used in this activity to determine how to use water as the coolant for a factory.

See TE margin for answer.

TABLE 1

	Mass (g)	T_i (°C)	T_f (°C)	c_{sp} (J/g·°C)
Metal (T_i for Metal)	55.20	99.6	28.5	0.14
Cups	32.42			
Cups and Water	82.27			
Water	49.85	25.8	28.5	

Name _______________________

Date _______________________

No Anchovies, Please!

Exploring Enthalpies of Solution and Reaction

Pizza is a favorite among students. Some college students practically live on pizza! Pizzas come in all shapes and sizes, and people love to pile on the toppings, such as pepperoni, mushrooms, and even anchovies. Pizza is a great source of energy for late-night study sessions. How do our bodies turn that food into energy? Through digestion. Digestion is a combined physical and chemical process by which humans and animals obtain energy. Chemical digestion uses chemical reactions to break chemical bonds, releasing energy. Our bodies use that energy for all sorts of metabolic functions. Any energy that is not used is stored for a later time.

Ideally we want to balance our energy intake with our energy usage because we don't need to be storing too much energy. To do that, we should eat balanced meals, so living

How can I measure the energy change during physical and chemical changes?

on pizza through college is probably not in our best interest. How can we measure the energy transfer during chemical and physical changes? We use calorimetry, the same process that you may recall using in Lab 15A.

In this lab activity, you will find the enthalpy of two different processes: the enthalpy of solution of potassium nitrate (KNO_3) and the enthalpy of reaction of hydrochloric acid (HCl) with magnesium. The thermal energy changes that you measure in your calorimeter are the enthalpy when the pressure is held constant.

$$Q = \Delta H$$

You'll need to do some calculations to make sure that your units are correct since enthalpy is measured in kJ/mol.

QUESTIONS

» How do I determine an enthalpy of solution?

» How can I find the enthalpy of reaction for a particular chemical reaction?

EQUIPMENT

- laboratory balance
- foam cups, 6–8 oz (2)
- beaker, 600 mL
- thermometer
- split rubber stopper, 1-hole, #4
- cardboard, 10 × 10 cm
- graduated cylinder, 100 mL
- room-temperature distilled water
- weighing dish
- potassium nitrate (KNO_3), solid
- hydrochloric acid (HCl), 1.00 M
- magnesium ribbon, 0.10–0.15 g
- goggles
- laboratory apron
- nitrile gloves
- puncture-proof gloves

» Determine the enthalpy of solution for potassium nitrate.

» Determine the enthalpy of reaction for chemical reactions.

A NOTE ON TIMING

Lab 15B should be done after students have worked through the material in Section 15.2 of the Student Edition.

EQUIPMENT NOTES

Lids

Lids with holes for the thermometers or probes dramatically decrease energy loss, producing more accurate results.

HCl Solution

The concentration of the HCl should be as close as possible to 1.00 M for the best results. To make 1 M HCl, pour approximately 70 mL of distilled water into a volumetric flask. Add 8.3 mL of concentrated HCl (12 M) and gently swirl to mix. Then add distilled water to dilute it up to 100 mL. These instructions can be scaled to make more HCl solution.

PRE-LAB CHECK

1. What will be used as the calorimeter in this experiment? *(two foam cups)*

2. What type of process does a negative ΔH indicate, endothermic or exothermic? *(exothermic)*

3. Do the products have more or less energy than the reactants in an endothermic process? *(more)*

Procedure

SETTING UP

A Nest the two foam cups together, and place them inside the 600 mL glass beaker for added stability as shown at right.

B While wearing the puncture-proof gloves, insert the thermometer into the split rubber stopper, using a drop or two of liquid soap to lubricate it. Adjust the position of the stopper on the thermometer so that the bulb of the thermometer does not touch the bottom of the inner cup when it is inserted through the hole in the cardboard lid (used in Step H). Set the thermometer, rubber stopper, and cardboard aside.

RUBBER STOPPERS AND BROKEN THERMOMETERS

Be sure that you wear gloves and exercise caution when inserting the thermometer into the split rubber stopper. Use soap as a lubricant, and don't put a force on the thermometer that could cause it to break.

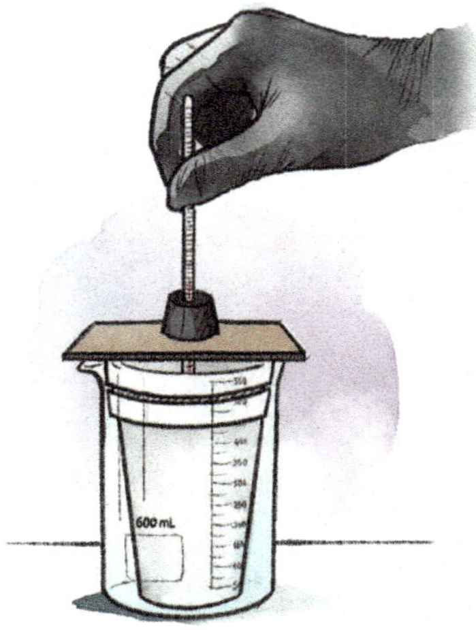

HEAT OF SOLUTION

C Measure the mass of the nested foam cups and record the value in Tables 1 and 2.

D Using the graduated cylinder, measure about 50 mL of distilled water into the inner cup and measure the mass of the cups and water. Record the mass in Table 1.

E Calculate the mass of water that you are using in your calorimeter and record it in Table 1.

F Determine the temperature of this water to the nearest 0.1 °C and record it in Table 1. Be sure that the thermometer bulb is in the center of the water sample.

G Using the weighing dish and laboratory balance, obtain 3–4 g of KNO_3. Record the actual mass of KNO_3 in Table 1.

WEIGHING WEIGHING PAPER?

If you have an electronic balance, tare or zero the balance with the weighing paper on it before adding KNO_3 so that you don't need to weigh the weighing paper and reweigh it with the KNO_3 on it.

1. Write a balanced chemical equation that shows the dissolving process of KNO_3. Don't include the water in the equation since it just makes the salt dissociate.

See TE margin for answer.

H Add the KNO_3 to the water and cover your calorimeter with the cardboard lid. Swirl the calorimeter gently while closely observing the temperature.

I Watch the temperature over a period of several minutes and record the most extreme temperature in Table 1.

2. Did the dissolving process release heat or absorb it?

Students should observe that the process absorbed heat.

Since ΔH_{soln} is expressed as J/mol (or kJ/mol), you must calculate the thermal energy and the moles of KNO_3 and then divide those answers to get J/mol.

3. Calculate the thermal energy change of the water. Show your work. Record your answer in Table 1.

Answers will vary. *Example:*

What we know: $m_{water} = 50.18$ g, $c_{sp\,water} = 4.18$ J/g·°C, $T_i = 24.4$ °C, $T_f = 19.5$ °C

Unknown: Q

Write the formula and solve for the unknown.

$$Q = c_{sp} m_{water} \Delta T$$

$$Q = c_{sp} m_{water}(T_f - T_i)$$

Evaluate.

$$Q = \left(4.18 \frac{J}{g \cdot °C}\right)(50.18\ g)(19.5\ °C - 24.4\ °C)$$

$$= \left(4.18 \frac{J}{g \cdot °C}\right)(50.18\ g)(-4.9\ °C)$$

$$= -1020\ J$$

4. Does the energy change of the water indicate that the dissolving of potassium chloride absorbs energy or releases energy? Explain.

Because the temperature of the water is decreasing, we know that the water is giving up energy to dissolve the salt. The dissolving process must be endothermic.

The temperature of the solution changed because the potassium nitrate broke up into ions. The change in temperature of the water also gives us the change in temperature of the KNO_3.

5. Now calculate the moles of KNO_3 that dissolved in the water.

Answers will vary. *Example:*

$$2.97\ g\ KNO_3 \left(\frac{1\ mol\ KNO_3}{101.11\ g\ KNO_3}\right) = 0.029\,37\ mol\ KNO_3$$

6. Using the appropriate sign—positive for endothermic or negative for exothermic—express the enthalpy in kJ/mol. Record your answer in Table 1.

Answers will vary. *Example:*

What we know: $Q = 1020$ J, $n_{KNO_3} = 0.029\,37$ mol KNO_3

Unknown: ΔH_{sol}

Convert to kJ/mol.

$$\left(\frac{1027.786\,76\ J}{0.029\,373\,9\ mol\ KNO_3}\right)\left(\frac{1\ kJ}{1000\ J}\right) = 34.7\ kJ/mol\ KNO_3$$

MATH IN LAB 15B

The symbol ΔT represents the change in temperature of the water, $T_1 - T_2$, and c_{water} is 4.18 J/g·°C. The calculations assume that no thermal energy is lost or gained by the calorimeter during the experiment.

Do not expect completely accurate results. Experimental error of 2%–5% can be expected because of inaccurate measurements and energy loss or gain from the environment.

7. Calculate a percent error on the basis of the standard value for the enthalpy of solution of KNO$_3$. This value is $\Delta H°_{sol}$ = +34.89 kJ/mol. Record your answer in Table 1.

Answers will vary. *Example:*

What we know: $\Delta H_{sol\,experimental}$ = 34.9897 kJ/mol, $\Delta H_{sol\,expect}$ = 34.89 kJ/mol

Unknown: %$_{error}$

Write the formula and solve for the unknown.

$$\%_{error} = \left(\frac{\Delta H_{sol\,experimental} - \Delta H_{sol\,accepted}}{\Delta H_{sol\,accepted}} \right) 100\%$$

Evaluate.

$$\%_{error} = \left(\frac{34.7 \text{ J/mol} - 34.89 \text{ J/mol}}{34.89 \text{ J/mol}} \right) 100\%$$

$$= \left(\frac{-0.1 \text{ J/mol}}{34.89 \text{ J/mol}} \right) 100\%$$

$$= 0\%$$

8. Is your value for enthalpy positive or negative? How does this relate to your observations in Question 3?

Students should observe a positive enthalpy, which corresponds to the absorption of thermal energy that they observed.

HEAT OF REACTION

The next reaction generates minimal hydrogen gas because you are using a fairly dilute hydrochloric acid. Nevertheless, make sure that there are no open flames nearby when doing this activity.

J. Empty the calorimeter, rinse well with water, and allow to drain.

K. Using the graduated cylinder, measure 75.0 mL of 1.00 M HCl solution and pour it into the inner cup. Measure the mass of the cups and hydrochloric acid, then calculate the mass of the acid alone. Record both masses in Table 2.

L. Measure the temperature of the HCl in the cup and record it in Table 2.

M. Measure the mass of the magnesium ribbon and record it in Table 2.

N. Roll the magnesium ribbon into a loose ball, drop it into the acid, and cover the cup.

This is an example of what happens when a metal reacts with a strong acid. This process usually liberates hydrogen gas and forms a salt with the anion of the acid and an ionized form of the metal, which will be a cation.

O. While gently swirling the cup, observe the temperature constantly. The metal should react completely. Record the most extreme temperature reached in Table 2.

9. What did you observe during this reaction?

The reaction formed bubbles of hydrogen gas and the temperature increased.

10. Write a balanced equation that shows the reaction of HCl with Mg to liberate hydrogen gas and form a salt.

    See TE margin for answer.

    Now calculate the theoretical enthalpy of the reaction.

11. Use the balanced equation and the standard values shown at right to calculate the enthalpy of reaction between HCl and Mg. Show your work.

ENTHALPIES OF FORMATION FOR QUESTION 11

HCl (aq): $\Delta H_f° = -167.2$ kJ/mol

Mg (s): $\Delta H_f° = -641.8$ kJ/mol

$$\Delta H = \Sigma \Delta H_{products} - \Sigma \Delta H_{reactants}$$

$$= \left[n_{H_2}\Delta H_{H_2} + n_{MgCl_2}\Delta H_{MgCl_2} \right] - \left[n_{HCl}\Delta H_{HCl} + n_{Mg}\Delta H_{Mg} \right]$$

$$= \left[(1\ \text{mol H}_2)\left(0.0\ \frac{kJ}{\text{mol HCl}}\right) + (1\ \text{mol MgCl}_2)\left(-641.8\ \frac{kJ}{\text{mol MgCl}_2}\right) \right] -$$

$$\left[(2\ \text{mol HCl})\left(-167.2\ \frac{kJ}{\text{mol HCl}}\right) + (1\ \text{mol Mg})\left(0.0\ \frac{kJ}{\text{mol Mg}}\right) \right]$$

$$= -641.8\ \text{kJ} - (-334.4\ \text{kJ})$$

$$= -307.4\ \text{kJ}$$

12. Calculate the thermal energy change of the HCl (assume that c_{sp} is the same as water) that actually happened. Record your answer in Table 2.

    Answers will vary. *Example:*

    What we know: $m_{HCl} = 76.26$ g, $c_{sp\ HCl} = 4.18$ J/g·°C, $T_i = 24.4$ °C, $T_f = 29.7$ °C

    Unknown: Q

    Write the formula and solve for the unknown.

    $$Q = c_{sp}m_{water}\Delta T$$

    $$Q = c_{sp}m_{water}(T_f - T_i)$$

    Evaluate.

    $$Q = \left(4.18\ \frac{J}{g \cdot °C}\right)(76.26\ g)(29.7\ °C - 24.4\ °C)$$

    $$= \left(4.18\ \frac{J}{g \cdot °C}\right)(76.26\ g)(5.3\ °C)$$

    $$= 1680\ J$$

13. Since the reaction occurred in the solution, does this change indicate that the reaction is endothermic or exothermic? Explain.

    Since the energy of the system increased during the reaction, the reaction must be exothermic.

Question 10 Answer

$$2HCl(aq) + Mg(s) \longrightarrow H_2(g) + MgCl_2(aq)$$

CALCULATING THERMAL ENERGY OF HCl

Since the HCl solution is mostly water, students should use the density of water to convert their volume of HCl to a mass. They should also use the specific heat of water (4.18 J/g·°C) in this equation.

The magnesium was the limiting reactant for this reaction.

14. Calculate the moles of Mg that reacted. Show your work. Record your answer in Table 2.

Convert grams of magnesium to moles.

$$0.11 \text{ g Mg} \left(\frac{1 \text{ mol Mg}}{24.31 \text{ g Mg}} \right) = 0.004\,52 \text{ mol Mg}$$

15. Express the enthalpy in kJ/mol. Record your answer in Table 2.

Answers will vary. *Example:*

What we know: $Q = -1680$ J, $n_{Mg} = 0.004\,52$ mol Mg

Unknown: ΔH_r

Convert energy per mole into kJ/mol.

$$\left(\frac{-1680 \text{ J}}{0.004\,52 \text{ mol Mg}} \right)\left(\frac{1 \text{ kJ}}{1000 \text{ J}} \right) = -371 \frac{\text{kJ}}{\text{mol Mg}}$$

$$= -3.7 \times 10^2 \text{ kJ/mol Mg}$$

16. Calculate a percent error using the enthalpy of reaction that you calculated in Question 12. Record your answer in Table 2.

What we know: $\Delta H_{r\text{ experimental}} = -371 \frac{\text{kJ}}{\text{mol Mg}}$,

$\Delta H_{r\text{ expected}} = -307.4 \frac{\text{kJ}}{\text{mol Mg}}$

Unknown: $\%_{\text{error}}$

Write the formula and solve for the unknown.

$$\%_{\text{error}} = \left(\frac{\Delta H_{r\text{ experimental}} - \Delta H_{r\text{ expected}}}{\Delta H_{r\text{ expected}}} \right) 100\%$$

Evaluate.

$$\%_{\text{error}} = \left(\frac{-371 \frac{\text{kJ}}{\text{mol Mg}} - \left(-307.4 \frac{\text{kJ}}{\text{mol Mg}} \right)}{-307.4 \frac{\text{kJ}}{\text{mol Mg}}} \right) 100\%$$

$$= \left(\frac{-63 \frac{\text{kJ}}{\text{mol Mg}}}{-307.4 \frac{\text{kJ}}{\text{mol Mg}}} \right) 100\%$$

$$= 20\%$$

17. Is your value for enthalpy positive or negative? How does this relate to your observations from Questions 9 and 11?

Students should observe a negative enthalpy, which corresponds to the release of thermal energy that they observed.

Going Further

On a small scale, people use the reaction of a metal such as copper with sulfuric acid and iron(III) chloride in a process called *photo etching*. This basically allows people to "print" a circuit that can be used in a circuit board. During this process, the acid is heated to speed up the reaction.

18. Why does heating the acid speed up the process?

Heating the acid makes its molecules move faster and react more quickly.

19. Suggest another way to use the process of etching.

Answers will vary. Artists also use etching to create prints and works of art that can be used to decorate metal pieces.

TABLE 1 *Enthalpy of Solution*

	Mass (g)	Temperature (°C)	Energy (J)	ΔH_{sol} (kJ/mol)	%$_{error}$
Cups	32.67				
Cups and Water	82.85				
Water (T_i)	50.18	24.4			
KNO_3	2.97				
KNO_3 Solution (T_f)		19.5	1.0×10^3	35	0

TABLE 2 *Enthalpy of Reaction*

	Mass (g)	Moles	Temperature (°C)	Energy (J)	ΔH_r (kJ/mol)	%$_{error}$
Cups	32.67					
Cups and HCl	108.93					
HCl (at T_i)	76.26		24.4			
Magnesium Ribbon	0.11	0.0045				
HCl (at T_f)			29.7	1.7×10^3	-3.7×10^2	20

SAMPLE DATA

The data in Tables 1 and 2 is sample data. Student data values may differ.

16A LAB

Chemistry—A Contact Sport?

Exploring Concentration's Effect on Reaction Rates

Rugby is clearly a contact sport. Two variants of the game are rugby sevens and rugby union. While both variants are played on the same-size field, the number of players differs significantly. In rugby union, each team has fifteen players on the field, while in rugby sevens there are only seven players per team. As you can imagine, there are a whole lot more collisions on the field in a rugby union game.

Chemistry is similar to a contact sport. According to the collision theory, a chemical reaction will occur only if particles collide, and when they collide they must do so with enough force and in the proper orientation. If there are more particles present, we would expect more reactions to occur.

The speed at which the reaction proceeds is its rate. *Rate* can also be defined as either the amount of product formed per unit of time or the amount of reactant consumed per unit of time. In this experiment, you will explore how varying the concentration of a reactant affects how fast a reaction takes place.

How does concentration affect reaction rate?

EQUIPMENT

- beakers, 100 mL (6)
- beakers, 200 mL (3)
- grease pencil or labeling tape
- graduated cylinders, 25 mL (3)
- stirring rod
- stopwatch
- sodium thiosulfate ($Na_2S_2O_3$), 0.20 M, 100 mL
- distilled water, 100 mL
- hydrochloric acid (HCl), 1.0 M, 175 mL
- sodium thiosulfate ($Na_2S_2O_3$), x M, 25 mL
- goggles
- laboratory apron
- nitrile gloves

QUESTIONS

» How do the concentrations of reactants affect the reaction rate?

» Once I know the rate law, can I determine the concentration of a reactant on the basis of the reaction time?

LAB 16A OBJECTIVES

» Determine the relationship between concentration and reaction rates on the basis of the collision theory.

» Predict the concentration of a solution on the basis of its reaction rate.

TIMING LAB 16A

Consider doing this lab activity to kick off Chapter 16. If you choose to do this, you may need to make your pre-lab discussion a bit longer to introduce students to the ideas of reaction rate and collision theory.

MAKING THE SOLUTIONS

0.20 M Sodium Thiosulfate

Pour approximately 50 mL of distilled water into a 100 mL volumetric flask. Add 3.16 g sodium thiosulfate anhydrous. Swirl gently until fully dissolved. Dilute to 100.0 mL.

1.0 M Hydrochloric Acid

Pour approximately 140 mL of distilled water into a 200 mL volumetric flask. Add 16.7 mL 12 M HCl. Swirl gently until fully mixed. Dilute to 200.0 mL.

Unknown Sodium Thiosulfate

Mix 12.5 mL of the 0.20 M sodium thiosulfate solution with 12.5 mL of distilled water. Swirl gently until fully mixed. This will make 25 mL of 0.10 M sodium thiosulfate solution for the unknown concentration.

You can scale these instructions to produce more solution.

PRE-LAB CHECK

1. Define *rate of a chemical reaction*. *(The rate of a chemical reaction is the speed at which a chemical reaction proceeds, the amount of reactant used per unit of time, or the amount of product formed per unit of time.)*

2. What factor will you test in this experiment? *(the effect of concentration on equilibrium)*

3. On a molecular basis, how does the factor listed above affect the rate of a reaction? *(It increases or decreases the number of collisions between molecules.)*

4. How many different concentrations will you test? *(five)*

5. How will you be able to tell when the reactions are completed? *(Reactions are completed when solutions turn cloudy.)*

6. How is the rate of a reaction related to the time that it takes for it to occur? *(Reactions with high rates occur in short time periods.)*

 SOLUBILITY

Students should recall that like dissolves like, so sulfur will not dissolve in water since water is polar. You may choose to review solubility with them at this point.

Question 1 Answer

The presence of water would decrease the concentration of the solutions. Some students might mention that if the test tubes had tap water in them, a dissolved substance in the water might react with the sodium thiosulfate.

Question 2 Answer

$$S_2O_3^{2-}(aq) + 2H^+(aq) \longrightarrow$$
$$S(s) + SO_2(aq) + H_2O(l)$$

MIXING SOLUTIONS

When solutions of the thiosulfate ion and the acid are first mixed, nothing seems to happen. Eventually, the solution will turn cloudy as sulfur forms. The sulfur is a colloidal precipitate. The rate of reaction will be the amount of time required for sulfur to be produced.

Procedure

SETTING UP

A Check that your beakers are clean and dry.

B Using the grease pencil or labeling tape, label five of the 100 mL beakers 1 through 5. Label a sixth beaker "Unknown."

C Label the 200 mL beakers "Na₂S₂O₃," "distilled water," and "HCl." Obtain 100 mL each of Na₂S₂O₃ and distilled water and 175 mL of HCl.

D In Beakers 1–5, prepare mixtures of varying concentration amounts by combining sodium thiosulfate and distilled water in the proportions shown in Table 1. Use a different graduated cylinder for the sodium thiosulfate and the water. Stir each mixture thoroughly, rinsing and drying the stirring rod before stirring subsequent mixtures.

1. When making the mixtures of the sodium thiosulfate solutions and water, why was it important that the test tubes have no water in them?

   See TE margin for answer.

   ____

The reaction that you will use is a redox reaction between the thiosulfate ion, $S_2O_3^{2-}$, supplied by the sodium thiosulfate, Na₂S₂O₃, and the hydrogen ion supplied by the hydrochloric acid, HCl. The sodium and chloride ions are spectator ions in this reaction. The two reactants produce elemental sulfur, sulfur dioxide, and water.

2. Write a balanced equation that shows this reaction.

   See TE margin for answer.

3. Elemental sulfur is a nonpolar substance. Do you think it is soluble in water? Explain.

   Nonpolar sulfur is insoluble in water, which is a polar solvent.

4. How do you think you will know when the reaction begins?

   The solution will turn cloudy as the insoluble sulfur is produced.

5. How do you think you can consistently measure the reaction time if the concentration changes the reaction rate?

   Answers will vary. Students should understand that they need to start timing the moment the reactants are mixed and stop timing when the same point of the reaction has been reached.

DETERMINING REACTION TIMES

E Set Beaker 1 on the image of the T on the right. You should be able to see the T as you look down through the solution.

F Use the remaining graduated cylinder to add 25.0 mL of 1.0 M hydrochloric acid to Beaker 1 and start the stopwatch. Stop timing once the T is no longer visible. Record the time in Table 1.

G Repeat Steps E and F for Beakers 2–5.

CREATING A MODEL

H Calculate the moles of thiosulfate ion, $S_2O_3^{2-}$, for Beakers 1–5. Show your work for Beaker 1 below and record all your results in Table 1.

What we know: $V_{solution} = 25$ mL $= 0.025$ L, $c_{solution} = 0.20$ mol $Na_2S_2O_3$/L

Unknown: n_{solute}

Write the formula and solve for the unknown.

$$c_{Na_2S_2O_3} = \frac{n_{Na_2S_2O_3}}{V_{solution}}$$

$$c_{Na_2S_2O_3} V_{solution} = \left(\frac{n_{Na_2S_2O_3}}{V_{solution}}\right) V_{solution}$$

$$n_{Na_2S_2O_3} = c_{Na_2S_2O_3} V_{solution}$$

Evaluate.

$$n_{Na_2S_2O_3} = (0.20 \text{ mol/L } Na_2S_2O_3)\, 0.025 \text{ L}$$

$$= 0.005 \text{ mol } Na_2S_2O_3$$

Convert moles $Na_2S_2O_3$ to moles $S_2O_3^{2-}$.

$$0.005 \text{ mol } Na_2S_2O_3 \left(\frac{1 \text{ mol } S_2O_3^{2-}}{1 \text{ mol } Na_2S_2O_3}\right) = 0.0050 \text{ mol } S_2O_3^{2-}$$

I Calculate the amount concentration of $S_2O_3^{2-}$ using the total volume in each beaker, including the volume of 1.0 M HCl. Show your work for Beaker 1 below and record all your results in Table 1.

What we know: $V_{solution} = 50$ mL $= 0.050$ L, $n_{solution} = 0.0050$ mol $S_2O_3^{2-}$

Unknown: c_{solute}

Write the formula and solve for the unknown.

$$c_{S_2O_3^{2-}} = \frac{n_{S_2O_3^{2-}}}{V_{solution}}$$

Evaluate.

$$c_{S_2O_3^{2-}} = \frac{0.0050 \text{ mol } S_2O_3^{2-}}{0.050 \text{ L}} = 0.10 \frac{\text{mol } S_2O_3^{2-}}{\text{L}}$$

J Create a scatterplot of the amount concentration of $S_2O_3^{2-}$ and reaction time. Include a curve of best fit.

SAMPLE CALCULATIONS

Sample calculations are based on the sample data. Your students' calculations will likely be different, but their processes should be similar.

6. What happens to the time required for the reaction and the rate of the reaction as the concentration of the thiosulfate ion increases?

As the concentration increases, the reaction rate increases, decreasing the time required for the reaction.

7. On the basis of the collision theory, explain why this relationship makes sense.

As the concentration increases, there are more particles available to collide. This will increase the number of effective collisions, causing the reaction rate to increase.

8. If the relationship between amount concentration and reaction time is correct, what is the relationship between amount concentration and reaction rate? Explain.

See TE margin for answer.

9. According to your scatterplot, how long would the reaction take if using a 0.15 M solution? if using a 0.01 M solution?

43 s; 654 s

TESTING AN UNKNOWN

K Four a 25.0 mL sample of a solution of $Na_2S_2O_3$ of unknown molarity into the clean, dry beaker marked "Unknown."

L Set the beaker on the image of the T.

M Using the graduated cylinder for the HCl, add 25 mL of HCl to the unknown and start timing. Stop timing once the T is no longer visible. Record the time in Table 2.

10. Using your scatterplot, estimate the concentration of thiosulfate in your unknown solution.

See TE margin for answer.

Question 8 Answer

The relationship between amount concentration and reaction time is an inverse one. Therefore, the relationship between amount concentration and reaction rate should be linear. Reaction time is the time per reaction, while reaction rate is the reactions per time. They are reciprocals of each other.

Question 10 Answer

Answers should be consistent with graphs. The concentration should be about 0.050 M if you produce the unknown solution as described in the note on page 157.

11. Get the actual amount concentration from your teacher. Record this value in Table 2 and calculate your percent error. Show your work below and record your results in Table 2.

Example:

What we know: $c_{predicted} = 0.054$ mol/L, $c_{actual} = 0.050$ mol/L

Unknown: $\%_{error}$

Write the formula and solve for the unknown.

$$\%_{error} = \left(\frac{value_{predicted} - value_{actual}}{value_{actual}} \right) 100\%$$

Evaluate.

$$\%_{error} = \left(\frac{0.054 \text{ mol/L} - 0.050 \text{ mol/L}}{0.050 \text{ mol/L}} \right) 100\%$$

$$= \left(\frac{0.004 \text{ mol/L}}{0.050 \text{ mol/L}} \right) 100\%$$

$$= 8\%$$

12. Look at the six beakers. Has any sulfur settled out? What does the answer to that question indicate?

No. The sulfur is likely a colloid.

13. As shown by your scatterplot, where is the reaction time more sensitive to changes in amount concentration? Why do you think this is true?

See TE margin for answer.

Going Further

In 2009 there was a fire at a nightclub in Perm, Russia. Over 100 people died as the burning insulation released cyanide. Many of the deaths and hospitalizations related to this accident were a result of cyanide poisoning.

Sodium thiosulfate can be administered intravenously to patients who experience both arsenic and cyanide poisoning. Increasing the concentration allows a patient's body to eliminate the cyanide as the sodium thiosulfate binds to the cyanide and helps it pass harmlessly through the body.

14. How does this example of chemical kinetics work to alleviate people's suffering?

Increasing sodium thiosulfate concentration saves lives, helping people recover from cyanide poisoning.

Question 13 Answer

The reaction time is more sensitive to changes in amount concentration at lower amounts concentration. When at 0.01 mol/L, it takes only a small amount of solute to double the concentration and therefore to cut the reaction time in half.

SAMPLE DATA

The data in Table 1 is similar to the data that students should produce. The scatterplot is based on the sample data. The students' plot will likely have different data points, but the curves should have the same basic shapes. The expectation is that students are using a spreadsheet program, but they may graph it on paper if desired.

TABLE 1 *Data*

Beaker	Volume $Na_2S_2O_3$ (mL)	Volume Water (mL)	Volume HCl (mL)	Total Volume (mL)	Reaction Time (s)	Moles $S_2O_3^{2-}$	Amount Concentration $S_2O_3^{2-}$ (mol/L)
1	25.0	0.0	25.0	50.0	65	0.0050	0.10
2	20.0	5.0	25.0	50.0	80	0.0040	0.080
3	15.0	10.0	25.0	50.0	108	0.0030	0.060
4	10.0	15.0	25.0	50.0	162	0.0020	0.040
5	5.0	20.0	25.0	50.0	326	0.0010	0.020

TABLE 2 *Testing an Unknown*

Reaction Time (s)	Predicted Amount Concentration (mol/L)	Actual Amount Concentration (mol/L)	Percent Error
120	0.054	0.050	8

Name ___________________

Date ___________________

Don't Overreact

Determining a Rate Law

We all understand that iron rusts slowly, while methane burns rapidly. Every reaction proceeds at its own rate. But we also understand that under different conditions, each individual reaction will proceed at different rates.

Chemists use reaction rate laws to describe how specific reactions respond to the concentration of the reactants. Rate laws can be determined only empirically.

How can we determine the rate law for a chemical reaction?

In this lab activity we will study the reaction between potassium iodide, potassium bromate, and hydrochloric acid. Iodide, bromate, and hydrogen ions react to form molecular iodine, free bromide ions, and water.

$$6KI(aq) + KBrO_3(aq) + 6HCl(aq) \longrightarrow$$
$$3I_2(aq) + 6KCl(aq) + KBr(aq) + 3H_2O(l)$$

Eliminating the spectator ions—K^+ and Cl^-—yields the net ionic equation.

$$6I^-(aq) + BrO_3^-(aq) + 6H^+(aq) \longrightarrow 3I_2(aq) + Br^-(aq) + 3H_2O(l)$$

To create a time delay in the reaction, we will also include thiosulfate ions, which react quickly with molecular iodine according to the following equation.

$$I_2(aq) + 2S_2O_3^{2-}(aq) \longrightarrow 2I^-(aq) + S_4O_6^{2-}(aq)$$

To indicate the "stop" point of our reaction, we will include some starch solution in the reaction vessel. Once all the thiosulfate ions are consumed, the molecular iodine will combine with the starch to form a black complex. The general form of the rate law for this reaction is shown below.

$$rate = k[I^-]^l\,[BrO_3^-]^m\,[H^+]^n$$

Note that l, m, and n are the reaction orders for the three ions.

In this lab activity you will vary the concentration of each ion to see its effect on reaction rate. From this information you will be able to determine the reaction rate, the reaction order for each ion, and the constant k for the temperature at which you conduct the trials.

QUESTION
» What affects the rate law?

EQUIPMENT

- beakers, 100 mL (7)
- graduated cylinders, 10 mL (2)
- graduated cylinders, 25 mL (4)
- stopwatch
- disposable pipette
- grease pencil or labeling tape
- potassium iodide solution, 0.010 M, 80 mL
- potassium bromate, 0.040 M, 80 mL
- sodium thiosulfate solution, 0.000 25 M, 80 mL
- hydrochloric acid, 0.10 M, 80 mL
- distilled water, 80 mL
- starch solution, 5 mL
- goggles
- laboratory apron
- nitrile gloves

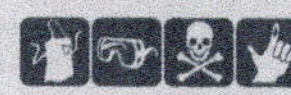

» Collect data related to changing reaction rates that are due to changes in concentration.

» Determine the reaction order for each ion empirically.

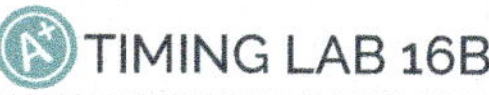 TIMING LAB 16B

Lab 16B should be done as a summative activity for Chapter 16.

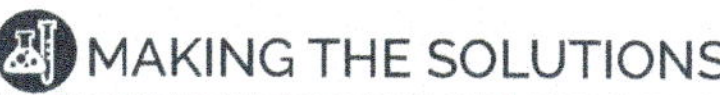 MAKING THE SOLUTIONS

0.010 M Potassium Iodide

Pour approximately 50 mL of distilled water into a 100 mL volumetric flask. Add 0.17 g KI. Swirl gently until fully dissolved. Dilute to 100.0 mL.

0.040 M Potassium Bromate

Pour approximately 50 mL of distilled water into a 100 mL volumetric flask. Add 0.67 g $KBrO_3$. Swirl gently until fully dissolved. Dilute to 100.0 mL.

0.000 25 M Sodium Thiosulfate

Pour approximately 250 mL of distilled water into a 500 mL volumetric flask. Add 0.20 g sodium thiosulfate anhydrous. Swirl gently until fully dissolved. Dilute to 500.0 mL. For each group, mix 8.0 mL of this solution with 72.0 mL of distilled water.

1.0 M Hydrochloric Acid

Pour approximately 70 mL of distilled water into a 100 mL volumetric flask. Add 8.3 mL 12 M HCl. Swirl gently until fully mixed. Dilute to 100.0 mL.

You can scale these instructions to produce more solution.

Starch Solution

Due to the fact that starch solution degrades fairly quickly, it is recommended that you purchase premade starch solution prior to doing this lab activity.

PRE-LAB CHECK

1. How will you know when the reaction is "done"? What causes this indication? *(The reaction is done when the solution changes color. This is caused by the starch combining with the iodine.)*

2. What is the purpose of including the thiosulfate in the reaction vessel? *(The thiosulfate reacts with the iodine so that there will be a delay in the color change.)*

3. Why is it important that the thiosulfate reacts quickly with the iodine? *(Since this reaction is quick, the iodine will not build up in the reaction vessel until all the thiosulfate is used up. Then the iodine can combine with the starch to indicate the end of the reaction.)*

4. How will you determine the reaction orders for the iodide, bromate, and hydrogen ions? *(by varying the concentration of only one ion at a time and measuring the change in the reaction rate)*

Procedure

A Label the beakers "KI," "KBrO₃," "Na₂S₂O₃," "HCl," "Water," "A," and "B."

B Label the 10 mL graduated cylinders "S" for starch and "T" for sodium thiosulfate.

C Label the 25 mL graduated cylinders "KI" for potassium iodide, "Br" for potassium bromate, "H" for hydrochloric acid, and "W" for water.

D Fill the appropriately labeled beakers with 80 mL of potassium iodide, potassium bromate, sodium thiosulfate, hydrochloric acid, and distilled water.

E Using Cylinder S, obtain approximately 5 mL of starch solution.

Table 1 indicates the volume of each solution needed in each trial. As you would expect, you will be changing the volume of the potassium iodide, potassium bromate, and hydrochloric acid. These changes allow you to check the effect that changing concentration has on reaction rate.

TABLE 1 *Volumes*

Beaker A			
Trial	Volume KI (mL)	Volume Na₂S₂O₃ (mL)	Volume H₂O (mL)
1	10.0	9.0	10.0
2	20.0	9.0	0.0
3	10.0	9.0	0.0
4	10.0	9.0	0.0

Beaker B			
Volume KBrO₃ (mL)	Volume HCl (mL)	Starch (mL)	Total Volume (mL)
10.0	10.0	1.0	50.0
10.0	10.0	1.0	50.0
20.0	10.0	1.0	50.0
10.0	20.0	1.0	50.0

1. Why do you think the water volume changes in the different trials?

See TE margin for answer.

Question 1 Answer

The water volume is changed in order to keep the total volume constant. By maintaining a constant volume, any change in the volume of a reactant results in an equal change in the concentration of that reactant. For example, doubling the volume of potassium iodide doubles the concentration of the iodide ions.

F Calculate the moles of each ion provided by the volume indicated for Trial 1. Show your work for iodide below.

What we know: $V_{solution} = 10$ mL $= 0.010$ L, $c_{KI} = 0.010$ mol/L

Unknown: n_{KI}

Write the formula and solve for the unknown.

$$c_{solute} = \frac{n_{solute}}{V_{solution}}$$

$$c_{solute}\, V_{solution} = \left(\frac{n_{solute}}{V_{solution}}\right) V_{solution}$$

$$n_{solute} = c_{solute}\, V_{solution}$$

Evaluate.

$$n_{KI} = (0.010 \text{ L})(0.010 \text{ mol/L KI})$$

$$= 0.000\,10 \text{ mol KI} \left(\frac{1 \text{ mol I}^-}{1 \text{ mol KI}}\right) = 0.000\,10 \text{ mol I}^-$$

G Using your answer from Step F, calculate the amount concentration of each ion in Trial 1. Remember that the amount concentration is affected by the total volume. Show your work for iodide below and record your answers in Table 2.

What we know: $V_{solution} = 50$ mL $= 0.050$ L, $n_{I^-} = 0.000\,10$ mol/L I$^-$

Unknown: c_{I^-}

Write the formula and solve for the unknown.

$$c_{solute} = \frac{n_{solute}}{V_{solution}}$$

Evaluate.

$$c_I = \frac{0.000\,10 \text{ mol I}^-}{0.0500 \text{ L}}$$

$$= 0.0020 \text{ mol/L I}^-$$

H Repeat Steps F–G for the remaining trials. Record these values in Table 2.

SAMPLE CALCULATIONS

Since calculations are based on the information in Table 1, students' calculations should match.

When solutions are first mixed, nothing seems to happen. The solution will eventually turn blue as the iodine-starch complex forms. The rate of reaction will be the amount of time required for iodide to be produced.

THE REACTION—TRIAL 1

I. In Beaker A, mix the appropriate volumes from Table 1 of the potassium iodide, sodium thiosulfate, and distilled water.

J. In Beaker B, mix the appropriate volumes (see Trial 1 of Table 2) of the potassium bromate, hydrochloric acid, and starch solution.

K. Pour the contents from Beaker B into Beaker A and start the stopwatch. Pour the entire mixture back into Beaker B to ensure complete mixing.

L. Stop timing when the solution turns blue and record the time in Table 3.

M. Clean Beakers A and B and dry completely in preparation for the remaining trials.

2. Why is it so important to dry the beakers between trials?

We are testing concentration in this lab activity. Any water in the beaker will alter the concentration of the solutions.

THE REACTION—REMAINING TRIALS

N. Repeat Steps I–M for the volumes of solutions from Table 1 for Trials 2–4.

3. Consider the values for the amount concentration of each substance. Explain what we are testing in each of Trials 2–4.

In Trial 2, we are testing the effect of doubling the amount concentration of iodide ions on the reaction rate. Trial 3 allows us to test the effect of doubling the amount concentration of bromate ions. Hydrogen ion concentration is tested in Trial 4.

4. Why does the concentration of thiosulfate ions not vary in any of the trials?

Thiosulfate is included to provide a time delay for the purpose of measuring the reaction. It is not a reactant in the reaction that we are studying. Therefore, it should not have any effect on the reaction rate. Also, in a controlled experiment we must keep all variables constant except for the ones that we are studying.

O. For Trial 2, calculate the change in concentration by determining the ratio of the new concentration to the original concentration for the iodide ions. Record this value in Table 3.

P For Trial 2, calculate the change in reaction rate by determining the ratio of the original reaction time to the new reaction time. Record this value in Table 3.

5. Notice that in Steps O and P we calculated a ratio. In Step O we found the change in concentration by dividing the new value by the original. But in Step P we found the change in rate by dividing the original by the new. Why do you think this change was made to calculate the change in rate?

See TE margin for answer.

____

____

Recall that the reaction orders are the exponents on each concentration in the rate law. For example, a reaction order of 0 is indicated by no change in the reaction rate when the concentration is changed. A reaction order of 1 (linear) means that the reaction rate changes by the same factor as the change in concentration. So if the concentration doubles, the reaction rate doubles also. Reaction orders of 2 (quadratic) indicate that the reaction rate will change by a square of the change in concentration. In other words, if we double the concentration, the reaction rate will be four times faster.

Q For Trial 2, determine the reaction order for iodide ions by comparing the change in concentration to the change in reaction rate. Record the reaction order in Table 3.

R Repeat Steps O–P for the bromate in Trial 3 and the hydrogen ions in Trial 4. Record these values in Table 3.

Going Further

Recall that the rate law for this reaction has the form shown below.

$$rate = k[\text{I}^-]^i[\text{BrO}_3]^m[\text{H}^+]^n$$

We have determined the reaction orders for each of the reactants. Now we want to determine the k value for the temperature at which you did your trials. To do this we are going to solve the rate law for k.

$$k = \frac{rate}{[\text{I}^-]^i[\text{BrO}_3]^m[\text{H}^+]^n}$$

In order to solve for k, we need to know the rate. Remember that we included thiosulfate to provide an indication of the reaction. We can use the concentration of thiosulfate to determine the rate of reaction.

$$rate = \frac{[\text{S}_2\text{O}_3^{2-}]}{6\,t_{\text{reaction}}}$$

The $[\text{S}_2\text{O}_3^{2-}]$ should make sense because our indication of reaction completion was when the thiosulfate was consumed. And the reaction time should also make sense as this is the time that the reaction took. The 6 in the equation comes from the two ionic equations at the beginning of the lab activity. Each occurrence of the first reaction produces three iodine molecules, and each of the iodine molecules consumes two thiosulfate ions. Therefore, the thiosulfate is consumed six times for each occurrence of the primary reaction.

Question 5 Answer

The second ratio is not the change in time but the change in rate. A rate is how something changes per unit time. So the reaction time and rate are inverses of each other. We are interested in rate, so we inverted the ratio.

 GOING FURTHER

The Going Further section is advanced. If time and interest permit, it will be interesting for students to work through determining the constant for the rate law.

6. How consistent were your values for the constant k?

Answers will vary. Most students will find the values relatively close to each other.

7. How can you explain the consistency of your values for the constant k?

Answers will vary.

8. Describe some situations where it may be important to understand reaction rates and what influences them.

See TE margin for answer.

Question 8 Answer

Answers will vary. Students may think of things undergoing decay or how we strive to prevent decay. They could mention applications such as adhesives setting or materials curing. In both of those applications we want the reaction to occur at a medium rate. Other applications involve digestion, ozone decomposition, and the growth of microorganisms. Also, an automotive engineer may want to determine ways to burn gasoline at a particular rate to improve efficiency in an engine.

 SAMPLE DATA

The values reported by students in Table 3 should be similar to those shown here but will likely show some variation according to their reported reaction times.

TABLE 2 *Concentrations*

Trial	$[\text{I}^-]$ (M)	$[\text{S}_2\text{O}_3^{2-}]$ (M)	$[\text{BrO}_3^-]$ (M)	$[\text{H}^+]$ (M)
1	0.0020	4.5×10^{-5}	0.0080	0.020
2	0.0040	4.5×10^{-5}	0.0080	0.020
3	0.0020	4.5×10^{-5}	0.0160	0.020
4	0.0020	4.5×10^{-5}	0.0080	0.040

TABLE 3 *Results*

Trial	Changed Substance X	t_{reaction} (s)	$\Delta[X]$	$\Delta\text{Rate}_{\text{reaction}}$	Reaction Order	$\text{Rate}_{\text{reaction}}$ ($\times 10^{-7}$)	k
1		98.03				0.8	12.0
2	I^-	48.74	2	2	1	1.5	12.0
3	BrO_3^-	50.42	2	2	1	1.5	11.6
4	H^+	24.15	2	4	2	3.1	12.1

Name

Date

Stressed Out

Inquiring into Le Châtelier's Principle

Personal space—people typically feel strongly about it. When someone moves into what we consider our personal space, we tend to feel uncomfortable and make adjustments to alleviate the stress that we feel.

Chemical systems do something similar. A reversible chemical reaction will reach equilibrium when the forward and reverse reactions are occurring simultaneously at the same rate. The amounts of reactants and products remain constant, a condition called the *equilibrium position*. But what happens when we interfere with a system at equilibrium? Just like when someone invades our personal space, a chemical system will adjust itself to reduce the change that was imposed upon it. This is what chemists call *Le Châtelier's principle*.

How do chemical systems respond to changing conditions?

In this inquiry lab activity you will investigate the effect of varying conditions on the equilibrium position of a chemical reaction involving solutions of iron(III) chloride and potassium thiocyanate.

Procedure

PLANNING/WRITING SCIENTIFIC QUESTIONS

A Label the test tubes "Fe" for iron(III) chloride and "K" for potassium thiocyanate.

B Obtain approximately 5 mL each of iron(III) chloride and potassium thiocyanate in the test tubes. Observe the two solutions. Put a pipette in each of the two tubes.

QUESTIONS

» What does a chemical reaction in equilibrium look like?

» How can I test the effect of different stresses on a chemical reaction in equilibrium?

» How can I predict the change in the equilibrium position of a reaction due to stresses that I put on the system?

EQUIPMENT

- test tubes (2)
- test tube rack
- Erlenmeyer flask, 250 mL
- pipettes (2)
- labeling tape or grease pencil
- iron(III) chloride (FeCl$_3$), 0.25 M, 5 mL
- potassium thiocyanate (KSCN), 0.25 M, 5 mL
- distilled water, 100 mL
- goggles
- laboratory apron
- nitrile gloves

LAB 17A OBJECTIVES

» Design experiments to test the effect of stresses placed on a chemical reaction in equilibrium.

» Analyze the effects of various inputs on the equilibrium positon of a reaction.

SCHEDULING LAB 17A

Lab 17A should be done after students have worked through the material in Section 17.2 of the Student Edition.

PREPARING SOLUTIONS

Iron(III) Chloride

One molar iron(III) chloride solution is available from science supply stores. Dilute this in a 1:3 ratio with distilled water to make your 0.25 M FeCl$_3$ solution. To make the solution from solid iron(III) chloride hexahydrate, add approximately 70 mL of distilled water to a 100 mL volumetric flask. Then add 6.76 g of FeCl$_3$·6H$_2$O. Swirl gently until fully dissolved and then dilute with distilled water to 100 mL.

Potassium Thiocyanate

One molar potassium thiocyanate solution is available from science supply stores. Dilute this in a 1:3 ratio with distilled water to make your 0.25 M KSCN solution. To make the solution from solid potassium thiocyanate, add approximately 70 mL of distilled water to a 100 mL volumetric flask. Then add 2.43 g KSCN. Swirl gently until fully dissolved and then dilute with distilled water to 100 mL.

Stressed Out | 169

PRE-LAB CHECK

1. Define *equilibrium*. *(Equilibrium is the point in a reaction at which the rate of the forward reaction equals the rate of the reverse reaction.)*

2. What happens to the concentration of each ion when the solution is at equilibrium? *(The concentration of each ion stays the same.)*

3. What stresses can affect an equilibrium? *(Changes in temperature, pressure, and concentration can all affect an equilibrium.)*

4. What is Le Châtelier's principle? *(Le Châtelier's principle states that a stress on a chemical system in equilibrium will shift the equilibrium position in the direction that reduces the effect of the stress.)*

5. Explain what is meant by the phrase "the equilibrium is shifted forward." *(When more products form as a result of stress on the system, the equilibrium is said to have shifted to the products—from left to right, or forward—as the reaction is written.)*

SOLUTION COLORS

If you have any students with impaired color vision, make sure to pair them with another student that can help them identify the colors of the indicators in this activity.

WRITING THE REACTION

It is crucial for students to correctly write the reaction in Question 1. You may choose to check their answers before progressing on to the rest of the lab activity. It may be helpful for them to write the colors of the chemicals below the reactants and products in this equation.

Question 1 Answer

$FeCl_3(aq) + 3KSCN(aq) \longrightarrow$
$Fe(SCN)_3(aq) + 3KCl(aq)$

C Fill the Erlenmeyer flask with 100 mL of distilled water, then add 5 drops of each of the solutions. Mix thoroughly and observe.

1. Write the balanced chemical equation between iron(III) chloride and potassium thiocyanate.

See TE margin for answer.

D Research the two products to determine which is producing the color observed.

E Brainstorm with your lab group about how you can disrupt the equilibrium of the system.

F Write specific questions related to chemical equilibrium that you could answer by collecting data.

DESIGNING SCIENTIFIC INVESTIGATIONS

G Write procedures to collect data that will allow you to answer the questions that you wrote in Step F above.

H Have your teacher approve your procedures.

CONDUCTING SCIENTIFIC INVESTIGATIONS

I Following the procedures that you have written, collect the data to answer the questions that you wrote.

DEVELOPING MODELS

J Develop a matrix of possible changes to the system and the effect of each of those changes.

SCIENTIFIC ARGUMENTATION

K Explain what you know about the reaction between iron(III) chloride and potassium thiocyanate. Support your claims with evidence from the data collected.

Teacher Guide

INQUIRING INTO LE CHÂTELIER'S PRINCIPLE

An inquiry lab allows students a degree of freedom, but the teacher must guide the process to achieve the desired educational outcomes. As teachers start using inquiry lab activities, they often struggle with this process of guiding students. The best method is to think through the goals that you want students to achieve and then use questions to guide them to reach those goals. You want to balance their freedom to investigate the question in the manner they choose with the need to meet your educational goals.

Procedure

PLANNING/WRITING SCIENTIFIC QUESTIONS

1. Each lab group starts with small samples of iron(III) chloride and potassium thiocyanate solutions. They will observe the two solutions and then combine them. They will observe changes in the mixture.

2. The groups will then brainstorm ways in which they could disturb the equilibrium of the system. They may think of the following:

   » adding more of the ions on either side of the equation

   » removing some of the ions from either side of the equation (difficult to do because all of the substances stay in solution)

   » changing the temperature of the reaction vessel

   » changing the pressure of the reaction vessel (In this case, the pressure change will have no effect because none of the substances are in gaseous form.)

   » adding a catalyst

   » adding water to decrease the concentration (should have no effect because all of the substances are equally affected)

3. Each lab group should write a series of questions related to chemical equilibrium that they could answer by collecting data.

DESIGNING SCIENTIFIC INVESTIGATIONS

1. Once the groups have written their questions, they need to write specific procedures to answer them. This may be a new task for some students, so you may want to do part of this task as a class discussion or review their procedures prior to them starting to collect data.

2. Help students think through how they will alter only one factor at a time. For example, if the students are testing the effect of temperature, they have to maintain constant initial concentrations.

CONDUCTING SCIENTIFIC INVESTIGATIONS

1. Once the groups have good procedures, allow them to collect data.

DEVELOPING MODELS

1. Students are to develop a matrix of possible changes to the system and the effect of each of those changes. Encourage them to work with other researchers to create a more comprehensive matrix.

2. A sample matrix is included on the facing page.

SCIENTIFIC ARGUMENTATION

1. Students should write a paragraph about what they can conclude about the reaction between iron(III) chloride and potassium thiocyanate. They should be able to explain the effect of varying conditions of the system at equilibrium. They should be able to state a claim about whether the reaction as written is exothermic or endothermic. As always, you should expect students to support their claims with evidence from the data collected.

EQUIPMENT

- hot plate
- test tubes (8)
- test tube rack
- Erlenmeyer flask with reference solution
- test tube filled with KSCN
- test tube filled with $FeCl_3$
- rubber stopper
- beakers, 250 mL (2)
- test tube clamp
- syringe without needle, 12 mL
- labeling tape or grease pencil
- potassium chloride (KCl), solid
- ice
- distilled water
- goggles
- laboratory apron
- nitrile gloves

Sample Procedure

A Label a test tube "R" for reference solution.

B Fill Tube R about half full with the solution in the Erlenmeyer flask to use as a reference for comparison with your experimental data. Record your observation about the color of the solution.

ADDING MORE OF A SUBSTANCE

C Label three additional test tubes "S1," "S2," and "S3," and fill each one about half full from the Erlenmeyer flask.

D To Tube S1, add 15 to 20 drops of the KSCN solution. Compare the tube with the solution in Tube R. Record any color change in Table 1.

E To Tube S2, add 15 to 20 drops of $FeCl_3$ solution. Compare the tube with the solution in Tube R. Record any color change in Table 1.

F To Tube S3, add approximately 1 g of solid KCl, close with a rubber stopper, and shake well. Compare the tube with the solution in Tube R. Record any color change in Table 1.

CHANGING TEMPERATURE

G Fill one of the 250 mL beakers with approximately 150 mL of tap water and place on the hot plate. Warm the water to approximately 60 °C.

H Fill the other 250 mL beaker half full with ice and then add tap water to approximately the 150 mL level.

I Label two additional test tubes "T1" and "T2" and fill each one about half full from the Erlenmeyer flask.

J Using the test tube clamp, place Tube T1 in the hot water bath and T2 in the ice water bath and allow their temperatures to stabilize.

K Compare the tubes with the solution in Tube R. Record any color changes in Table 1.

CHANGING PRESSURE

L Draw approximately 5 mL of solution from the Erlenmeyer flask into the syringe.

M Cover the inlet with your fingertip and then press the syringe plunger to increase the pressure on the fluid. Compare the tubes with the solution in Tube R. Record any color change in Table 1.

N With the inlet still covered with your fingertip, pull the syringe plunger to decrease the pressure on the fluid. Compare the tubes with the solution in Tube R. Record any color change in Table 1.

DILUTING THE SUBSTANCES

O Label two additional test tubes "D1" and "D2" and fill each one about half full from the Erlenmeyer flask.

P To Tube D1, add distilled water so that the test tube is about three-quarters full. Compare the tube with the solution in Tube R. Record any color change in Table 1.

Q To Tube D2, add distilled water so that the test tube is full. Compare the tube with the solution in Tube R. Record any color change in Table 1.

TABLE 1 *Sample Matrix*

Tube	Change	Observation	Equilibrium Shifts	[Products]	[Reactants]	K_{eq}
R	none	cherry red				
S1	added KSCN	deep red	right (forward)	increase	decrease	none
S2	added $FeCl_3$	deep red	right (forward)	increase	decrease	none
S3	added KCl	pale yellow	left (reverse)	decrease	increase	none
T1	heated reaction	pale yellow	left (reverse)	decrease	increase	decrease
T2	cooled reaction	deep red	right (forward)	increase	decrease	increase
Syringe	increased pressure	cherry red	none	none	none	none
Syringe	decreased pressure	cherry red	none	none	none	none
D1	diluted by 50%	red	none	none	none	none
D2	diluted by 100%	red	none	none	none	none

17B LAB

Precipitous Changes

Exploring Solubility Products

Water is a precious commodity. It is also a key medium for many industrial processes, during which it may become contaminated. How do we clean the water used in these processes to protect this vital resource?

How can we change the substance that precipitates from a chemical reaction?

Chemists often use chemical precipitation to clean the water used in industrial processes. By chemically treating the water, they can force hazardous compounds out of solution. These compounds can then be filtered from the water.

In this lab activity you will investigate solubility products and how they are related to solubility of solutes in solution and to the formation of precipitates.

Procedure

SETTING UP

A Label three beakers and three 10 mL graduated cylinders each for silver nitrate, sodium chloride, and sodium sulfide.

B Pour approximately 25 mL of the sodium sulfide solution into its beaker.

C Use one of the 25 mL graduated cylinders to add 25.0 mL of distilled water to each of the other two beakers.

D Using one of the weighing dishes, add 0.43 g of $AgNO_3$ to its beaker and stir until all the solid is dissolved. Using the other weighing dish, add 0.15 g of NaCl to its beaker and stir until all the solid is dissolved. Be sure to clean and dry the stirring rod between stirring the two solutions.

E Label the two test tubes "A" and "B."

1. Write the balanced equations for the dissolving of $AgNO_3$ and NaCl.

See TE margin for answer.

QUESTIONS

» How can I make some of a precipitate go away when the solution is saturated?

» How can I determine the concentration of a saturated solution?

EQUIPMENT

- **beakers, 50 mL (3)**
- **graduated cylinders, 10 mL (3)**
- **graduated cylinders, 25 mL (3)**
- **weighing dishes (2)**
- **stirring rod**
- **test tubes (2)**
- **test tube rack**
- **sodium sulfide (Na_2S) solution, 0.1 M**
- **distilled water**
- **silver nitrate ($AgNO_3$), 0.43 g**
- **sodium chloride (NaCl), 0.15 g**
- **labeling tape or grease pencil**
- **goggles**
- **laboratory apron**

LAB 17B OBJECTIVES

» Shift the equilibrium position of a precipitate reaction by adding a less soluble substance.

» Calculate the amount concentration of saturated solutions on the basis of their solubility constants.

SCHEDULING LAB 17B

Lab 17B should be done after students have worked through the material in Section 17.3 of the Student Edition.

PREPARING SOLUTIONS

Sodium Sulfide

You can purchase sodium sulfide solution from science supply companies. To make the solution from solid sodium sulfide nonahydrate, pour approximately 70 mL of distilled water into a 100 mL volumetric flask. Then add 2.40 g of $Na_2S \cdot 9H_2O$. Swirl gently until fully dissolved and then dilute with distilled water to 100 mL.

Question 1 Answer

$AgNO_3(s) \longrightarrow Ag^{3+}(aq) + NO_3^-(aq)$

$NaCl(s) \longrightarrow Na^+(aq) + Cl^-(aq)$

PRE-LAB CHECK

1. Compare the terms *concentration* and *solubility*. (Concentration is a measure of how much solute is dissolved in a solution, while solubility refers to the maximum amount of solute that can be dissolved at a particular temperature.)

2. What does a solubility product measure? (Solubility product indicates the solubility of a salt.)

3. Solubility product is a special case of what calculation? Explain. (Solubility product is a special case of the equilibrium constant. It is the equilibrium constant for dissolving a solute.)

4. What numerical values indicate soluble salts? insoluble salts? (values greater than 1; values less than 1)

5. If the equation for a solubility product includes only the ions, how can we use it to calculate the solubility of the solute? (The concentrations of the ions are related to the concentration of the solute according to the balanced equation for the dissolving of the solute. Therefore, the solubility of the solute can be determined through substitution in the solubility product equation.)

F Calculate the amount concentration of the $AgNO_3$ and NaCl solutions that you have made. Show your work for $AgNO_3$ below and record your answers for both calculations in Table 1.

What we know: $m_{AgNO_3} = 0.43$ g , $V_{solution} = 25.0$ mL $= 0.0250$ L

Unknown: n_{AgNO_3}, c_{AgNO_3}

Convert mass silver nitrate to moles of silver nitrate.

$$0.43 \text{ g AgNO}_3 \left(\frac{1 \text{ mol AgNO}_3}{169.88 \text{ g AgNO}_3} \right) = 0.002\,53 \text{ mol AgNO}_3$$

Write the formula and solve for the unknown.

$$c_{AgNO_3} = \frac{n_{AgNO_3}}{V_{solution}}$$

Evaluate.

$$c_{AgNO_3} = \frac{0.002\,53 \text{ mol AgNO}_3}{0.0250 \text{ L}} = 0.10 \text{ mol AgNO}_3/\text{L}$$

G Using the appropriate graduated cylinders, transfer 10 mL each of the silver nitrate and sodium chloride solutions to both Tube A and Tube B.

2. What are your observations of the reaction in Tube A?

   When the chemicals in Tube A mixed, a white precipitate formed and settled to the bottom of the tube.

3. Write the balanced chemical equation for combining the silver nitrate and the sodium chloride solutions.

   See TE margin for answer.

4. Write ionic and net ionic equations for combining the silver nitrate and sodium chloride solutions.

   $Ag^+ + NO_3^-(aq) + Na^+ + Cl^-(aq) \longrightarrow AgCl(s) + Na^+ + NO_3^-(aq)$

   $Ag^+ + Cl^-(aq) \longrightarrow AgCl(s)$

5. Why did a precipitate form?

   The combination of silver and chloride ions exceeded the solubility of silver chloride. Therefore, any dissolved ions above this concentration precipitated out of solution.

6. Without looking up the solubility or K_{sp} values for silver chloride, what can you conclude about them?

   See TE margin for answer.

Question 3 Answer

$AgNO_3(aq) + NaCl(aq) \longrightarrow$
$\quad AgCl(s) + NaNO_3(aq)$

Question 6 Answer

If the combination of solutions had made a saturated solution, the concentration of the silver chloride would be 0.050 M. Therefore, we know that the solubility of silver chloride must be less than 0.050 M. If the solubility of silver chloride is less than 0.0500 M, its K_{sp} must be less than 0.0025. The combination of silver and chloride ions exceeded the solubility of silver chloride. ($K_{sp} = [Ag+][Cl-]$, $K_{sp} < (0.050)(0.050) < .0025$)

Name ______________________

H Look up the K_{sp} value for silver chloride and record it in Table 1.

7. Was your prediction in Question 5 correct?

   Yes

8. On the basis of the K_{sp} value for silver chloride, calculate the solubility of the solution in Tube A.

   What we know: $K_{sp} = 1.3 \times 10^{-9}$, $[Ag^+] = [Cl^-] = [AgCl]$

   Write the formula and solve for the unknown.

   $$K_{sp} = [Ag^+][Cl^-]$$

   $$K_{sp} = [AgCl][AgCl]$$

   $$K_{sp} = [AgCl]^2$$

   $$\sqrt{K_{sp}} = \sqrt{[AgCl]^2}$$

   $$[AgCl] = \sqrt{K_{sp}}$$

   Evaluate.

   $$[AgCl] = \sqrt{1.3 \times 10^{-9}}$$

   $$= 3.6 \times 10^{-5}\ M$$

9. Look back at the ionic equation in Question 3. What will be the effect on the equilibrium of the system if you were to add sodium sulfide solution to Tube B? Explain.

   See TE margin for answer.

   ____

   ____

   ____

   ____

   ____

I Using the appropriate graduated cylinder, transfer 10 mL of sodium sulfide solution to Tube B.

10. What are your observations of the reaction in Tube B?

   When the sodium sulfide was added to Tube B, some of the white precipitate disappeared and seemed to be replaced with a black precipitate. We ended up with a layer of white precipitate on the bottom and a layer of black precipitate on the top.

11. Did the reaction behave as expected? Explain.

   Answers will vary. Students should see a decrease in the amount of the white precipitate. They will likely be surprised by the formation of a black precipitate.

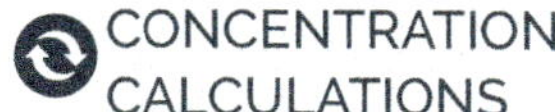 CONCENTRATION CALCULATIONS

You may need to review using the K_{sp} formula for calculating the concentration of the solute.

Question 9 Answer

Answers will vary. The addition of sodium sulfide allows sulfide ions to react with silver ions, removing silver ions from solution and shifting the reaction to the left. As the reaction shifts left, there is a decrease in silver chloride precipitate as silver sulfide precipitate forms. Some students may think that adding sodium sulfide solution increases the concentration of sodium ions, causing the reaction to shift left. However, the sodium ions are spectator ions and therefore don't affect the equilibrium position.

12. On the basis of your observations, how do you expect the K_{sp} value for silver sulfide to compare with the value for silver chloride? Explain.

Because the silver sulfide precipitate "replaces" some of the silver chloride, students should expect that silver sulfide's K_{sp} is lower than that of silver chloride. If the K_{sp} for silver chloride were the lower value, there would be very few silver ions available to react.

J. Look up the K_{sp} value for silver sulfide and record it in Table 1.

13. On the basis of the K_{sp} value for silver sulfide, calculate the solubility of the solution in Tube A.

What we know: $K_{sp} = 6.3 \times 10^{-50}$, $[S^{2-}] = [Ag_2S]$, $[Ag^+] = 2[Ag_2S]$

Write the formula and solve for the unknown.

$$K_{sp} = [Ag^+]^2[S^{2-}]$$

$$K_{sp} = (2[Ag_2S])^2[Ag_2S]$$

$$K_{sp} = 4[Ag_2S]^2[Ag_2S]$$

$$\frac{K_{sp}}{4} = \frac{4[Ag_2S]^3}{4}$$

$$\sqrt[3]{\frac{K_{sp}}{4}} = \sqrt[3]{[Ag_2S]^3}$$

$$[Ag_2S] = \sqrt[3]{\frac{K_{sp}}{4}}$$

Evaluate.

$$[Ag_2S] = \sqrt[3]{\frac{6.3 \times 10^{-50}}{4}}$$

$$= \sqrt[3]{2.0 \times 10^{-48}}$$

$$= 2.5 \times 10^{-17}\ M$$

Going Further

14. How could we use solubility and solubility products to benefit people?

Solubility and solubility products could be used to remove unwanted compounds from solutions. Scientists can select the reactants so that the substance they want to remove has the lowest solubility product, thus allowing that substance to precipitate out of solution. The scientist could then filter the liquid to remove the compound.

TABLE 1

Initial [AgNO$_3$]	0.10
Initial [NaCl]	0.10
K_{sp} Silver Chloride	1.3×10^{-9}
K_{sp} Silver Sulfide	6.3×10^{-50}
Final [AgCl]	3.62×10^{-5}
Final [Ag$_2$S]	2.5×10^{-17}

✔ SAMPLE DATA

The data shown in Table 1 is sample data. Student data may differ slightly.

18A LAB

Name ___________________________

Date ___________________________

Colorful Chemistry

Exploring Acid-Base Indicators

Indicators are chemical substances that are used to indicate the presence or concentration of another chemical. You probably think that indicators exist only in laboratories. But they're actually everywhere! Many acid-base indicators occur naturally in the world around us. The leaves of red cabbage, the flowers of geraniums and poppies, the stems of rhubarb, and the fruit of blueberries, cherries, and black currants all contain chemicals that can indicate acids or bases. Litmus, a very common laboratory indicator, comes from lichen! With the right combination of indicators, you can cover the whole pH spectrum. Of course, pH meters give you a more accurate and objective reading of pH, but indicators are useful because they are more widely available.

Acid-base indicators (identified by "In" in the equation below) are generally weak organic acids or bases that exist in equilibrium between the conjugate acid-base pair. Each member of the pair has a different color. The equation below shows this.

$$HIn(aq) + H_2O(l) \rightleftharpoons H_3O^+(aq) + In^-(aq)$$
$$\text{color 1} \qquad\qquad \text{color 2}$$

According to Le Châtelier's principle, changes in the concentration of H_3O^+ will shift the equilibrium, affecting the color of the indicator. When the color changes, we can know what's happening to the hydronium concentration of the solution.

In this experiment, you'll try your hand at making a rainbow of chemistry. You will make three sets of standards. Each set will cover a range of five concentrations and will contain an acid-base indicator. You'll use these colors to estimate the hydronium ion concentration and pH of some solutions.

How does amount concentration of weak acids affect pH?

QUESTIONS
» What do indicators do?
» How can I estimate how acidic a solution is?
» How can I measure the concentration of an acid?

EQUIPMENT
• pH meter
• test tubes (21)
• test tube racks (2)
• graduated cylinder, 100 mL
• graduated cylinder, 10 mL
• transfer pipettes (3)
• labeling tape or grease pencil
• hydrochloric acid (HCl), 0.1 M
• distilled water
• thymol blue solution
• methyl orange solution
• methyl red solution
• acetic acid $(HC_2H_3O_2)$, 0.1 M
• unknown weak acid
• goggles
• laboratory apron
• nitrile gloves

SCHEDULING LAB 18A

This lab activity works well after pH is discussed in Section 18.2 of the Student Edition.

NATURAL INDICATORS

Certain plant pigments belong to a class of molecules known as *anthocyanins*, which are different colors in solutions of different acidity; that is, they can serve as acid-base indicators. This introduction mentions plants that contain anthocyanins. Other examples include turmeric, cranberries, carrots, strawberries, tea, and pansies. Onions and vanilla extract are olfactory indicators, in which smell is determined by pH.

WORKING WITH INDICATORS

If you have a student with impaired color vision, make sure to pair him with another student who can help identify the colors of the indicators in this activity.

EQUIPMENT NOTES

pH Meter

The use of the pH meter is optional. If you have one available, it is good for students to see the numerical value in addition to the visual indications from the acid-base indicators.

PRE-LAB CHECK

1. What is the difference between the two acids that you will be working with, other than their identity? *(Hydrochloric acid is a strong acid, and acetic acid is a weak acid.)*

2. How will you find the $[H_3O^+]$ of the $HC_2H_3O_2$ solution and of the unknown weak acid? *(These will be determined by testing the solutions with indicators to estimate the pH and then calculating the $[H_3O^+]$ from the pH equation.)*

3. How would you expect the pH of 0.1 M $HC_2H_3O_2$, a weak acid, to compare with the pH of 0.1 M HCl? Explain. *(The pH of $HC_2H_3O_2$ will be higher—less acidic—because it does not ionize as much as HCl and so produces fewer H_3O^+ ions.)*

4. In general, what chemical composition does an acid-base indicator have? What is its purpose? *(An acid-base indicator is a weak organic acid or base whose molecules appear as one color and whose ions appear as another. Since an indicator changes color over a specific pH range, it can be used to estimate the pH of a solution.)*

5. How many different sets of indicators will you make? How many samples will you test with each indicator? *(three; five)*

If you prefer, use microchemistry well plates instead of the test tubes and racks described in this experiment. Doing so will also have the benefit of producing less waste acid.

Solutions

You can purchase 1 M hydrochloric acid and acetic acid from science supply stores.

To make the 0.1 M HCl, pour approximately 70 mL of distilled water into a 100 mL volumetric flask. Add 10 mL of 1 M HCl. Gently swirl to mix thoroughly and then dilute with distilled water to 100 mL.

You can use any weak acid, such as 1.0 M acetic acid, as the unknown acid solution.

To make the 0.1 M $HC_2H_3O_2$, pour 70 mL of distilled water into a 100 mL volumetric flask. Add 10 mL of 1 M $HC_2H_3O_2$. Gently swirl to mix thoroughly and then dilute with distilled water to 100 mL.

EQUILIBRIUM AND INDICATORS

The more the concentration of H_3O^+ is increased, the more the equilibrium will shift toward the left; the more it is decreased, the greater the shift toward the right. The degree of acidity that is necessary to cause this shift will depend on the specific indicator and its K_a value. In any case, Color 1 will predominate in solutions that are more acidic than the indicator's K_a, and Color 2 will predominate in solutions that are more basic. The pH range in which both colors are present in significant amounts will exhibit a color that is a mixture of them both.

DATA COLLECTION HINT

You may choose to have colored pencils available so that students can color their observations rather than describe the color in words. This would allow them to shade the same color lighter or heavier, allowing for better record keeping.

Your students could also photograph their test tubes for easy reference.

Procedure

SETTING UP INDICATORS

A Label five test tubes "1A" through "1E," five test tubes "2A" through "2E," and five test tubes "3A" through "3E." Place them in the test tube racks.

B Measure 15 mL of 0.1 M HCl solution in the 100 mL graduated cylinder.

C Using the 10 mL graduated cylinder, transfer 3 mL portions of the HCl into Tubes 1A, 2A, and 3A. You should have 6 mL of 0.1 M HCl remaining in the 100 mL graduated cylinder.

D Add distilled water to the 100 mL graduated cylinder up to a total of 60 mL and mix well. This dilutes the acid to 0.01 M HCl. *Normally we don't add water to acid because the acid can splatter. But in this setting, you are using extremely dilute acids, which eliminates the risk.*

E Rinse the 10 mL graduated cylinder thoroughly and then pour 3 mL portions of the 0.01 M HCl into Tubes 1B, 2B, and 3B.

F Repeat Steps D and E to make the solutions listed in Table 1, pouring 3 mL portions of the HCl solutions into the appropriate test tubes as specified below.

0.001 M HCl in Tubes 1C, 2C, and 3C

0.0001 M HCl in Tubes 1D, 2D, and 3D

0.00001 M HCl in Tubes 1E, 2E, and 3E

G For each solution in Tubes 1A–1E, use one of the pipettes to add three drops of thymol blue solution. Swirl the solutions well and record the color of each in Table 2.

TABLE 1

HCl	Thymol Blue	Methyl Orange	Methyl Red
0.1 M	1A	2A	3A
0.01 M	1B	2B	3B
0.001 M	1C	2C	3C
0.0001 M	1D	2D	3D
0.000 01 M	1E	2E	3E

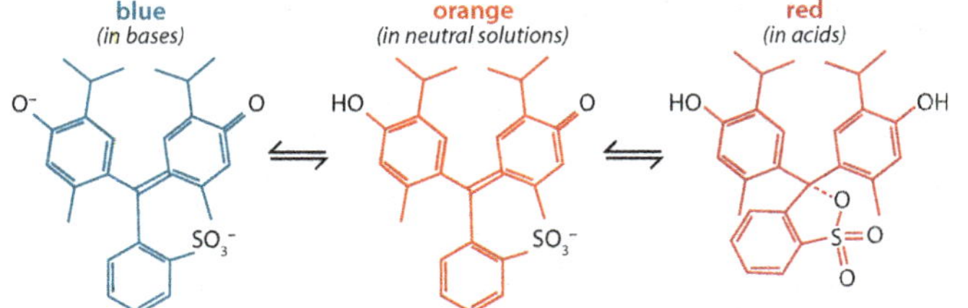

Thymol blue is an artificial indicator that can undergo two color changes. From left to right, these structures produce blue in basic solutions, orange in neutral solutions, and red in acids. Study these three structures to see how the structure changes with each color change.

1. In the image shown above, how does the structure of thymol blue change, as indicated by a color change?

   Each color change involves adding a hydrogen to the structure of thymol blue.

2. Which structure do you have in your test tubes right now? How can you tell?

   The rightmost structure exists in Tubes 1A–E, indicated by a red color and an acidic pH.

H. For Tubes 2A–2E, use a second pipette to add three drops of methyl orange solution. Swirl the solutions well and record the color of each in Table 2.

I. For Tubes 3A–3E, use the third pipette to add three drops of methyl red solution. Swirl the solutions well and record the color of each in Table 2.

J. Use a pH probe to measure the exact pH of your dilutions in all fifteen tubes to the nearest 0.1. Make sure that you rinse the probe between tests. If a pH probe is not available, calculate the pH from $[H_3O^+]$. Record these values in Table 2.

3. What effect does a dilution by a factor of 10 have on the pH of an HCl solution?

   It raises the pH by 1 pH unit.

4. On the basis of your answer to Question 3, describe how easy it is to change the pH of a solution?

   The pH of a solution is difficult to change by adding water alone.

5. Estimate the pH range over which each of the indicators changes color.

   thymol blue: 1–3; methyl orange: 3–4; methyl red: 4–5

Now you are going to use the indicator standards that you just created to learn more about the pH range in which they work best.

COMPARING $HC_2H_3O_2$ TO STANDARDS

K. Label three test tubes "4," "5," and "6." Using the 10 mL graduated cylinder, measure 3.0 mL of 0.1 M $HC_2H_3O_2$ solution into these test tubes.

L. Add three drops of thymol blue solution to Tube 4, three drops of methyl orange solution to Tube 5, and three drops of methyl red solution to Tube 6.

M. Compare the colors in Tubes 4–6 with the appropriate indicator standards in Tubes 1A–E, 2A–E, and 3A–E, and record your observations in Table 2.

N. Use these observations to estimate the $[H_3O^+]$ in the $HC_2H_3O_2$ solution and record your estimate in Table 2.

6. How do the strengths of HCl and the $HC_2H_3O_2$ compare? Explain how this affects their behavior.

   HCl is a strong acid, so it ionizes completely. This is different from

   acetic acid, which is a weak acid. Equally concentrated solutions of HCl

   and acetic acid will have different pH values and amounts of

   hydronium ions. HCl will have more hydronium ions, resulting in a

   lower pH and more acidic characteristics.

More Tubes!

Tube 4: 3.0 mL of 0.1 M $HC_2H_3O_2$, 3 drops of thymol blue

Tube 5: 3.0 mL of 0.1 M $HC_2H_3O_2$, 3 drops of methyl orange

Tube 6: 3.0 mL of 0.1 M $HC_2H_3O_2$, 3 drops of methyl red

 **pH METER**

If you have access to probeware, have students use it here to measure the pH of their dilutions.

O Using the estimated [H_3O^+], record the pH of the 0.1 M $HC_2H_3O_2$ solution in Table 2.

7. How did you arrive at this estimate?

ts colors best matched the Group C set of standards. Students may also mention using the pH formula.

8. Compare concentrations and pH values of 0.1 M $HC_2H_3O_2$ and 0.001 M HCl. How do they compare? What does this mean?

The acetic acid is 100 times more concentrated than the hydrochloric acid. They have similar concentrations of hydronium ions and therefore similar pH values. This means that HCl, the stronger acid, must release many more—100 times more—hydronium ions.

COMPARING AN UNKNOWN ACID TO STANDARDS

P Label the last three test tubes "7," "8," and "9." Using the 10 mL graduated cylinder, measure 3.0 mL of the unknown weak acid into these test tubes.

Q Add three drops of thymol blue solution to Tube 7, three drops of methyl orange solution to Tube 8, and three drops of methyl red solution to Tube 9.

R Compare the colors in Tubes 7–10 with the appropriate indicator standards in Tubes 1A–E, 2A–E, and 3A–E, and record your observations in Table 2.

S Use these observations to estimate the [H_3O^+] and pH of the unknown weak acid and record your estimate in Table 2.

Let's practice writing some equations and acid-ionization constant expressions for a weak acid, such as hypochlorous acid (HClO). The dissociation equation for dissolving HClO in water is shown below.

$$HClO(aq) + H_2O(l) \longrightarrow H_3O^+(aq) + ClO^-(aq)$$

From this equation we write the K_a expression.

$$K_a = \frac{[ClO^-][H_3O^+]}{[HClO]}$$

A 1.0 M hypochlorous acid solution has a pH of 3.8 because it produces a hydronium ion concentration of 1.73×10^{-4} M. We can therefore conclude two things about the concentrations in the solution. Because the coefficients of the hydronium ions and hypochlorite ions are both one, we know that their concentrations will be equal. Therefore, [ClO^-] = 1.73×10^{-4} M too. We also know that since very little of a weak acid dissociates, we can assume that the concentration of HClO is still 1.0 M. We can substitute these values into the K_a expression to calculate the acid-ionization constant for HClO.

Even More Tubes!

Tube 7: 3.0 mL of 1.0 M unknown acid, 3 drops of thymol blue

Tube 8: 3.0 mL of 1.0 M unknown acid, 3 drops of methyl orange

Tube 9: 3.0 mL of 1.0 M unknown acid, 3 drops of methyl red

$$K_a = \frac{[H_3O^+][ClO^-]}{[HClO]}$$

$$= \frac{(1.73 \times 10^{-4})(1.73 \times 10^{-4})}{(1.0)}$$

$$= 3.0 \times 10^{-8}$$

Now you can use a similar process to analyze the unknown acid. Ask your teacher for the identity of the acid.

9. Write the equilibrium equation for the dissociation of the acid.

   See TE margin for answer.

10. Using this equation, write the K_a expression.

$$K_a = \frac{[C_2H_3O_2^-][H_3O^+]}{[HC_2H_3O_2]}$$

11. Find the K_a value for this acid in Appendix I of your textbook. Use the K_a value and your estimated $[H_3O^+]$ to determine the amount concentration of the unknown weak acid.

Answers will vary. *Example:*

What we know: $K_{a\,HC_2H_3O_2} = 1.7 \times 10^{-5}$,
   $[C_2H_3O_2^-] = [H_3O^+] = 0.0001$ mol/L

Unknown: $[HC_2H_3O_2]$

Write the formula and solve for the unknown.

$$K_a = \frac{[C_2H_3O_2^-][H_3O^+]}{[HC_2H_3O_2]}$$

$$K_a[HC_2H_3O_2] = \frac{[C_2H_3O_2^-][H_3O^+]}{[HC_2H_3O_2]}[HC_2H_3O_2]$$

$$\frac{K_a[HC_2H_3O_2]}{K_a} = \frac{[C_2H_3O_2^-][H_3O^+]}{K_a}$$

$$[HC_2H_3O_2] = \frac{[C_2H_3O_2^-][H_3O^+]}{K_a}$$

Evaluate.

$$[HC_2H_3O_2] = \frac{(0.0001)(0.0001)}{1.7 \times 10^{-5}}$$

$$= .0006 \text{ M}$$

Question 9 Answer

Answers will vary. Example:
$$HC_2H_3O_2(aq) + H_2O(l) \longrightarrow$$
$$H_3O^+(aq) + C_2H_3O_2^-(aq)$$

IONIZATION CONSTANT

The ionization constant is simply a special case of the equilibrium constant. Remind students that the expression [HA] stands for the concentration of the acid in moles per liter. Because the acid has ionized slightly, the amount concentration cannot exactly equal the concentration of the weak acid.

12. Ask your teacher for the amount concentration of the unknown acid.
How well were you able to determine the amount concentration?

Answers will vary.

13. How did the concentrations and pH values of the unknown acid and the
0.1 M acetic acid compare?

While the pH values differed only by about 1, the 0.1 M acetic acid
concentration is about 200 times higher.

Going Further

14. On the basis of what you learned in this lab activity, why do you think
buffer systems use a weak acid and its conjugate base?

With a weak acid, there is little change in pH with significant changes
in concentration.

15. How do you think knowing the strength of an acid would make a
difference?

The strength of an acid determines its use and necessary
concentration in manufacturing, diet, and medicine, for example.

✓ SAMPLE DATA

The data shown in Table 2 is sample data.
Student data values may differ.

TABLE 2

Test Tubes	Thymol Blue	Methyl Orange	Methyl Red	$[H_3O^+]$	pH
Group A	deep red	red	red	0.1	1.0
Group B	lighter red	red	red	0.01	2.0
Group C	reddish orange	red	red	0.001	3.0
Group D	yellow	reddish orange	red	0.0001	4.0
Group E	yellow	yellow	reddish orange	0.000 01	5.0
$HC_2H_3O_2$ (4, 5, and 6)	reddish orange	red	red	0.001	3.0
Unknown Weak Acid (7, 8, and 9)	varies	varies	varies	0.0001	4.0

18B LAB

Name _______________

Date _______________

» Standardize a solution of sodium hydroxide using a primary standard.

» Determine the mass fraction of white vinegar.

» Evaluate the reported mass fraction of acetic acid contained in vinegar.

QUESTIONS

» What does it mean to standardize a solution?

» How do I do titration?

» How can I determine the concentration of vinegar by titration?

Say Cheese!

Measuring Concentration by Titration

At Shelburne Farms in Vermont, blocks of cheddar cheese are set to age in a cooler. But how do the cheese makers at Shelburne Farms know when the cheese is ready? This is where chemistry comes in. Farmers do a titration of the whey from cheese to determine its acidity. The acid in cheese is lactic acid, which is what helps cheese to develop taste and texture.

We can also determine when a particular level of acidity has been reached by using a pH meter or an indicator. In this experiment, you will find the concentration of a vinegar by doing an acid-base titration.

How do chemists determine amount concentration of acids and bases?

Procedure

Before we can titrate the vinegar, we must first determine the concentration of a standardized solution by titrating it with a primary standard. A *primary standard* is a chemical substance that is easy to work with and of such purity that it can be used as a reference.

SETTING UP

A Add about 100 mL of NaOH solution to a clean, dry 150 mL beaker.

B Observe the burettes set up by your teacher. Burette A is for the acid and Burette B is for the base.

C Check that the tip of Burette B is filled with liquid (no air in the tip). If necessary, drain some of the NaOH solution out of the burette into the waste beaker provided by your teacher. If there is not enough solution, use a funnel to carefully fill Burette B with NaOH solution until the level of the liquid is near but not above 0.00 mL. Record this initial volume of NaOH under Trial 1 in Table 1.

EQUIPMENT

- laboratory balance
- beakers, 150 mL (2)
- ring stand
- burette clamp
- burettes, 50 ml or 100 mL (2)
- funnel
- Erlenmeyer flasks, 250 mL (2)
- wash bottle
- pipette
- sodium hydroxide (NaOH) solution
- potassium hydrogen phthalate ($KHC_8H_4O_4$)
- distilled water
- phenolphthalein solution
- white vinegar
- goggles
- laboratory apron
- nitrile gloves

SCHEDULING LAB 18B

This lab activity works well as a concluding activity for the chapter. It is best to do this after students have worked through the material in Section 18.3 of the Student Edition.

WORKING WITH INDICATORS

If you have a student with impaired color vision, make sure to pair him with another student who can help identify the colors of the indicators in this activity.

EQUIPMENT NOTES

Burettes

The procedures as written assume that you have the burettes set up prior to student arrival. Label your burettes "A" for the acid and "B" for the base.

Solutions

You will need about 150 mL of NaOH solution for each lab group. Use approximately 0.4 M NaOH for the titration. To make this solution, add approximately 140 mL of distilled water to a 200 mL volumetric flask. Add 3.20 g of NaOH to the flask and gently swirl until dissolved. Then dilute with distilled water to 200 mL.

You can scale these instructions to make more of the solution.

Burette for Vinegar

To save time and to eliminate the need for each group to have two burettes, students may obtain their vinegar sample from a single burette located at a central location where you can monitor its use and refill as needed. In this case, each group would bring a flask to this central location to get vinegar at Steps P and Q.

PRE-LAB CHECK

1. What is the main purpose of the first titration of this activity? *(The purpose is to standardize the NaOH solution, that is, to find its concentration by using the primary standard, KHP.)*

2. Once you have weighed out the KHP, why is it *not* important to measure the volume of water that you add? *(You are titrating the KHP, not the water, and you already know how much of it is present. Adding more or less water will not change how much of the measured reacting substance is present.)*

3. What is amount concentration? *(a measure of concentration: moles of solute per liter of solution)*

4. Define *titration*. *(Titration is a laboratory technique used to measure the concentration of a solution by reacting it with a known amount of another substance.)*

SHOWING A TITRATION

It may be a good idea to show students a video of a titration being done using phenolphthalein as an indicator. Do an internet search using the keywords "acid base titration video."

END POINT

Review with students the definition of the end point of a titration. The end point is the moment when the indicator color change occurs and remains because the two reagents are chemically or stoichiometrically equivalent; that is, they have reacted according to a balanced chemical reaction.

Question 1 Answer

The top marking is 0.00 mL; volumes increase from top to bottom down the burette. Also, the burette is a more precise instrument, allowing volumes to be read to the nearest 0.01 mL, rather than just the 0.1 mL seen in many graduated cylinders.

Question 2 Answer

$$KHC_8H_4O_4(aq) + NaOH(aq) \longrightarrow$$
$$H_2O(l) + NaKC_8H_4O_4(aq)$$

Question 3 Answer

The amount of water does not affect the amount of KHP in the flask. Rather, the amount of KHP is the information used to determine the concentration of the NaOH. The water is primarily used to dissolve the KHP.

Question 4 Answer

Students should recognize that the structure of phenolphthalein is nonpolar, indicating that it is probably insoluble in water. They should indicate that the solvent must be nonpolar and may suggest alcohol or acetone, especially if they smell the solution.

STANDARDIZATION TIPS

Students should observe phenolphthalein to be colorless in an acid (KHP) and pink or magenta in a base. The white paper makes the color change easier to see. If students go beyond the end point, they will have to start over by making a new primary standard solution to titrate.

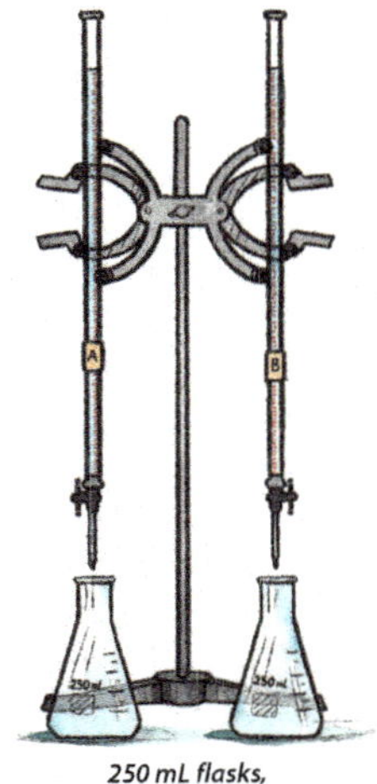

Acid-base titration setup

1. How are the calibrations of a burette different from that of a graduated cylinder?

   See TE margin for answer.

D Measure the mass of one of the Erlenmeyer flasks and record this value under Trial 1 in Table 1.

STANDARDIZING THE NaOH SOLUTION

E Add 1.0–1.5 g of potassium hydrogen phthalate (KHP) to the flask on the balance. Measure and record the combined mass under Trial 1 in Table 1.

F Calculate the mass of KHP used. Record this result under Trial 1 in Table 1.

2. Write a balanced chemical equation for the reaction of NaOH with KHP to form water and sodium potassium phthalate.

   See TE margin for answer.

G Add at least 30 mL of distilled water to the flask and swirl the flask until all the solid dissolves. Wash any crystals that cling to the wall of the flask down into the solution with a few milliliters of water from a wash bottle.

3. Why does the water not need to be measured accurately?

   See TE margin for answer.

H Add two drops of phenolphthalein indicator to the KHP solution in the flask.

4. The structure of phenolphthalein is shown in the image at right. Phenolphthalein is a white solid. But you are using drops of solution. Suggest what it might be dissolved in. (*Hint*: Think about the polarity of phenolphthalein.)

   See TE margin for answer.

I Place the flask on a piece of white paper under Burette B. Lower the burette until the tip extends into the flask.

5. What do you think is the purpose of the white paper?

   Since phenolphthalein is an indicator, we are expecting a color change. The color change will be easier to see against a white background.

J Titrate the KHP solution by adding a few milliliters of NaOH solution from Burette B as you swirl the flask to mix the solutions (see image at right). Stop adding NaOH when the light pink color doesn't go away with swirling. Read the final volume on the burette and record it in Table 1.

K Calculate the volume of NaOH used in this trial and record it in Table 1.

L Dispose of the waste in accordance with your teacher's instructions, then rinse out your flask well. Repeat Steps D–J to do Trial 2 for standardizing the NaOH solution.

6. Describe the color change that you observed for phenolphthalein in the trials on the basis of the acidity or basicity of the solution.

   Phenolphthalein is pink in basic solutions and colorless in acidic
   solutions.

7. Calculate the moles of NaOH used in each trial. Show your work for Trial 1 below. Record your answers for both trials in Table 1.

   Answers will vary.

   What we know: $m_{KHC_8H_4O_4} = 1.03$ g

   Unknown: n_{NaOH}

   Convert mass $KHC_8H_4O_4$ to moles NaOH:

   $$1.03 \text{ g } KHC_8H_4O_4 \left(\frac{1 \text{ mol } KHC_8H_4O_4}{204.23 \text{ g } KHC_8H_4O_4} \right) \left(\frac{1 \text{ mol NaOH}}{1 \text{ mol } KHC_8H_4O_4} \right)$$

   $$= 0.005\ 0\underline{4}3 \text{ mol NaOH}$$

8. Calculate the amount concentration of the NaOH for each trial. Show your work for Trial 1 below. Record your answers for both trials in Table 1.

   Answers will vary.

   What we know: $n_{NaOH} = 0.005\ 0\underline{4}3$ mol, $V_{NaOH} = 12.38$ mL $= 0.012\ 38$ L

   Unknown: c_{NaOH}

   Write the formula and solve for the unknown.

   $$c_{solute} = \frac{n_{solute}}{V_{solution}}$$

   Evaluate.

   $$c_{solute} = \frac{0.005\ 0\underline{4}3 \text{ mol NaOH}}{0.012\ 38 \text{ L}}$$

   $$= 0.407 \text{ mol/L NaOH}$$

M Calculate the average amount concentration on the basis of the values that you calculated for each trial and record this as the amount concentration for NaOH in Table 2 (enter twice).

N Refill Burette B with NaOH solution for the next part of the activity.

Titrating the unknown solution

HOW TO TITRATE

1. Control the stopcock with one hand while you swirl the flask with the other hand.

2. Continue to titrate by adding the NaOH solution slowly until the light pink color lingers before disappearing; then add the NaOH by drops.

3. Stop titrating when the light pink color remains for at least 30 seconds; you have reached the end point.

4. If the pink color fades, add one drop at a time until there is a change from colorless to a permanent pink.

MORE ON PHENOLPHTHALEIN

Phenolphthalein undergoes a second color change from colorless to cherry red in solutions that are extremely acidic (pH less than 0).

O. Pour about 55 mL of white vinegar into the second clean, dry 150 mL beaker.

P. Check that the tip of Burette A is filled with liquid (no air in the tip). If necessary, drain some of the vinegar solution out of the burette into the appropriate waste beaker provided by your teacher. If there is not enough solution, use a funnel to carefully fill the burette with vinegar solution until the level is near but not above 35.00 mL. Record the initial volume of vinegar in Table 2.

Q. Allow about 15 mL of the vinegar to drain into the other clean 250 mL Erlenmeyer flask. Add two drops of phenolphthalein indicator to the vinegar in the flask.

R. Record the initial volume of the NaOH in Table 2.

S. Titrate the vinegar in the flask with the NaOH solution from Burette B.

T. If you feel like you've added more than barely the amount needed to titrate the solution, add a few drops of vinegar from the burette and then carefully add NaOH until one drop causes the color to change to pink. Read the final volumes of the NaOH and vinegar and record both in Table 2.

U. Calculate the volumes of NaOH and vinegar used in each trial and record them in Table 2.

9. Write the balanced chemical equation between sodium hydroxide and acetic acid.

See TE margin for answer.

V. If needed, refill your burettes and repeat Steps O–T to accomplish a second trial.

10. Calculate the moles of acetic acid neutralized in each trial. Show your work for Trial 1 below. Record your answers for both trials in Table 2.

Answers will vary. *Example:*

What we know: $c_{NaOH} = 0.39$ mol/L, $V_{NaOH} = 34.16$ mL

Unknown: $n_{NaOH}, n_{HC_2H_3O_2}$

Convert mL of NaOH to moles of $HC_2H_3O_2$.

$$34.16 \text{ mL}\left(\frac{1 \text{ L}}{1000 \text{ mL}}\right)\left(\frac{0.39 \text{ mol NaOH}}{1 \text{ L}}\right)\left(\frac{1 \text{ mol } HC_2H_3O_2}{1 \text{ mol NaOH}}\right)$$

$$= 0.0133 \text{ mol } HC_2H_3O_2$$

Question 9 Answer

$$HC_2H_3O_2(aq) + NaOH(aq) \longrightarrow H_2O(l) + NaC_2H_3O_2(aq)$$

11. Calculate the amount concentration of the vinegar for each trial.
    Show your work for Trial 1 below. Record your answers for both trials in
    Table 2.

    Answers will vary. *Example:*

    What we know: $n_{HC_2H_3O_2} = 0.0133$ mol, $V_{HC_2H_3O_2} = 15.34$ mL $= 0.01534$ L

    Unknown: $c_{HC_2H_3O_2}$

    Write the formula and solve for the unknown:

    $$c_{solute} = \frac{n_{solute}}{V_{solution}}$$

    Evaluate.

    $$c_{solute} = \frac{0.0133 \text{ mol } HC_2H_3O_2}{0.01534 \text{ L}}$$

    $$= 0.867 \text{ mol/L } HC_2H_3O_2$$

12. Report the average amount concentration of acetic acid on the basis of
    the values that you calculated for each trial. Show your work in the mar-
    gin if needed.

    Answers will vary. *Example:* 0.90 M

Going Further

13. Use dimensional analysis with the amount concentration, molar mass,
    and density (1.01 g/mL) of acetic acid to calculate the mass fraction of
    acetic acid in your vinegar.

    Answers will vary. See below for worked-out solution.

14. Check your bottle of vinegar. How does this compare with what the label
    says?

    See TE margin for answer.

    Scientists titrate more than just the things you eat, like the lactic acid in
    the whey of cheese. Titration is used during the production of biodiesel fuel,
    for aquarium water testing, and in medical analysis such as medication levels
    in IV drips and the glucose levels of diabetic patients.

15. How do these uses of titration play a role in helping people solve real-
    world problems?

    Titrations can protect and improve people's health and way of life in

    these applications of chemistry. Students may mention more specific

    examples, such as cheese making.

Question 13 Answer

A sample worked-out calculation is shown
in the bottom margin.

Question 14 Answer

Students should find that their value is
similar. The vinegar used in the sample data
was 5% acetic acid, so 5.4% is a match for
this value.

Question 13 Answer

What we know: $c_{HC_2H_3O_2} = 0.90$ mol/L, $\rho_{HC_2H_3O_2} = 1.01$ g/mL

Unknown: $w_{HC_2H_3O_2}$

Convert amount concentration to mass fraction.

$$0.90 \frac{\text{mol } HC_2H_3O_2}{\text{L solution}} \left(\frac{1 \text{ L}}{1000 \text{ mL}}\right)\left(\frac{1 \text{ mL}}{1.01 \text{ g}}\right)\left(\frac{60.06 \text{ g } HC_2H_3O_2}{1 \text{ mol } HC_2H_3O_2}\right) = 0.0535 \frac{\text{g } HC_2H_3O_2}{\text{g solution}}$$

$$= (0.0535)100\% = 5.4\%$$

SAMPLE DATA

The data shown in Tables 1 and 2 is sample data. Student data values may differ.

TABLE 1

	Trial 1				Trial 2			
	Mass (g)	Volume (mL)	Moles	M (mol/L)	Mass (g)	Volume (mL)	Moles	M (mol/L)
Flask	257.40				257.40			
Flask and $KHC_3H_4O_4$	258.43				258.51			
$KHC_3H_4O_4$	1.03				1.11			
NaOH Initial		1.50				1.14		
NaOH Final		13.88		0.41		15.71		0.37
NaOH Used		12.38	0.0050			14.57	0.0054	

TABLE 2

	Trial 1			Trial 2		
	Volume (mL)	Moles	M (mol/L)	Volume (mL)	Moles	M (mol/L)
NaOH Initial	1.75			3.20		
NaOH Final	35.91			42.34		
NaOH Used	34.16		0.39	39.14		0.39
Vinegar Initial	1.40			6.70		
Vinegar Final	16.74			22.93		
Vinegar Used	15.34	0.013	0.87	16.23	0.015	0.94

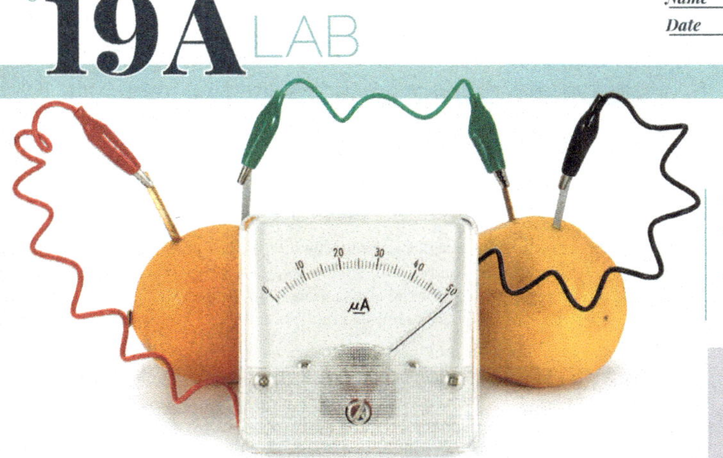

The Dead, Twitching Frog Mystery

Investigating a Voltaic Cell

A whole new world opened to chemists in the late 1790s with the discovery of electrochemistry. When Alessandro Volta built the first battery, he probably didn't realize the significance of his invention. Volta created the battery to settle a dispute that he was having with fellow scientist Luigi Galvani. Galvani had noticed that a freshly dissected frog leg would twitch when he touched it with two strips of different metals joined at one end. He believed that this experiment demonstrated the existence of what he called "animal electricity"—electricity generated by a biological system.

Where does the electricity in a battery come from?

Volta disagreed with Galvani's interpretation of his experiment. Volta believed that the electricity originated because of the dissimilar metals. He constructed a number of simple batteries to confirm that biological material was not needed. In an example of one of Volta's batteries shown below, a battery is made from a series of cups containing an electrolytic solution and strips of two different metals. In reality, both Volta and Galvani were partially correct. Volta was right in that the metals were the key to what Galvani had observed. But Galvani was correct in believing that biological tissues *can* generate electricity. Neither, however, understood the essential role that an electrolyte played in the overall picture.

Let's take a look at Volta's battery to expand our understanding of redox reactions—the chemistry that makes a battery work.

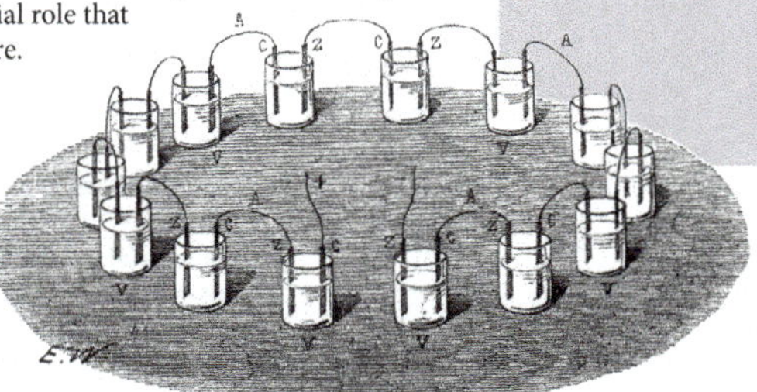

One of Volta's batteries, known as a "crown of cups"

QUESTIONS

» How do chemical reactions produce electricity?

» How does the choice of material for electrodes affect a redox reaction?

EQUIPMENT

- digital multimeter
- beaker, 50 mL
- alligator clip leads (2)
- white vinegar (5%, acetic acid)
- mechanical pencil leads (4)
- tape
- sandpaper
- zinc strip or galvanized nail
- magnesium ribbon
- goggles
- laboratory apron
- nitrile gloves

LAB 19A OBJECTIVES

» Analyze the redox reaction occurring in a voltaic cell.

» Analyze output voltage of voltaic cells constructed with different electrodes.

SETTING UP LAB 19A

While a digital multimeter is preferred due to its ease of use and its high internal resistance, a mechanical analog voltmeter may be used, provided it can read low voltages.

ANOTHER VOLTA BATTERY

Students may have seen pictures of another Volta battery, better known as a *voltaic pile*—a stack of copper and silver discs separated by brine-soaked felt. But it was his crown of cups battery that led to practical batteries, the ancestors of modern ones. It is also closer to the voltaic cells that students will examine in this lab activity.

The Dead, Twitching Frog Mystery | 187

PRE-LAB CHECK

1. Draw a picture with labels of the voltaic cell that you will be building. *(Students should draw a picture similar to the image on page 189.)*

2. Batteries involve the flow of electrons, so where are the electrons coming from in your battery? *(the zinc electrode)*

3. What kind of chemical reaction is releasing the electrons? *(a redox reaction)*

4. How do the electrons get from the zinc electrode to the carbon electrode? *(They travel through the voltmeter.)*

5. Describe voltage in your own words. *(Voltage is the electrical difference between two charged areas. Voltage is the "push" that moves electrons.)*

HANDLING PENCIL LEAD

Warn students to be gentle with the pencil lead because it will snap easily. As long as they are gentle, the bundles of pencil lead will work well for this activity.

FOOD BATTERIES

The voltaic cell explored in this lab activity is very similar to the lemon battery or potato battery commonly demonstrated in earlier science courses. You may want to mention this fact to your students. These batteries rely on the natural acids found in foods to provide the electrolyte. In the case of citrus fruits, the most significant acid is citric acid. For an interesting discussion of the chemistry behind simple batteries, see the paper "Observations on Lemon Cells" by Jerry Goodisman, published in the *Journal of Chemical Education*, Vol. 78, No. 4 (April 2001).

DIFFERENCES IN VOLTAIC CELLS

Students may be confused if they try to use the image on page 475 of the Student Edition to understand the voltaic cells explored in this lab activity. The cell shown there is significantly different and more complicated than the cell being constructed in this activity. In its case, *both* electrodes are undergoing redox reactions, two electrolytes are involved, and the salt bridge is necessary to keep the electrolytes separated. The cell's output voltage is the difference between the standard electrode potentials of zinc (−0.76 V) and copper (+0.34)—about 1.1 V.

In the cells explored in this activity, only one electrode is involved in a redox reaction (the metal), there is just one electrolyte, and the output voltage is based on the difference between the metal electrode's standard electrode potential and that of hydrogen (0 V). One of the purposes for the Daniell cell depicted in the Student Edition was to eliminate the hydrogen bubbles produced by cells like the one that students create in this activity. Be sure to make these distinctions clear.

Procedure

SETTING UP

A Fill the beaker with approximately 40 mL of acetic acid.

B Connect the alligator clip leads to the multimeter's probes. Switch the meter on and set it to measure DC voltage.

C Form the mechanical pencil leads into a bundle, using a small piece of tape in the middle to secure them. They will act as a carbon electrode. Gently connect the alligator clip lead coming from the positive (+) multimeter probe (usually red) to one end of the bundle.

TESTING METALS

D Using the sandpaper, rub the zinc strip so that it brightens, and then clip the alligator lead connected to the negative (−) multimeter probe (usually black) to one end of the zinc strip.

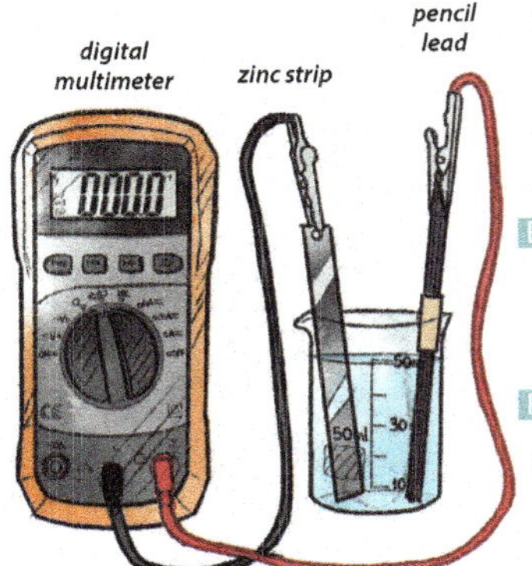

1. Why is it important to clean the metal with sandpaper before performing the experiment?

   The metal may have corrosion products on the surface that would interfere with the chemical reaction being studied.

E Dip the carbon and zinc electrodes in the acetic acid, making sure that they don't touch. As soon as the multimeter gives a definite voltage reading, record this value in Table 1. Generally, the voltage will start to drop within just a few seconds, so read the meter quickly.

F Repeat Steps D and E using the magnesium strip in place of the zinc.

2. What did you observe around the magnesium strip during the procedure?

   Bubbles formed around the magnesium strip.

3. How could you experimentally determine whether your observations for Question 2 were due to a redox reaction with the carbon or to the magnesium reacting with the acid?

   Answers may vary. The best way to decide is to immerse the magnesium strip in acetic acid alone without it being connected to anything else.

G Use the procedure that you outlined in Question 3 to determine whether the observations in Question 2 were due to a redox reaction. When sanding the magnesium ribbon, lay the ribbon on the table and sand the flat surface. *The edge of the ribbon can cause significant cuts.*

---

4. What did you conclude in Step G? Explain.

See TE margin for answer.

____

____

____

H Connect the carbon and zinc electrodes with a single alligator clip lead between them (see image at right). The meter will not be used.

I Immerse the two electrodes in the acetic acid, making sure that they don't touch. Leave the setup undisturbed for 10 minutes. Read on and answer Questions 5–12 while you wait.

Voltaic cells involve redox reactions, but what is being oxidized and what is being reduced? The following equations describe the reaction for the first experiment (zinc electrode).

$$Zn(s) \longrightarrow Zn^{2+}(aq) + 2e$$

$$2H^+(aq) + 2e^- \longrightarrow H_2(g)$$

5. What's happening in the first reaction? How do you know?

The zinc is being oxidized. The zinc atoms are losing two electrons and turning into zinc ions. By definition, losing electrons in this way is known as oxidation.

6. Where is the zinc going? How do you know?

The zinc ions are going into solution. The (aq) notation in the equation indicates that the ions are in solution.

7. Where are the electrons going? How do you know?

They must be staying in the zinc, wire, and carbon system since there is no (aq) in the equation to show that they are in solution.

8. What is happening electrically to the zinc electrode?

It is becoming negatively charged since it has an excess of free electrons.

9. If this process goes on long enough, what do you expect will happen to the zinc electrode?

Assuming that there is sufficient acetic acid in solution, the zinc will dissolve into the solution and disappear.

The setup for the voltaic cell

QUESTION 4: WHEN METAL MEETS ACID

The reaction going on here is the same reaction that occurs when many metals come in contact with strong acids. Since magnesium is very reactive, it undergoes a single-replacement reaction, even in a weak acid like vinegar, liberating hydrogen gas. The zinc strip would do the same thing if a stronger acid, such as hydrochloric, were used.

Question 4 Answer

The magnesium strip bubbled when it was immersed in the acid while not being connected to anything else. Therefore, the bubbles must have been caused by a different reaction (the single-replacement reaction of the metal with the hydrogen from the acetic acid) from the redox reaction being studied in this activity.

TRACKING ELECTRONS (QUESTION 7)

Some students may have difficulty answering Question 7 since it seems logical that the electrons would also enter the solution. Remind them that the equation notation makes it clear that only the ions are in the solution. The electrons remain in the electrodes.

You may also need to review the concept of electrical charge if your students have difficulty with Question 8. Since the zinc ions are leaving the metal, positive charges (protons) are also leaving. Consequently, there is an overall excess of electrons, leading to a negative charge.

10. What's happening in the second reaction? How do you know?

Hydrogen ions are reducing to gaseous hydrogen. The hydrogen ions are gaining electrons. By definition, this process is known as reduction. This is evidenced by the formation of hydrogen bubbles.

11. Where are the hydrogen ions in the second equation coming from?

The hydrogen ions are produced when acetic acid dissociates in water.

12. What do you expect to see happening as a consequence of the second reaction?

Hydrogen bubbles should begin to appear somewhere in the solution.

J. After 10 minutes has elapsed, carefully examine both electrodes in a good light. Do not touch or jiggle them!

13. What do you observe?

Tiny bubbles have accumulated on the carbon electrode.

14. Describe in detail what you think is happening when you connect the electrodes together and immerse them in the acid.

See TE margin for answer.

15. Does the carbon electrode undergo a redox reaction? Explain.

No. Carbon doesn't appear in either equation, so it doesn't appear to be involved in a redox reaction of its own. Students may mention that it's a source of electrons and a point of electrical contact, so it is important to the zinc-hydrogen redox reaction. But it's not actually reacting itself since it is maintaining a consistent number of electrons.

16. What role do you think the carbon electrode plays in the voltaic cell?

The carbon electrode completes the electrical circuit. It is also a place for the hydrogen to reduce since it provides a source of electrons.

Question 14 Answer

Answers will vary. Zinc is oxidized, forming zinc ions. The zinc ions go into solution. The hydrogen ions in solution are reduced at the carbon electrode, producing hydrogen gas. As the carbon electrode loses electrons, the excess electrons in the zinc electrode flow through the clip lead to the carbon rod. This process goes on continuously, causing an electric current to flow through the wire and ions to flow through the solution.

⁉️ UNREACTIVE ELECTRODE

Some students may think that the carbon electrode is involved in a redox reaction since it provides the electrons that reduce the hydrogen ions. Emphasize that the carbon is nothing more than a conduit for the electrons in the zinc electrode, just like the wires in the battery. Carbon atoms are neither losing nor gaining electrons. Consequently, the carbon electrode isn't being changed in any way. Given enough time, the zinc electrode will dissolve away, but the carbon will remain unchanged.

Any other material of low reactivity could be substituted for the carbon electrode. Copper, platinum, and gold would all work just as well as carbon since they don't react in acetic acid. Carbon is used in this activity instead of copper to avoid confusion with the cell shown on page 475 of the Student Edition. Gold and platinum obviously are not practical!

COMPARING VOLTAGES

When you measured the voltages developed by the zinc and magnesium, you may have wondered why they were different. Volta and other scientists also observed that batteries made from different metals performed differently. Eventually scientists figured out a way to classify different materials to predict how they would behave in a battery.

Each substance to be tested is used as an electrode in a voltaic cell containing a 1 M electrolyte solution at STP (see image below). The second electrode is made from a glass tube containing hydrogen gas. This electrode is treated as a *standard reference electrode*, which is assumed to have a value of 0 V. The voltage between the two electrodes is measured. Since the reference electrode is assumed to be 0, the voltage reading is considered to be the tested material's specific voltage value.

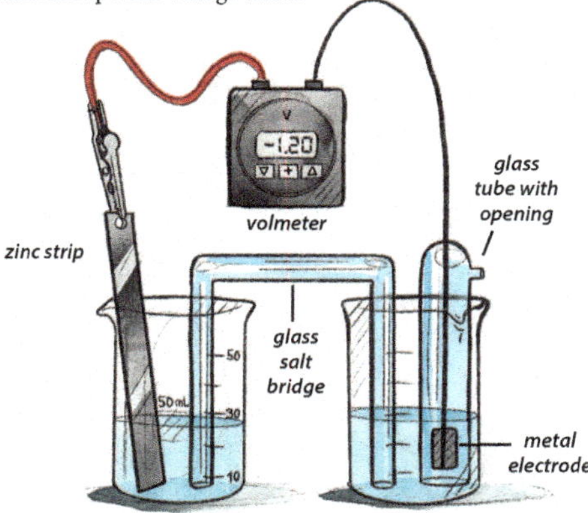

Apparatus used to determine standard electrode potentials

17. Ignoring the sign on the standard electrode potentials for zinc and magnesium given in Table 1, compare those standards to your measured values. Do the measured values reflect the general trend shown by the standard values?

    Yes. Magnesium produces a larger voltage than zinc in both the standard and the measured values.

18. Are the measured values reasonably close to the standard values?

    Answers will vary. Generally, they will not be very close, although they will be reasonably so.

19. If they are *not* reasonably close, can you think of possible reasons to account for the difference?

    See TE margin for answer.

!?! STANDARD CONDITIONS?

Thoughtful students might question whether it is legitimate to compare the standard electrode potentials with their measured results. Remind them that their voltages are also measured relative to hydrogen, so the comparison, while a bit crude, is reasonable. As Question 19 hopefully leads students to realize, differences will arise due to nonstandard conditions and other poorly controlled factors that are not present in the conditions used to determine the standard electrode potentials.

Question 19 Answer

Answers will vary. The test conditions (molarity, temperature, pressure) are different from the standard conditions used to determine the "official" values. The acidic solution is likely not a 1 M solution. The metals may have contaminants or corrosion products on them, or the meter may not be accurate. Students may also mention that the value for magnesium may be affected by the second chemical reaction that is simultaneously going on (indicated by bubbling around the metal).

SAMPLE DATA

The voltages shown in Table 1 are representative. Your students' data may be different due to variations in the experimental conditions, but should show a similar tendency.

20. So where do you think the electricity comes from in a nickel-zinc battery?

Answers will vary. In a nickel-zinc battery, the zinc is oxidized, releasing electrons (two, in this case). The electrons flow through the circuit to the nickel, which is being reduced.

21. If zinc and nickel have a voltage of −1.2 V and +0.5 V respectively when testing in the voltaic cell described above in "Comparing Voltages," what voltage would you expect from a nickel-zinc battery?

Answers will vary. Students should expect a voltage of 1.7 V from a nickel-zinc battery.

TABLE 1

Material	Measured Voltage (V)	Standard Electrode Potential (V)
zinc (Zn)	0.95	−0.76
magnesium (Mg)	1.87	−2.37

19B LAB

Essential Medicine

Using Redox Titration

The World Health Organization (WHO) includes potassium permanganate on its list of essential medicines. As a medicine, it is used to treat a number of skin diseases. It is also used in processes of water purification to remove iron and hydrogen sulfide. It can also combat invasive species such as zebra mussels. These applications all involve redox reactions.

In this lab activity you will do a redox titration to find out the concentration of a potassium permanganate solution ($KMnO_4$). You won't need an indicator for this titration—the potassium permanganate acts as its own indicator! As this redox titration reaches its end point, you will watch the solution change color, similar to the color change of the indicator in acid-base titrations.

How can I use titration for redox reactions?

QUESTIONS

» How can I model the movement of electrons in a redox reaction?

» How is the concentration of an oxidizer determined by titration?

Procedure

SETTING UP

A Add about 60 mL of $KMnO_4$ solution to a clean, dry 150 mL beaker.

1. Record your observations of the $KMnO_4$ solution.

   Students should observe that the solution is dark purple in color.

B Make sure that your titration setup looks like the image shown at left. Check that the burette stopcock is closed and, with the aid of a funnel, carefully fill the burette with $KMnO_4$ solution. Open the stopcock and drain out some of the solution into the 50 mL beaker until the tip of the burette is filled. Check that the level of the $KMnO_4$ solution is near but not above 0 mL. Record the starting volume in the Trial 1 column of Table 1.

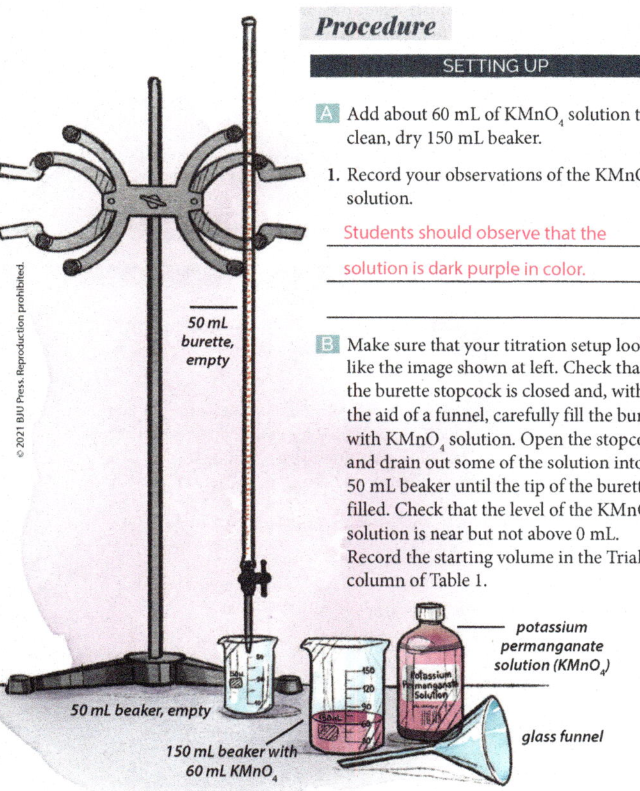

EQUIPMENT

- laboratory balance
- beaker, 150 mL
- burette, 50 mL
- burette clamp
- ring stand
- filtering funnel
- beaker, 50 mL
- weighing dish
- Erlenmeyer flask, 125 mL
- wash bottle with distilled water
- graduated cylinder, 10 mL
- graduated cylinder, 100 mL
- potassium permanganate solution ($KMnO_4$)
- iron(II) sulfate heptahydrate ($FeSO_4 \cdot 7H_2O$)
- distilled water
- sulfuric acid (H_2SO_4), 6 M
- goggles
- laboratory apron
- nitrile gloves

LAB 19B OBJECTIVES

» Analyze a redox reaction to determine the movement of electrons.

» Determine the concentration of an oxidizer by titration.

EQUIPMENT NOTES

Potassium Permanganate

To make this solution, add 3.16 g of $KMnO_4$ to 1 L of water. There is no need to standardize this solution since students will be calculating its concentration. To confirm their answers, consider comparing the results of the lab groups.

Iron(II) Sulfate

You should try this lab activity ahead of time to check the purity of your iron sulfate. If you see a precipitate form, it's possible that your iron sulfate is contaminated with iron oxide. Using nitric acid instead of sulfuric acid will make the precipitates disappear and enter solution, since more nitrates are soluble. If you have a 1 M solution of either sulfuric or nitric acid, add 5 mL to the Erlenmeyer flask before titration.

If you have access to anhydrous iron(II) sulfate, have your students use about 0.4 g for Step C.

Burette

While you can have students set up the burette, it will save time to have a burette set up for each lab group before they arrive.

PRE-LAB CHECK

1. What distinguishes a redox reaction from other types of reactions? *(A redox reaction is a reaction in which the oxidation numbers of some of the elements change.)*

2. Why don't you need an indicator for this titration? *(The titrant acts as its own indicator.)*

3. What color change will you observe in the titration that will signal the end point? *(The appearance will change from colorless to purple.)*

QUESTION 2:
WHERE DID THE OXYGEN GO?

Your students may be uncomfortable to see oxygen atoms on the reactant side but not on the product side of this reaction. You can use the two half-reactions shown below to help them track the oxygen.

$$Fe^{3+} + 1\ e^- \longrightarrow Fe^{2+}$$

$$MnO_4^- + 8H^+ + 5e^- \longrightarrow Mn^{2+} + 4H_2O$$

Question 2 Answer

$$MnO_4^{2-}(aq) + Fe^{2+}(aq) \longrightarrow$$
$$Mn^{2+}(aq) + Fe^{3+}(aq)$$

QUESTION 3:
NET IONIC EQUATIONS

You may want to review ionic equations, net ionic equations, and spectator ions (Ch. 10) while studying this chapter on redox chemistry. Ions are often used to track changing oxidation numbers in redox reactions.

QUESTION 4: REVIEWING
REDOX TERMINOLOGY

Use Questions 4–6 to review the terms of redox reactions. Students may benefit from the mnemonic device, "LEO the lion goes GER": Loss of Electrons means Oxidation, and Gain of Electrons means Reduction.

⚠ STEP E: SULFURIC ACID

As a safety precaution, you may want to have one location for the sulfuric acid. Students can bring their flask to that location and you can dispense the sulfuric acid.

QUESTION 6:
ACIDS AND REDOX REACTIONS

Review the definition of an acid from the previous chapter. Acids often take part in redox reactions because the flow of electrons will result in the development of areas of charge. In some redox equation balancing techniques, hydrogen ions are used to balance the charge differences in a redox chemical equation. Reminding students of this role of acids in redox equations will prepare them to answer Question 6.

C Using a weighing dish, obtain approximately 0.70 g of iron(II) sulfate ($FeSO_4$). Record the mass of the iron(II) sulfate in Table 1. Transfer the iron(II) sulfate to the flask.

2. Write the chemical equation for the reaction of permanganate ions with iron(II) ions to form manganese and iron(III) ions. Don't be concerned with balancing the elements in this equation for now.

See TE margin for answer.

3. Why are the potassium and sulfate ions not included in this reaction?

They are spectator ions and don't participate in the reaction.

4. Which element is being reduced? Which is being oxidized?

manganese; iron

5. Which element is the reducing agent? Which is the oxidizing agent?

iron; manganese

D Add approximately 10 mL of distilled water to the flask and swirl the flask until all the solid iron(II) sulfate dissolves. Use a few milliliters of water from a wash bottle to wash any crystals that cling to the wall of the flask down into the solution.

E Using the 10 mL graduated cylinder, add 1 mL of 6 M sulfuric acid (H_2SO_4) to the flask.

6. What does the sulfuric acid add to the solution that affects the redox reaction?

hydrogen ions

TITRATING KMnO₄

F Place the flask on a piece of white paper under the burette. This will help you see when the color in the flask starts changing. Lower the burette until the tip extends into the flask.

G Titrate the $FeSO_4$ solution by adding a few milliliters of $KMnO_4$ from the burette as you swirl the flask to mix the solutions (see image at left). Stop adding $KMnO_4$ when the light pink color doesn't go away with swirling. Read the final volume in the burette and record it in Trial 1 of Table 1.

H Calculate the amount of $KMnO_4$ used in the titration and record it in Table 1.

I Now do a second trial. Refill the burette with $KMnO_4$ solution and record the initial volume in the Trial 2 column of Table 1.

J Repeat Steps D–H for a second sample of iron(II) sulfate.

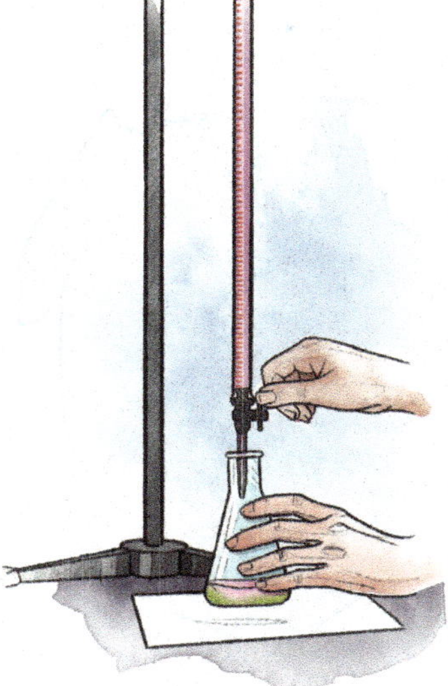

K Drain any unreacted potassium permanganate solution from your burette into the waste beaker provided by your teacher.

7. Describe the color change that you observed as you did the titration.

As the potassium permanganate solution entered the flask, the purple color disappeared with swirling. The purple color stayed longer until it got to the end point of the titration, at which time the solution remained permanently light purple or even just pink.

The balanced redox equation for the reaction of potassium permanganate and iron(II) sulfate is shown below.

$$2KMnO_4(aq) + 8H_2SO_4(aq) + 10FeSO_4(aq) \longrightarrow$$
$$K_2SO_4(aq) + 2MnSO_4(aq) + 5Fe_2(SO_4)_3(aq) + 8H_2O(aq)$$

This might make a little more sense if you look at what is happening to the ions.

$$2MnO_4^-(aq) + 16H^+(aq) + 10Fe^{2+}(aq) \longrightarrow$$
$$2Mn^{2+}(aq) + 10Fe^{3+}(aq) + 8H_2O(l)$$

8. What happens to the oxygens in the permanganate ion?

They combine with the hydrogens from the sulfuric acid to form water molecules.

L Calculate the moles of $KMnO_4$ reacted in each titration. Show your work for Trial 1 below and record your answers for both trials in Table 1.

Answers will vary. *Example:*

What we know: $m_{FeSO_4} = 0.73$ g

Unknown: n_{KMnO_4}

Convert from mass of iron(II) sulfate to moles potassium permanganate.

$$0.73 \text{ g FeSO}_4 \cdot 7H_2O \left(\frac{1 \text{ mol FeSO}_4 \cdot 7H_2O}{278.05 \text{ g FeSO}_4 \cdot 7H_2O} \right) \left(\frac{1 \text{ mol FeSO}_4}{1 \text{ mol FeSO}_4 \cdot 7H_2O} \right) \left(\frac{2 \text{ mol KMnO}_4}{10 \text{ mol FeSO}_4} \right)$$

$$= 5.25 \times 10^{-4} \text{ mol KMnO}_4$$

TITRATION TIPS

Unless your water is quite hard, students should observe potassium permanganate to be pink or magenta and to turn colorless as it reacts. A piece of white paper makes the color change easier to see. If students go beyond the end point, they will have to start over by weighing another sample of iron(II) sulfate to titrate.

At the end point of this reaction, there should be about five times as much iron sulfate used as potassium permanganate.

 ## POTASSIUM PERMANGANATE WASTE DISPOSAL

In most places, you may flush small amounts of dilute potassium permanganate solution down the drain with plenty of water. Large amounts of potassium permanganate should be disposed of at chemical waste facilities.

You may consider saving the solution to be used in future redox titration lab activities or react remaining solutions with iron(II) sulfate in the presence of sulfuric acid. Of course, check waste disposal regulations in your area to make sure that you comply.

 ## SAMPLE CALCULATIONS

The sample calculations are for the data shown in Table 1. Your students' calculations may be different due to variations in their data, but the process should be similar.

M Calculate the amount concentration of the $KMnO_4$ solution. Show your work for Trial 1 below and record your answers for both trials in Table 1.

Answers will vary. *Example:*

What we know: $n_{KMnO_4} = 5.25 \times 10^{-4}$ mol,
$V_{KMnO_4} = 27.84$ mL $= 0.027\ 84$ L/Unknown: c_{KMnO_4}

Write the formula and solve for the unknown.

$$c_{solute} = \frac{n_{solute}}{V_{solution}}$$

Evaluate.

$$c_{solute} = \frac{5.25 \times 10^{-4}\ \text{mol } KMnO_4}{0.027\ 84\ \text{L}}$$

$$= 0.0188\ \frac{\text{mol}}{\text{L}}\ KMnO_4$$

N Obtain the expected concentration of the potassium permanganate solution from your teacher and calculate a percent error. Show your work for Trial 1 below and record your answers for both trials in Table 1.

Answers will vary. *Example:*

What we know: $c_{measured} = 0.0188$ mol/L, $c_{expected} = 0.020$ mol/L

Unknown: $\%_{error}$

Write the formula and solve for the unknown.

$$\%_{error} = \left(\frac{value_{measured} - value_{expected}}{value_{expected}}\right)100\%$$

Evaluate.

$$\%_{error} = \left(\frac{0.0188\ \text{mol/L} - 0.020\ \text{mol/L}}{0.020\ \text{mol/L}}\right)100\%$$

$$= \left(\frac{-0.0012\ \text{mol/L}}{0.020\ \text{mol/L}}\right)100\%$$

$$= -6\%$$

Name ________________________

Going Further

9. Potassium permanganate is used for medical purposes such as treating skin diseases and infections. On the basis of what you have observed in this activity, suggest some reasons why.

Students may suggest its oxidizing properties and the safe byproducts of its reactions.

TABLE 1

	Trial 1	Trial 2
Initial Volume $KMnO_4$ (mL)	2.21	0.09
Mass of $FeSO_4$ (g)	0.73	0.75
Final Volume $KMnO_4$ (mL)	30.05	29.49
Volume $KMnO_4$ used (mL)	27.84	29.40
Moles of $KMnO_4$	5.3×10^{-4}	5.4×10^{-4}
Amount Concentration $KMnO_4$ (mol/L)	0.019	0.018
Percent Error	−6%	−9%

POTASSIUM PERMANGANATE AS A DISINFECTANT

When potassium permanganate reacts with water, it produces safe byproducts, but it also produces free-radical oxygen, which is what kills microbes. The reaction is shown below.

$$KMnO_4 + H_2O \longrightarrow$$
$$MnO_2 + K^+ + 2OH^- + O^-$$

You may want to extend the discussion to talk about free radicals and why potassium permanganate shouldn't be used on deep cuts. Free radicals in the body are a bad thing!

SAMPLE DATA

The data shown in this table is representative. Your students' data may be different due to variations in the experimental conditions but should show a similar tendency.

20A LAB

» Predict the esters produced by reacting carboxylic acids and alcohols.

» Synthesize esters from carboxylic acids and alcohols.

Makes Scents!

Synthesizing Esters

The smell that we associate with bananas is due to a naturally produced ester called *3-methylbutyl acetate*. This ester can be synthesized in laboratories to be used for everything from banana-flavored chewing gum to solvents for varnishes and lacquers. This chemical also has the very convenient ability to attract large numbers of honey bees—an amazing design feature of banana trees!

Where do artificial scents come from?

Smaller esters, such as 3-methylbutyl acetate, have an attractive scent. They form when an alcohol reacts with a carboxylic acid. The hydrogen from the alcohol reacts with the hydroxyl group on the carboxylic acid to form water.

$$R-C\underset{O-H}{\overset{O}{\diagdown}} + R'\,OH \rightleftharpoons R-C\underset{O-R'}{\overset{O}{\diagdown}} + H_2O$$

In this lab activity you'll actually get to make some chemicals that smell good! You'll mix three different combinations of carboxylic acids and alcohols to form three different esters.

EQUIPMENT

- microwave oven
- beakers, 50 mL (3)
- test tubes (3)
- test tube rack
- beaker, 250 mL
- laboratory thermometer
- pipettes (3)
- test tube clamp
- distilled water
- 2-hydroxybenzoic acid ($C_7H_6O_3$)
- methanol (CH_3OH)
- ethanoic acid (CH_3COOH)
- ethanol (CH_3CH_2OH)
- propan-2-ol ($CH_3CHOHCH_3$)
- sulfuric acid, concentrated (H_2SO_4)
- sodium bicarbonate ($NaHCO_3$) solution
- labeling tape or grease pencil
- goggles
- laboratory apron
- nitrile gloves

QUESTIONS

» What esters can be formed from different carboxylic acids and alcohols?

» How can I produce an ester?

EQUIPMENT NOTES

Set up four 100 mL beakers in a place that all lab groups can access. Label them and fill them with sulfuric acid, methanol, ethanol, and ethanoic acid. Equip each beaker with a dropper so that students can easily access each solution. If you have a large class, consider making multiple stations.

Use solutions that are as concentrated as possible. For best results, use concentrated sulfuric acid, 90% ethanol, and glacial ethanoic acid. These will produce a maximum amount of ester, making it easier for students to smell their products.

After adding sodium bicarbonate solution to their test tubes to consume excess carboxylic acids, students may find it helpful to dip a strip of paper towel in their ester while it is in the test tube to observe the scents of their esters.

Consider having coffee beans or grinds available for students to smell. This aroma "resets" the sense of smell, making it easier for students to distinguish the smells of the esters and the reactants used to produce them.

PRE-LAB CHECK

1. What is a key characteristic of esters? *(They are very fragrant.)*

2. What kind of structures do esters have? *(They contain an ether and a carbonyl group attached to the same carbon.)*

3. What are the two carboxylic acids that you will be using? *(2-hydroxybenzoic acid and ethanoic acid)*

4. What are the three alcohols that you will be using? *(methanol, ethanol, and propan-2-ol)*

5. What will you do to help speed up the reaction between the carboxylic acid and alcohol? *(Heat up the reaction and add an acid catalyst.)*

Procedure

A Label three 50 mL beakers "1," "2," and "3," and fill each half full with distilled water. You will be using these beakers of water to help you observe the smell of the esters that you produce.

B Label three clean, dry test tubes "1," "2," and "3."

1. In the reaction shown on page 199, what do the R and R′ represent?

   carbon chains (or alkyl groups)

Later in this activity, you will add sulfuric acid to the mixtures of carboxylic acids and alcohols and then heat these mixtures.

2. Keeping in mind that the reaction is an endothermic process, what do you think is the purpose for heating the reaction and adding the acid?

   Heating the reaction and adding the acid shifts the equilibrium to the

   right to produce more esters.

You will put the following carboxylic acid and alcohol pairs into the test tubes as follows:

   Tube 1: 2-hydroxybenzoic acid and methanol

   Tube 2: ethanoic acid and ethanol

   Tube 3: ethanoic acid and propan-2-ol

3. What is R and R′ for 2-hydroxybenzoic acid and methanol? See the structure of 2-hydroxybenzoic acid below.

   R is 2-hydrobenzoate, and R′ is a methyl group.

4. Write a reaction predicting the result of combining 2-hydroxybenzoic acid and methanol.

   salicylic acid   +   methanol   ⇌   methyl salicylate

C Name the ester produced in this reaction, and record it in the Tube 1 row of Table 1.

5. What is R and R′ for ethanoic acid and ethanol?

R is acetate, and R′ is an ethyl group.

6. Write a reaction predicting the result of combining ethanoic acid and ethanol.

$$CH_3 - \overset{\overset{O}{\|}}{C} - O - H \ + \ CH_3CH_2OH \ \rightleftharpoons \ CH_3 - \overset{\overset{O}{\|}}{C} - O - CH_2CH_3 \ + \ H_2O$$

acetic acid   +   ethanol $\rightleftharpoons$ ethyl acetate

D Name the ester produced in this reaction, and record it in the Tube 2 row of Table 1.

7. What is R and R′ for ethanoic acid and propan-2-ol?

R is ethanoate, and R′ is a propan-2-yl group.

8. Write a reaction predicting the result of combining ethanoic acid and propan-2-ol.

$$CH_3 - \overset{\overset{O}{\|}}{C} - O - H \ + \ CH_3 - \overset{\overset{OH}{|}}{CH} - CH_3 \ \rightleftharpoons \ CH_3 - \overset{\overset{O}{\|}}{C} - O \nearrow \overset{\overset{CH_3}{|}}{\underset{CH_3}{CH}}CH_3 \ + \ H_2O$$

acetic acid   +   isopropanol $\rightleftharpoons$ isopropyl acetate

E Name the ester produced in this reaction, and record it in the Tube 3 row of Table 1.

Now you're ready to make some esters!

COOKING UP SOME ESTERS

F Fill the 250 mL beaker half full with hot tap water. Heat it in a microwave to about 70 °C, *but no warmer*.

G In Tube 1, add 10 drops of 2-hydroxybenzoic acid and 20 drops of methanol.

H In Tube 2, add 10 drops of ethanoic acid and 20 drops of ethanol.

I In Tube 3, add 10 drops of ethanoic acid and 20 drops of propan-2-ol.

J Now add 2 drops of concentrated sulfuric acid to each tube. The sulfuric acid acts as a catalyst to help speed up the reaction between the alcohol and the carboxylic acid. Swirl to mix the contents of each tube.

NO FLAMES!

The chemicals you are working with in this activity are highly flammable, so make sure that there are no laboratory burners turned on as you work.

MAKING MORE ESTER

This method of esterification is called *Fischer esterification*. If you want to increase the amount of ester produced, make sure to use concentrated alcohols and carboxylic acids. An alternative to putting the ester in the beaker of water is to add sodium carbonate or calcium carbonate. This will effervesce as it reacts with any remnant carboxylic acid. You can add the water and ester mix in the beaker to an extraction flask to aid in observing the scents of the esters if desired.

To show students the mechanism by which esters form during Fischer esterification, do an internet search for videos using the keywords "Fischer esterification."

ADDING SODIUM BICARBONATE

If you choose to use glacial ethanoic acid, students may need to add more drops of sodium bicarbonate to neutralize it.

K Put all three test tubes into the hot water. Let them sit in the water for about five minutes. If the contents of a test tube begin to boil, raise the test tube out of the hot water with a test tube holder, and then return it to the hot water bath when boiling stops.

L Let the test tubes cool. Add 10–20 drops of sodium bicarbonate ($NaHCO_3$) solution, about 1 g, until the contents of the tube stop fizzing.

M Pour the contents of each test tube into the beaker that corresponds to it. Tube 1 gets poured into Beaker 1, for example.

N Waft the aroma of each liquid toward your nose (see left), and describe what you observe in Table 1. You may want to waft the aromas of the carboxylic acids and alcohols to contrast the smells. *Do not put your nose directly over the containers while smelling.*

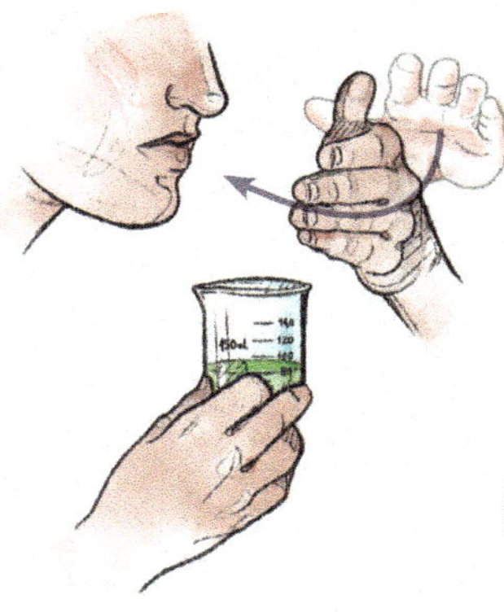

This is how you observe the aroma of the solutions in the beakers.

Going Further

9. What kinds of products would esters be useful for? Explain.

Answers will vary. Esters may be used for the food and perfume industries, for pharmaceuticals, and for cosmetics.

10. Where would the esters to make these products come from, and how could chemistry help?

The esters could come from natural sources, such as plants, or they could be synthesized in laboratories. Chemistry could help us to isolate esters from natural sources and find ways to safely and economically synthesize them in laboratories.

TABLE 1

	Carboxylic Acid	Alcohol	Ester	Observations
Tube 1	2-hydroxybenzoic acid	methanol	methyl 2-hydrobenzoate	wintergreen smell
Tube 2	ethanoic acid	ethanol	ethyl acetate	fruity smell
Tube 3	ethanoic acid	propan-2-ol	propan-2-yl ethanoate	pear-like smell

20B LAB

Name ___________

Date ___________

LAB 20B OBJECTIVES

» Investigate the chemical action of soaps and detergents in hard water.

» Determine which of three given substances is the best emulsifying agent.

» Demonstrate the use of a precipitation reaction to soften hard water.

EQUIPMENT NOTES

To make sodium carbonate, heat approximately 5 g of sodium bicarbonate strongly for 10 to 15 minutes. If time is a concern, the hard water solution can be prepared before the lab activity.

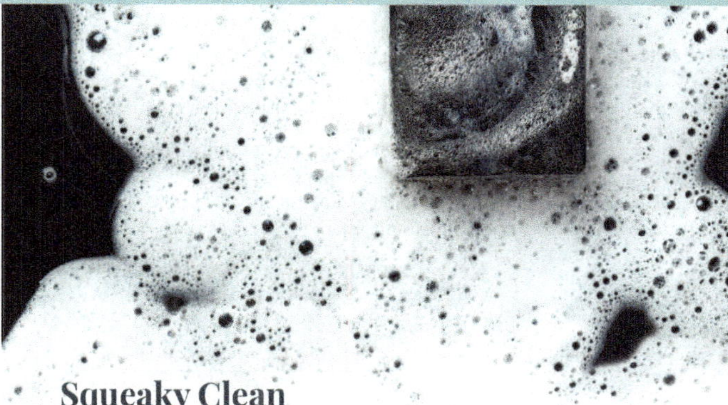

Squeaky Clean

Investigating Soaps and Detergents

The soap-making process has been known for over 2000 years. Prior to AD 100, soap was used as a medicine, but since that time it has been used primarily for washing and cleaning. Early soap makers produced soap by the same process that is still used today. During this process, called *saponification*, a metal hydroxide is heated with a lipid to produce a fatty acid salt and glycerol (together called soap). Originally, the potash (potassium compounds, e.g., KOH) in wood and plant ashes provided the hydroxide for the reaction; today NaOH is used. The following reaction shows this saponification process.

How do soaps and detergents work?

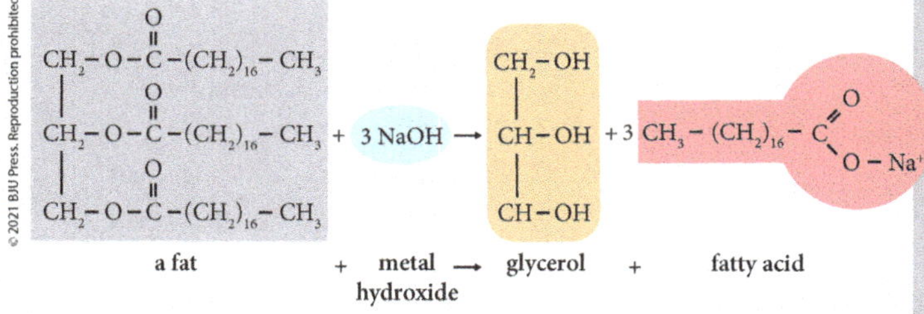

a fat + metal hydroxide → glycerol + fatty acid

When dissolved in water, soap can remove grease and dirt from surfaces as a result of its structure. A soap molecule has a nonpolar end that dissolves in oil or grease, and it has a polar end that dissolves in water. As the soap molecules dissolve in the oil or grease, the polar ends remain protruding from the droplet that forms. Thus, the soap forms a coating on the oil or grease that appears negatively charged to water molecules and results in clusters known as *micelles*. Once coated, the oil or grease is said to be *emulsified*. It can then mix with water and be rinsed away. In this lab activity, you will conduct a series of three tests to investigate how soaps and detergents work.

EQUIPMENT

- laboratory balance
- graduated cylinder, 100 mL
- beaker, 250 mL
- weighing dish (2)
- stirring rod
- test tubes (6)
- test tube brush
- test tube rack
- graduated cylinder, 10 mL
- spatula
- corks (6)
- metric ruler
- beaker, 50 mL
- ring stand with iron ring
- filtering funnel
- filter paper
- distilled water
- magnesium sulfate heptahydrate ($MgSO_4 \cdot 7H_2O$)
- bar of soap
- liquid detergent
- shampoo
- cooking oil
- trisodium phosphate (Na_3PO_4)
- goggles
- laboratory apron
- nitrile gloves

QUESTIONS

» How do soaps and detergents act in hard water?

» Is soap, detergent, or shampoo better at removing oils?

» How can I soften hard water?

Squeaky Clean | 203

PRE-LAB CHECK

1. What two kinds of chemicals react together in saponification? *(a metal hydroxide and a lipid)*

2. What term describes oil or grease coated by soap? *(emulsified)*

3. On what basis will you determine the hardness of your water samples? *(the height of the column of suds produced by each)*

4. Why do laundry detergents contain water-softening agents? *(Detergents clean better in soft water.)*

Procedure

HARD WATER TEST

Hard water is water that contains a significant amount of dissolved minerals. The solubility and cleaning action of soap are greatly reduced in hard water due to the presence of calcium and magnesium ions, which form insoluble carboxylates with the soap. For this reason, soap is often replaced by a synthetic detergent, which has a structure and a function similar to soap but does not form the insoluble compounds in hard water that soap does.

While the presence of suds is not a requirement for soap or detergent to clean well, the loss of suds may indicate the loss of cleansing power.

1. What ion is responsible for creating the hard water in this lab activity?

   magnesium ion

A Using a 100 mL graduated cylinder, measure 150 mL of distilled water and place it in the 250 mL beaker. Using the weighing dish and laboratory balance, measure 3.0 g $MgSO_4 \cdot 7H_2O$, then mix it with distilled water. Stir well. This will be your hard water stock solution.

B Clean three test tubes, and drain as much water from them as possible. Label the tubes "1," "2," and "3," then place them in the test tube rack.

C Use the 10 mL graduated cylinder to add 10 mL of distilled water to Tube 1, 10 mL of tap water to Tube 2, and 10 mL of the prepared hard water stock solution to Tube 3.

D Using a spatula, add a dime-sized scraping of bar soap to each of the test tubes.

E Cork the test tubes and shake each one individually up and down vigorously ten times. Return each tube to the rack and wait one minute. Then use the metric ruler to measure the height of the suds in each tube. (Wherever you are asked in this lab activity to measure the suds, consistently measure only the suds that span the tube, not where they are thin and consist of only a few bubbles.) Record your measurements in Table 1.

F Dispose of the contents of the three test tubes. Clean the tubes thoroughly with the brush, rinse them with tap water, and then do a final rinse with distilled water.

2. Why do you think the bar soap produced the different suds heights in the different samples of water?

   The dissolved minerals interfere with the activity of soap, so the suds height was the smallest in the hard water. The bar soap produced the highest suds in the distilled water because there are no dissolved minerals in the distilled water.

G Repeat Steps C through F, but substitute 10 drops of liquid detergent for the bar soap.

H Repeat Steps C through F again, but substitute 10 drops of shampoo for the bar soap.

3. Which cleanser produced the most suds in each type of water?

   See TE margin for answer.

4. Why does a detergent form suds in hard water when soap doesn't?

   A detergent does not form insoluble precipitates with magnesium or calcium ions the way that soap does.

EMULSIFYING ACTION TEST

I Clean a fourth test tube, label it "4," and allow it to drain. Place it with the other three test tubes in the test tube rack. Add 10 mL of the prepared hard water stock solution to Tubes 1–4.

J Add 20 drops of cooking oil to each test tube.

K Using your spatula, add a nickel-sized scraping of bar soap to Tube 2. Add 20 drops of liquid detergent to Tube 3 and 20 drops of shampoo to Tube 4.

5. What is the purpose of Tube 1?

   Since there is no cleanser added to Tube 1, it

   must be a control.

L Cork the tubes and individually shake each test tube vigorously up and down twenty times. Return each tube to the rack and wait ten minutes. While waiting, continue on to Steps N and O in the Precipitation Test below. After ten minutes, use the metric ruler to measure the height of the suds in each test tube and record your measurements in Table 1.

M Since the goal of removing grease or oil is to emulsify it, or break it up and suspend it in the water, you should also note whether there is a difference in the amount of visible oil droplets at the surface, just under the suds. If the oily layer is not obvious or is milky, you have evidence that it has been emulsified.

6. How did each cleanser perform in the emulsification action test?

   See TE margin for answer.

   ____

   ____

   ____

7. Which substance is the best emulsifying agent in hard water: soap, detergent, or shampoo? Explain your reasoning.

   Shampoo (or shampoo and detergent equally) is

   the best because more of the oil was

   emulsified, and because there were more suds

   than there were with soap.

PRECIPITATION TEST

Commercial laundry and dishwasher detergents also contain substances that soften the water and thus aid their cleaning ability. In this part of the activity, we will use trisodium phosphate to soften the water.

8. Write the balanced equation for the reaction of magnesium sulfate with trisodium phosphate.

   See TE margin for answer.

N Using the weighing dish and laboratory balance, measure 2.0 g Na_3PO_4, then mix it with 20 mL of the hard water stock solution in the 50 mL beaker.

O Stir a couple of minutes, or until all the Na_3PO_4 is dissolved. Note that since a precipitate is forming as the Na_3PO_4 dissolves, it may be difficult to determine when the Na_3PO_4 is completely dissolved.

P Set up the filtration equipment as shown at right. Filter the white precipitate. Dispose of the filtered precipitate in accordance with your teacher's instructions and save the clear filtrate.

Q Clean the last two test tubes, label them "5" and "6," and allow them to drain. Place them in the test tube rack. Add 10 mL of the clear filtrate to Tube 5 and 10 mL of stock hard water solution to Tube 6.

R Add a dime-sized scraping of bar soap to each of the tubes.

S Cork the tubes and shake them ten times. After one minute, use the metric ruler to measure the height of the suds in each tube. Record your measurements in Table 1.

Question 6 Answer

Answers will vary. Results will depend on the types of shampoo and detergent used; results may be similar for the shampoo and the detergent. In any case, the detergent and shampoo should emulsify the oil better than the bar soap does.

Question 8 Answer

$$3MgSO_4\,(aq) + 2Na_3PO_4\,(aq) \longrightarrow$$
$$Mg_3(PO_4)_2\,(s) + 3Na_2SO_4\,(aq)$$

9. According to your data, do you think shampoo is more like a detergent or a soap? Explain your reasoning.

Shampoo is more like a detergent because it forms suds in either soft or hard water.

10. Did Na_3PO_4 affect the hardness of the water? What is the basis for your answer?

Yes. The phosphate ions formed a precipitate with the magnesium ions, removing them from solution, thereby softening the water. The ability of the soap to form suds proves that the hard water was softened.

Going Further

11. One of the key defenses against many diseases (flu, coronavirus, and other viruses) is hand washing. Why do you think that emphasis is put on using soap and washing for at least twenty seconds?

By washing with soap, you allow the soap molecules to adhere to the viruses. Twenty seconds allows ample time for the soap to be well distributed and to be worked into all parts of the hands (e.g., fingernail cuticles).

SAMPLE DATA

The data shown in Table 1 is sample data. Student data may differ but should show similar trends.

TABLE 1

Hard Water Test		
Water	**Cleanser**	**Suds (cm)**
distilled	bar soap	2.4
tap	2.1	
hard	0.4	
distilled	liquid detergent	2.7
tap	2.4	
hard	1.6	
distilled	shampoo	2.8
tap	2.4	
hard	1.5	
Emulsifying Action Test		
Cleanser	**Suds (cm)**	
none	0.0	
bar soap	1.8	
liquid detergent	2.6	
shampoo	2.9	
Precipitation Test		
Additive	**Suds (cm)**	
none	0.5	
Na_3PO_4	1.7	

21A LAB

Name ___________________

Date ___________________

» Determine which macronutrients are in the components of a common sandwich.

» Evaluate whether a sandwich is a balanced meal.

» Identify macronutrients that would need to be added to a sandwich to provide a balanced meal.

 PLANNING FOR LAB 21A

This lab activity is fairly involved, but students will love doing real chemistry. The activity will probably take two days.

Two Options for Saving Time

1. In the Testing Controls section, have students test only the substance that will have a positive result *(lipid—oil; protein —egg white; starch—starch solution; and sugar—sugar solution)* and water as a negative control.

2. Do the Testing Controls section as a teacher demonstration or group procedure with the whole class.

 EQUIPMENT NOTES

Sandwich Fixings

You may either have all the lab groups test the same sandwich components or allow each group to test the components of their choice.

If students are going to choose, give them enough lead time to bring the items to class. Decide in advance whether you will allow condiments (e.g., mustard, ketchup, mayonnaise) to count as food items. If you do allow condiments, the mortar and pestle instructions in Step AA will not be necessary. Also, it would be good to have a limited supply of ingredients on hand to meet the needs of groups that forget a component.

Allergen Warning

If students are bringing their own sandwich components, consider not allowing peanut butter due to the high risk of allergic reaction for some students.

Glucose Test Strips

Be sure to use glucose test strips that indicate glucose levels by a color change, not the type that must be used with a glucose meter. Such strips are available from science supply companies such as Flinn Scientific.

EQUIPMENT

- laboratory balance
- labeling tape or grease pencil
- beakers, 50 mL (7)
- disposable pipettes (7)
- test tubes (5)
- test tube rack
- rubber stoppers (10)
- graduated cylinder, 10 mL
- mortar and pestle
- weighing dishes (3)
- vegetable oil, 10 mL
- egg white, 10 mL
- cornstarch solution, 10 mL
- glucose solution, 10 mL
- distilled water, 10 mL
- biuret solution, 20 mL
- iodine solution, 20 mL
- brown lunch bag
- glucose test strips (8)
- food for testing
- goggles
- laboratory apron
- nitrile gloves

QUESTIONS

» What is a balanced meal?

» What macronutrients do the components of a sandwich provide?

» What can I add to my lunch to make it more balanced?

Balancing Act

Testing Macronutrients in Food

What should I have for lunch? This is a question that most people ask each day. We like to think about what we enjoy eating and what tastes good, but we should also think about what macronutrients are in our food. All the energy we need comes from the food we eat. While we don't need to eat a completely balanced meal every time, we should seek to understand the sources of the building blocks that we need and the foods that provide them to us.

Am I eating balanced meals?

In this lab activity you will test the ingredients of a common sandwich to see which macronutrients are in each ingredient. We will test for the presence of carbohydrates (both starches and sugars), proteins, and lipids (fats).

Procedure

A balanced meal should include carbohydrates, proteins, and lipids. The appropriate proportions of these depend on the energy needs of the individual, but the general rule of thumb is that we need 15%–35% of our calories to come from proteins, 25%–35% from lipids, and 35%–65% from carbohydrates.

PRE-LAB CHECK

1. What four substances will you be testing for in each food item? *(lipids, proteins, starches, and sugars)*

2. What will you use to test for each substance? *(lipids—brown paper bag; proteins—biuret solution; starches—iodine solution; sugars—glucose test strip)*

3. Why must you handle biuret solution carefully? *(It is a strong base and is corrosive.)*

⚠ BIURET SOLUTION

Biuret solution contains sodium hydroxide, which is a strong base. As with all bases, there is a risk due to its corrosive nature. Most students will be able to handle this material safely. If you prefer, you can do the protein test as a teacher demonstration or simply skip that part of the procedure.

Biuret solution must also be neutralized with an acid prior to disposal. Label a beaker "Used Biuret Solution" in which your students may dispose of their used solution. A teacher note at the end of the Protein Test section outlines the proper disposal process. All other liquids in this activity can be safely put down the drain with plenty of water.

THE TESTING CONTROLS

To begin, you will test five substances to determine your baseline data for positive and negative results for each macronutrient—lipid, protein, and carbohydrate (both starch and sugar). You will test vegetable oil, egg whites, cornstarch solution, glucose solution, and water.

A Label five beakers from 1 to 5.

B Pour 10 mL of each of the following samples into the labeled beakers as follows—vegetable oil (1), egg white (2), cornstarch solution (3), glucose solution (4), and water (5).

1. Identify the macronutrient for which each solution will provide the positive control indication.

   vegetable oil: lipid; *egg whites:* protein; *cornstarch:* carbohydrate (starch); *glucose:* carbohydrate (sugar)

C Label two additional beakers "Biuret Solution" and "Iodine Solution."

D Pour 20 mL of the appropriate solution in these two labeled beakers.

   Biuret solution contains sodium hydroxide (NaOH), which is a strong base. Bases are corrosive and must be handled with care. If you spill any biuret solution on your skin, wash immediately and notify your teacher. Iodine is less hazardous, but it will stain your skin and clothes.

E Place a pipette in each of the seven beakers.

LIPID TEST

You will test for lipids by putting some of each sample on a brown paper bag.

F Cut the brown lunch bag so that you have a single layer of paper. Section your brown paper bag into six sections and label five of them from 1 to 5 corresponding to the solutions in Beakers 1–5. Save the sixth section for the Testing Lunch Food portion of this activity.

G In each section, rub the bag with some of the appropriate liquid. Wipe off any excess sample without spreading it into other sections.

H Allow the material on the paper to dry, about 10–15 minutes. (While you wait, continue with the Protein Test below.)

I Once the bag is dry, hold the bag up to a bright light. Record in the Lipid Test column of Table 1 whether the bag in each test area is opaque or translucent.

J After collecting your data, set the bag aside until Step DD.

2. What is the indication of a positive test for lipids according to your observations?

   Lipids will cause the paper bag to become translucent.

PROTEIN TEST

The usual test for the presence of a protein in food is the biuret test. In this test, a liquid sample is mixed with biuret solution. A color change takes place if peptides are present.

K Label five test tubes from 1 to 5 corresponding to the solutions in Beakers 1–5.

L Using the pipettes, transfer 3 mL each from Beakers 1–5 to the appropriate test tubes.

M Using a pipette, add 3 mL of biuret solution to each of the test tubes. Stopper the tubes, shake well, and let sit.

N Allow time for a color change to occur—about 3–5 minutes. Record your observations in the Protein Test column of Table 1.

3. What are the indications of positive and negative results for proteins according to your observations?

The presence of protein turned the solution violet, while the solution remained blue if no protein was present.

O Empty the test tubes into the waste collection beaker provided by your teacher. Rinse and dry the test tubes for use in the Starch Test.

STARCH TEST

Iodine solution is used to test for the presence of starch in foods.

P Label five test tubes from 1 to 5 corresponding to the solutions in Beakers 1–5.

Q Using the pipettes, transfer 3 mL each of Beakers 1–5 to the appropriate test tubes.

R Using a pipette, add 3 mL of iodine solution to each of the test tubes. Stopper the tubes, shake well, and let sit.

S After a couple of minutes, observe any color change. Record the color in the Starch Test column of Table 1.

4. What are the indications of positive and negative results for starches according to your observations?

The presence of starch turned the solution dark blue or black, while the mixture remained yellow-brown if no starch was present.

T Empty the test tubes down the drain with excess water. Rinse and dry the test tubes.

⚠ DISPOSAL OF BIURET SOLUTION

When you are ready to dispose of the biuret solution, you will need water, 1 M hydrochloric acid, and a pH meter (or pH paper).

1. Dilute the biuret solution with eight times as much water (for instance, if you have 100 mL of solution, add 800 mL of water).

2. Add hydrochloric acid in 200 mL increments until the solution reaches neutral (pH = 7). It will take about equal volumes of acid and base to neutralize.

3. Once the solution is neutralized, you may pour it down the drain with plenty of water.

A glucose test strip is used to test for the presence of sugar in foods.

U Divide a piece of paper into five sections and label them from 1 to 5 corresponding to the solutions in Beakers 1–5. Place a glucose test strip in each section.

V Take each test strip from its section, dip it in the appropriate sample, and return it to the paper.

W After three minutes, use the color code provided with the test strips to evaluate each strip and record the results in Table 1.

5. What color indicates the presence of glucose?

green

X After you have collected your data, dispose of the test strips and paper in the trash.

Y Dispose of any unused sample materials down the drain with excess water. Then clean and dry Beakers 1–5 for use in the next section. You also still need the biuret and iodine solutions, so don't dispose of them.

6. Why did we test the water?

Water contains none of the macromolecules and therefore acts as a universal negative control.

Now that you know the process for testing each of the macromolecules, it's time to test some food. Select three ingredients of a typical sandwich, perhaps your favorite! You will now test each of these ingredients for the four different macromolecules.

Z Label three of the beakers appropriately for the foods that you will be testing.

AA Using the laboratory balance and a weighing dish, obtain an approximately 2 g sample of the first food item for testing. Since the food samples are solid, you need to crush each sample with a mortar and pestle. Once sufficiently crushed, transfer the sample to one of the beakers.

BB Using the 10 mL graduated cylinder, add 10 mL of water to the beaker and stir.

CC Repeat Steps AA and BB for each sandwich ingredient. Be sure to clean the mortar and pestle between foods.

DD Repeat the macromolecule tests on each ingredient and record your data in Table 2.

EE When finished, dispose of all biuret solution (used and unused) in the disposal container. Dispose of all other liquids down the drain with excess water. Dispose of all solids in the trash.

7. Which ingredients contained proteins? starches? sugars? lipids?

   Answers will vary: *protein*: ham, cheese; *starch*: bread; *sugar*: bread; *lipid*: ham, cheese.

Going Further

8. Did your lunch contain a balance of macronutrients?

   See TE margin for answer.

9. If your meal was not balanced, what food would be a good choice to make it balanced?

   Answers will vary.

TABLE 1 *Test Control Data*

Test Tube	Material	Positive for	Lipid Test	Protein Test	Starch Test	Glucose Test
1	vegetable oil	lipid	translucent	blue	yellow-brown	yellow
2	egg whites	protein	opaque	violet	yellow-brown	yellow
3	corn starch solution	starch	opaque	blue	dark blue or black	yellow
4	glucose solution	glucose	opaque	blue	yellow-brown	green
5	water	none	opaque	blue	yellow-brown	yellow

TABLE 2 *Lunch Data*

Item	Lipids		Protein		Carbohydrates			
					Starch		Glucose	
	Translucent?	Present?	Color	Present?	Color	Present?	Color	Present?
bread	no	no	blue	no	black	yes	green	yes
ham	yes	yes	violet	yes	brown	no	yellow	no
cheese	yes	yes	violet	yes	brown	no	yellow	no

TRACKING A HEALTHY DIET

Interested students may monitor their diets and count calories via any of a number of online resources. Do an internet search using the keywords "online diet and calorie tracker" to find examples.

Question 8 Answer

Answers will vary. A ham and cheese sandwich would include a balance of macronutrients. But students may also recognize that they need some fruits and vegetables in their meal.

SAMPLE DATA

The Glucose Test results in Table 1 may be different depending on the brand of strips used. The data in Table 2 is sample data; students' data may differ but should show similar trends.

21B LAB

Name ___________

Date ___________

The Proof Is in the Jell-O

Investigating Enzymes

Enzymes are polymers of amino acids that act as catalysts in nearly every biochemical reaction in living organisms. They regulate the speed of a reaction by providing an alternative reaction mechanism with a lower activation energy. Like all catalysts, they are not consumed by the reaction. The amino acid sequence and the three-dimensional structure of the enzyme molecule determine how specific the action of an enzyme will be. It is thought that the various folds of the amino acid chains bring certain areas of the chain together in the same region, forming pockets, or active sites, into which only a specific substrate will fit.

Enzymes are required for most biochemical reactions to proceed at an organism's core temperature. Without them, large quantities of reactants, extremely high temperatures, and long intervals of time would be required for reactions to occur. Several factors affect the activity of an enzyme, including temperature, pH, inhibitors, and activators, all of which are specific for particular enzymes. For example, most human enzymes exhibit optimum activity between 35 °C and 40 °C, but the bacteria that inhabit hot springs have enzyme systems that can effectively operate at 70 °C or higher.

Bromelain is an enzyme found in pineapple. It is classified in the group of enzymes called *proteases*—enzymes that separate proteins into amino acids. Gelatin is a protein from animal bones and connective tissues. In its pure form, gelatin is colorless, odorless, tasteless, brittle, and transparent. Although you may be more familiar with gelatin as a dessert dish, its uses are far broader than the production of food. Gelatin is also used in the manufacture of photographic film, in bacteriology as culture media, and in medicine as capsule shells and surgical dressings.

How do enzymes affect biochemical reactions?

QUESTIONS

» How do enzymes affect gelatin?

» How does temperature affect the activity of an enzyme?

» Which foods contain enzymes that affect gelatin the most?

EQUIPMENT

- microwave oven
- freezer
- beakers, 50 mL (11)
- beaker, 600 mL
- labeling tape or grease pencil
- prepared gelatin
- pineapple, fresh, canned, and frozen (but thawed)
- apple
- banana
- fig
- grape
- kiwi
- papaya
- goggles
- laboratory apron

LAB 21B OBJECTIVES

» Demonstrate the effect of bromelain on a gelatin substrate.

» Observe the effect of heat on the activity of an enzyme.

» Observe the effect of food enzymes on gelatin.

EQUIPMENT NOTES

You can use either flavored or unflavored gelatin for this lab activity. It usually takes one to two hours for the gelatin to fully set.

You can save a great deal of time by placing the pineapple sample in the freezer (see Step E) and preparing the gelatin the night before you plan for students to perform this activity. The sample should be left in the freezer until frozen hard.

Each of the pineapple samples should be the same size, approximately 2.5 cm square.

All materials other than the gelatin should be at room temperature when performing the activity except where indicated as frozen or microwaved.

PRE-LAB CHECK

1. What are enzymes? *(organic catalysts)*

2. What is the name of the enzyme that you will be testing for? *(bromelain)*

3. What evidence will be considered a positive test for the enzyme? *(dissolved gelatin)*

In this lab activity, you will determine the effect of the enzyme bromelain on gelatin and the effect of temperature on the activity of this enzyme. Then you will determine whether other fruits affect gelatin in a manner similar to the way that pineapple does.

1. Why would the necessary conditions for biochemical reactions without enzymes, as described in the introduction, be a problem for living organisms?

   Large amounts of reactants would require more resources than are likely available. Cells need a constant supply of energy; long time intervals for providing this resource would likely result in the death of the cells. Most organisms would not survive the high temperatures required for these processes to occur without enzymes.

Procedure

TESTING ENZYMES

A Label the 50 mL beakers from 1 to 11.

B Obtain a sample of prepared gelatin from your teacher and divide it equally between the beakers.

C Microwave a sample of fresh pineapple for 1 minute on high. Set it aside to cool.

2. Hypothesize about what effect you think heating pineapple will have on the activity of the enzyme bromelain.

   Answers will vary. Assess students' answers on the basis of whether their individual hypotheses are testable.

Once you have placed fruit samples as described in the steps below, you will observe each fruit's effect on gelatin after 10 minutes, 20 minutes, and 30 minutes. Therefore, you will want to get all your samples prepared in advance so that you can place them on the gelatin at approximately the same time.

D Place a sample of fresh pineapple on the surface of the gelatin in Beaker 2.

E Once the pineapple from Step C has cooled, place it on the surface of the gelatin in Beaker 3.

F Place a sample of the thawed pineapple on the surface of the gelatin in Beaker 4.

G Dry any excess juice from a sample of canned pineapple and then place the pineapple on the surface of the gelatin in Beaker 5.

H After 10, 20, and 30 minutes, examine the sample in each beaker and record your observations in Table 1.

TESTING OTHER FRUIT

I. Place a sample of the following fruits into the labeled beakers as follows: fresh apple (6), banana (7), fig (8), grape (9), kiwi (10), and papaya (11).

J. After 10, 20, and 30 minutes, examine the sample in each beaker and record your observations in Table 2.

3. What do you think you are testing with the samples of pineapple that are heated and frozen? What effect do you think heating and freezing will have on the enzyme? Explain.

   See TE margin for answer.

4. What is the purpose of Beaker 1?

   Because Beaker 1 contains only the gelatin, it will serve as a point of comparison for the rest of the beakers. It is the control group.

5. Which sample(s) of pineapple did *not* dissolve the gelatin?

   the canned and microwaved samples

6. Which sample(s) of pineapple caused the gelatin to dissolve?

   the fresh and frozen samples

7. How are the canned and the microwaved samples similar?

   Both have been heated to a temperature high enough to destroy, or *denature*, the enzyme.

8. What conclusion can you draw about the effect of freezing on this particular enzyme?

   Freezing the pineapple does not denature the enzyme.

9. Are the effects on the bromelain that you determined experimentally different from your predictions in Question 2?

   Answers will vary according to students' predictions.

10. Of the fruits that you tested, which one(s) dissolved the gelatin? What does this observation indicate?

    See TE margin for answer.

Question 3 Answer

Answers will vary. Evaluate students' answers on the basis of their reasoning. Many students will correctly assume that heating the pineapple will damage the enzyme, resulting in it having no effect on the gelatin. Some students may think incorrectly that heating the pineapple will speed the reaction, associating this situation with reaction rates. Many students will believe that freezing will have the same effect on the enzyme as heating it did. However, freezing it should have no effect on the enzyme.

Question 10 Answer

Fig, kiwi, and papaya dissolved the gelatin. Each one contains proteases. Students may incorrectly assume that bromelain is the protease in each fruit, but this is not the case.

Question 11 Answer

The proteases in fresh or frozen pineapple, fig, kiwi, and papaya will prevent gelatin from setting. You could use these fruits if they were heated or canned first.

🧪 TESTING OTHER FRUIT

If you choose to test more foods, you can continue to use Beaker 1 from the previous section as the control in the optional section. The samples should be fresh, not canned, preserved, or cooked. Be sure to place the peeled surface of the food on the gelatin. For example, do not place the skin side of a grape on the gelatin—cut the grape so that the inside of the fruit contacts the gelatin. Figs, papayas, and kiwi contain protease enzymes—ficin, papain, and actinidain, respectively. The effect of these enzymes on the gelatin will be similar to that of the bromelain.

Question 13 Answer

The main focus for treatment would be providing the enzymes that the body is not producing. This would allow the body to properly digest food, resulting in vitamin and mineral levels returning to normal.

✔️ OBSERVATIONS

Students' observations for this lab activity will vary. There should be no dissolution of the gelatin by either the microwaved or the canned pineapple specimen. After twenty minutes, there may be little observable difference between the samples. Usually, the first difference is a narrow "rim" of liquid gelatin around the edges of the fresh and frozen samples. As more time elapses, the differences will become more obvious.

11. Have you ever made Jell-O salad? You have to be careful about which fruit you put in your salad! Explain which fruits will and will not work.

See TE margin for answer.

Going Further

12. Exocrine pancreatic insufficiency (EPI) is a condition in which the pancreas fails to produce key digestive enzymes. What would you expect might be some symptoms of EPI?

Because EPI results in insufficient digestive enzymes, the symptoms will probably include those associated with low energy, vitamin, and mineral levels.

13. What would you expect for common treatments of EPI?

See TE margin for answer.

TABLE 1

	Beaker 1	Beaker 2	Beaker 3	Beaker 4	Beaker 5
	No Pineapple	Fresh Pineapple	Microwaved Pineapple	Thawed Pineapple	Canned Pineapple
10 min	no change	dissolved	no change	dissolved	no change
20 min	no change	dissolved	no change	dissolved	no change
30 min	no change	dissolved	no change	dissolved	no change

TABLE 2

	Beaker 6	Beaker 7	Beaker 8	Beaker 9	Beaker 10	Beaker 11
	Apple	Banana	Fig	Grape	Kiwi	Papaya
10 min	no change	no change	dissolved	no change	dissolved	dissolved
20 min	no change	no change	dissolved	no change	dissolved	dissolved
30 min	no change	no change	dissolved	no change	dissolved	dissolved

» Model a simple radioactive decay.

» Model a complex radioactive decay.

22A LAB

Name

Date

It's Only a Matter of Time

Investigating Half-Life

The word *radioactive* makes many people nervous. They associate radioactivity with the negative aspects of this fascinating field of chemistry. In reality, our discovery of radioactivity has led to many benefits for people.

There are many uses for radioactivity in a variety of fields. If you walk through a hospital, you can find many applications that use nuclear chemistry. In addition to the obvious x-rays, there are other imaging applications, such as CT scans and PET scans. You can find radioactivity being used in the form of radiotracers in nuclear pharmacology.

Nuclear chemistry is used to provide food and water around the world. A large portion of our produce is irradiated to destroy bacteria and insects. Nuclear chemistry is also applied in the desalination of water. Even everyday items such as watches, clocks, and smoke detectors operate using radioactive isotopes.

While there are many benefits of nuclear radiation, there are also hazards. As with many technologies, we have to take advantage of the beneficial aspects while minimizing the negative aspects. For example, how long will a material remain radioactive? How do we store it while it remains radioactive? In this lab activity, you will investigate radioactive decay and decay series.

How can I model radioactive decay?

QUESTIONS

» What determines how long a sample will remain radioactive?

» How is the decay model affected if the decay products are also radioactive?

EQUIPMENT
• three different color paper "isotopes" (red, white, and blue; 100 each)

1. What is radioactive decay?

See TE margin for answer.

2. What is half-life?

Half-life is the time required for half a sample of radioactive material to decay into a more stable isotope.

RADIOACTIVITY

Many students will have a negative view of radioactivity. Help them understand that radioactivity has many uses, though we do have to avoid unnecessary exposure.

ISOTOPES

Many items can be used in lieu of the paper "isotopes," as long as the two sides are easily recognized. Some teachers may choose to use coins or other objects. You will still need three different isotopes to do the decay series (100 each).

There are also online coin-flipping simulations. Do an internet search using the keywords "coin-flip simulator."

Question 1 Answer

Radioactive decay is the naturally occurring emission of particles and energy from an unstable nucleus as it progresses toward nuclear stability.

PRE-LAB CHECK

1. What are some benefits of radioactivity? *(food and water production, desalination, smoke detectors, clocks, medical technology)*

2. What is a half-life? *(A half-life is the time for one half of the atoms in a sample of a radioactive sample to decay.)*

3. What fraction of a radioactive sample would remain after three half-lives? *(one-eighth)*

4. How is radioactive decay similar to flipping a coin? *(Both are 50-50 probability events.)*

3. You have a sample that contains 137.6 g of copper-64. Copper-64 is unstable and has a half-life of 12.70 h. It undergoes beta decay to become stable zinc-64. After 1.75 d, how much of your sample will still be copper-64 and how much will be zinc-64?

What we know: $N_0 = 137.6$ g, $t_{1/2} = 12.70$ h,

$t = 1.75$ d $= 42.0$ h

Unknown: N, m_{Zn}

Write the formula and solve for the unknown.

$N = N_0(0.5)^{t/t_{1/2}}$

Evaluate.

$N = (137.6 \text{ g Cu})(0.5)^{42.0 \text{ h}/12.70 \text{ h}}$

$= (137.6 \text{ g Cu})(0.5)^{3.307\,087}$

$= (137.6 \text{ g Cu})(0.101\,034)$

$= 13.9$ g Cu

Determine the mass of zinc-64.

Since we know that copper-64 turns into zinc-64,

$m_{Zn} = N_0 - N$

$= 137.6 \text{ g} - 13.9 \text{ g} = 123.7 \text{ g Zn}$

Procedure

SIMPLE DECAY

In these simple decay trials we are assuming that each of our radioactive isotopes undergoes a single decay event to become a stable atom.

A Start with 40 paper "isotopes," which are blank on one side and have an X on the other.

4. What do you think each paper represents?

an unstable nucleus

5. What do you think the two different sides of the paper represent?

See TE margin for answer.

B Predict how many of the atoms you expect to decay during a first drop of the papers. Record your prediction in Table 1.

6. What does each drop of the papers represent?

the passing of one half-life

C Drop the papers onto the table or floor from at least 1 m up. All subsequent drops should also be made from at least 1 m up.

D Separate all the decayed atoms—the ones with the marked side facing up—from the undecayed. Count the number of decayed atoms, record this value in Table 1, and then set these decayed atoms aside.

E Count the number of undecayed atoms and record this value in Table 1.

F Using only the undecayed atoms, repeat Steps B–E until all the atoms are decayed.

G Repeat Steps A–F for two more trials.

7. What effect would adding more papers have on the number of drops needed to decay all the papers?

It will take more drops, on average, to decay all the atoms.

H Repeat Steps B–G, starting with 60 papers. Record the data from these trials in Table 2.

Question 5 Answer

The two sides represent whether the isotope has decayed or not decayed.

8. Were there challenges to making predictions? Explain. How did you solve any challenges? Why did you choose any solutions that you used?

Answers will vary. Most students will say that it was challenging to make a prediction when there was an odd number of atoms. Groups will have had to round up, round down, or do a combination of both.

9. Were your predictions correct for all the numbers of decayed atoms?

No. Almost all the predictions were off.

10. Could you have made better predictions? Explain.

Assuming that the students always predict half the atoms will decay, there is no way for them to make better predictions.

11. Were the numbers of half-lives exactly the same each trial? Why or why not?

No. Because of statistical fluctuation, we would not expect the number of half-lives to be exactly the same.

12. On average, how many additional half-lives were required to decay the 60 atom samples versus the 40 atom samples? Does this make sense? Explain.

On average it took one additional half-life to decay the extra atoms. This would be expected since each doubling of sample size should take one additional half-life. We increased the sample size by only 50%.

13. On the basis of your results, how much time would it take for 60 atoms of radium-226 ($t_{1/2}$ = 1599 y) to decay?

What we know: n (number of half-lives) = 6

Unknown: t

Write the formula and solve for the unknown.

$$n = \frac{t}{t_{1/2}}$$

$$nt_{1/2} = \left(\frac{t}{t_{1/2}}\right)t_{1/2}$$

$$t = nt_{1/2}$$

Evaluate.

$$t = (6 \ \cancel{\text{half-lives}})\left(\frac{1599 \text{ y}}{\cancel{\text{half-life}}}\right)$$

$$= 9594 \text{ y}$$

✓ SAMPLE CALCULATIONS

The sample calculations are for the data shown in Table 1. Student calculations may be different due to variations in their data, but the process should be similar.

14. Using the data from Trials 4–6, compare the average numbers of atoms that decayed during the first half-life to the average number that decayed in the fifth half-life. Do these results make sense? Explain.

Yes. The averages of sample data in Table 2 show that 30 atoms decayed during the first half-life, but only 2 during the fifth. This makes sense because there were so few atoms remaining by the fifth half-life.

[T] Create a scatterplot of undecayed atoms in Trials 1–3 on a single graph. Make each trial its own color. Create a second scatterplot of undecayed atoms in Trials 4–6. Make each trial its own color.

15. How do the trends of the two scatterplots compare? What type of mathematical relationship do the scatterplots represent?

All the curves on the two scatterplots show exponential decay.

© 2021 BJU Press. Reproduction prohibited.

DECAY SERIES

In the second part of this lab activity we will model a decay series in which each original atom has to go through three decay events before it is stable.

[J] Start with 100 red papers.

[K] Drop the papers onto the table.

[L] Count all the undecayed red atoms and record this value in Table 3.

[M] Count all the decayed atoms and replace each decayed red paper with a white paper. Record the number of white papers in Table 3.

[N] Drop the papers onto the table again.

[O] Separate all the decayed atoms from the undecayed. Replace the decayed red papers with white papers. Replace the decayed white papers with blue papers.

[P] Record how many red, white, and blue papers you now have in Table 3.

[Q] Drop the papers onto the table again.

[R] Separate all the decayed atoms from the undecayed. Replace the decayed red papers with white papers. Replace the decayed white papers with blue papers. Remove the decayed blue papers.

[S] Record how many red, white, and blue papers you now have.

[T] Repeat Steps Q–S until all the blue papers are removed.

16. How do you expect a scatterplot of the decay series will compare with the first two scatterplots?

See TE margin for answer.

[U] Create a scatterplot of undecayed atoms in the decay series on a third graph. Each color will be its own curve.

17. How did the decay series differ from the simple decay? Was this as you expected?

Answers will vary depending on students' answers to Question 16.

Question 16 Answer

Answers will vary. Some students will expect that the scatterplots will look the same, only with the third scatterplot starting at a higher initial value (100 atoms). Other students will recognize that the curve of the red atoms will match the first two scatterplots but that the curves for the white and blue atoms will have to grow first and then decay.

Name ___________________

Going Further

18. Many radioactive isotopes need a multistep decay series before they become a stable isotope. What does what you observed in this lab activity imply about the issue of nuclear waste?

   See TE margin for answer.

   ____

   ____

   ____

   ____

19. Does your answer to Question 18 imply that we should abandon the use of nuclear materials?

   See TE margin for answer.

   ____

   ____

   ____

   ____

20. From your study of biology, you may recall a concept called *genetic drift*, which describes how the frequency of gene variants within a population of organisms, called *allele frequency*, changes over time. Do an internet search using the keywords "genetic drift." Then compare genetic drift and nuclear decay as examples of random processes.

   See TE margin for answer.

   ____

   ____

   ____

   ____

   ____

   ____

PROBLEM OF WASTE

Questions 18 and 19 highlight the challenge of using radioactive material. Help students understand the balance between using things to benefit people and protecting people and the environment from the hazards of nuclear energy.

Question 18 Answer

Answers will vary. *Example:* This lab activity demonstrates the difficulty of dealing with nuclear waste. As one isotope decays, the next one in the series is formed. Each step of the decay series for each atom in the sample represents the emission of particles and energy. This coupled with the long half-lives of some of these isotopes demonstrates that we need a safe, long-term solution for storing this waste.

Question 19 Answer

Answers will vary. *Example:* No. Nuclear materials can benefit humans in many ways. As with most things that benefit people, there can be negative consequences. It is our challenge to discover ways to put the benefits to good use while mitigating the negative aspects.

Question 20 Answer

Genetic drift and nuclear decay both consist of events (a generation of organisms or a half-life of isotopes) whose outcomes are random: an offspring inherits either one or the other variant of a gene (an allele) and an isotope either does or doesn't decay. Over time, both processes drive a sample toward having only a single variant. But unlike nuclear decay, the rate of genetic drift is affected by factors other than the probability of inheritance alone, such as sample size. Genetic drift happens more slowly in larger populations.

SAMPLE DATA

The data shown in these tables are representative. Student data may be different due to variations in the experimental conditions but should show similar trends.

TABLE 1 *Simple Decay*

Half-Life	Trial 1			Trial 2			Trial 3		
	Predicted Decayed	Total Decayed	Undecayed (remaining)	Predicted Decayed	Total Decayed	Undecayed (remaining)	Predicted Decayed	Total Decayed	Undecayed (remaining)
0			40			40			40
1	20	18	22	20	24	16	20	14	26
2	11	12	10	8	5	11	13	8	14
3	5	4	6	5	8	3	7	7	7
4	3	4	2	2	3	0	4	4	3
5	1	2	0				2	2	1
6							1	1	0
7									

TABLE 2 *Simple Decay*

Half-Life	Trial 4			Trial 5			Trial 6		
	Predicted Decayed	Total Decayed	Undecayed (remaining)	Predicted Decayed	Total Decayed	Undecayed (remaining)	Predicted Decayed	Total Decayed	Undecayed (remaining)
0			60			60			60
1	30	30	30	30	30	30	30	29	31
2	15	9	21	15	15	15	16	17	14
3	11	10	11	8	8	7	7	10	4
4	5	3	8	4	3	4	2	3	1
5	4	3	5	2	2	2	1	1	0
6	3	4	1	1	2	0			
7	1	0	1						
8	1	1	0						

TABLE 3 *Decay Series*

Half-Life	Red Papers Remaining	White Papers Remaining	Blue Papers Remaining	Blue Papers Removed
0	100	0	0	0
1	57	43	0	0
2	26	50	24	0
3	16	33	38	13
4	7	27	34	32
5	5	15	35	45
6	2	11	28	59
7	1	9	21	69
8	0	4	18	78
9	0	4	11	85
10	0	1	9	90
11	0	1	5	94
12	0	1	2	97
13	0	1	1	98
14	0	1	1	98
15	0	0	1	99
16	0	0	0	100

Name ___________________

Date ___________________

» Relate the mass defect of nuclides to the binding energy per nucleon.

» Analyze nuclear reactions to determine energy released.

Atomic Asteroids

Determining Mass Defect and Binding Energy

There are some asteroids that are essentially collections of rocks held together by the force of gravity alone. What would it take to break these asteroids apart, especially if one was on a collision course with Earth? The energy required would need to overcome the force of gravity.

Nuclides are like those asteroids in some ways. They are collections of subatomic particles held together by nuclear binding energy. Just like the asteroids, it would take enough energy to overcome the binding energy for nuclides to undergo fission.

A fundamental idea of modern nuclear chemistry is that mass and energy are equivalent. Mass can be changed into energy, and energy can be changed into mass. The equivalence of mass and energy seems to be responsible for the force that holds a nucleus together. The equation $E = mc^2$ relates the quantities of mass and energy. In nuclear reactions, E is nuclear binding energy, m is the mass defect, and c is the speed of light. When a nuclear reaction occurs, a small amount of mass is converted into a large amount of energy.

In this lab activity you will calculate the binding energy per nuclear particle for He-4, Fe-56, and U-232 nuclides so that you can do a direct comparison of the energy packed into these nuclides. You will need to make several preliminary calculations for each atom. You might be surprised to know which nuclides are the most stable!

Where does the mass lost in a nuclear reaction go?

QUESTIONS

» How do I determine mass defect?

» How much energy holds nucleons together in a nucleus?

» How much energy is released in a nuclear reaction?

EQUIPMENT

• reference source (with precise masses of nuclides) or access to the internet

✔ ATOMIC MASS DATA

For a complete list of atomic masses of different nuclides along with their abundances, visit the National Institute of Standards and Technology (NIST) website and do a keyword search for "Atomic Weights and Isotopic Compositions with Relative Atomic Masses." When the page opens up, click on "All Elements" or "All Nuclides." You may choose to have students pick three isotopes of their own to do the activity with. You could then create a class graph of all the data and see how it compares to the graph on page 542 of the Student Edition.

PRE-LAB CHECK

1. What does Einstein's equation, $E = mc^2$, indicate about mass and energy? *(Mass and energy are equivalent, and we can transform mass to energy and energy to mass.)*

2. What are the proper units for energy, mass, and the speed of light in the equation $E = mc^2$? *(joules, kilograms, and meters per second, respectively)*

3. How will you determine the total mass of nucleons for a nuclide? *(by adding the masses of all the protons, neutrons, and electrons contained in it)*

4. How will you determine the actual mass of a nuclide? *(by looking it up in a reference book or online)*

5. How will you find the mass defect? *(by calculating the difference between the actual and total masses of nucleons)*

Procedure

You will start the activity by finding the total mass of all the electrons, protons, and neutrons for He-4, Fe-56, and U-232. You will then add these masses together to find the total expected mass of nucleons for each nuclide. Assume that a proton is 1.0073 u, a neutron is 1.0087 u, and an electron is 0.000 55 u. You will then determine the mass defect—the difference between the expected mass and the actual mass of the nuclide. Once you know the mass defect, you can use the mass-energy equation to calculate the binding energy of the nuclide. You can then calculate the binding energy per nucleon (see the graph on page 542 of your textbook). Let's see how this is done by working through these calculations for sodium-23.

Na-23 nuclides have 11 protons, 11 electrons, and 12 neutrons. Use the calculations below as a sample. Make sure that you follow the rules for significant digits.

SAMPLE CALCULATION (Na-23)

Total Mass of Particles:

$$11 \text{ protons}\left(\frac{1.0073 \text{ u}}{\text{proton}}\right) = 11.0803 \text{ u}$$

$$12 \text{ neutrons}\left(\frac{1.0087 \text{ u}}{\text{neutron}}\right) = 12.1044 \text{ u}$$

$$11 \text{ electrons}\left(\frac{0.000 \, 55 \text{ u}}{\text{electron}}\right) = 0.006 \, 05 \text{ u}$$

$$\text{Total} = 23.190 \, \underline{7}5 \text{ u}$$

Mass Defect:
For the Na-23 atom, the mass defect is the difference between the total mass of nucleons and the nuclide's actual mass.

$$23.190 \, \underline{7}5 \text{ u} - 22.989 \, 77 \text{ u} = 0.200 \, \underline{9}8 \text{ u}$$

$$1 \text{ u} = 1.66 \times 10^{-27} \text{ kg}$$

The mass defect of the Na-23 atom is

$$0.200 \, 98 \text{ u}\left(\frac{1.66 \times 10^{-27} \text{ kg}}{1 \text{ u}}\right) = 3.33\underline{6} \, 268 \times 10^{-28} \text{ kg}.$$

Binding Energy:
$$E = mc^2$$
$$= (3.33\underline{6} \, 268 \, 4 \times 10^{-28} \text{ kg})(3.00 \times 10^8 \text{ m/s})^2$$
$$= 3.00\underline{2} \, 641 \, 2 \times 10^{-11} \text{ J}$$

Binding Energy per Nucleon:
The Na-23 atom has 23 nucleons, so its binding energy per nucleon is

$$\frac{3.00\underline{2} \, 641 \, 2 \times 10^{-11} \text{ J}}{23 \text{ nucleons}}, \text{ or } 1.305 \times 10^{-12} \text{ J/nucleon}.$$

DATA COLLECTION

A Find the total mass of all the protons, neutrons, and electrons for the three nuclides in Table 1. Show your work for He-4 below. Record the total masses in Table 1.

$$2 \text{ protons}\left(\frac{1.0073 \text{ u}}{\text{proton}}\right) = 2.0146 \text{ u}$$

$$2 \text{ neutrons}\left(\frac{1.0087 \text{ u}}{\text{neutron}}\right) = 2.0174 \text{ u}$$

$$2 \text{ electrons}\left(\frac{0.000\,55 \text{ u}}{\text{electron}}\right) = 0.001\,10 \text{ u}$$

Total = 4.033 10 u

1. How do these masses compare to the atomic masses on the periodic table? Explain.

   These masses are close but not exact because the atomic masses on the periodic table are weighted averages of the masses of the various nuclides according to their abundance.

B Use reference books or the internet to look up the actual masses of the nuclides. Be aware that these are different from the atomic masses on the periodic table. Record these in Table 1.

C Find the mass defect of each isotope. Convert the mass defect to kilograms. Show your work for He-4 below. Record the mass defects in Table 1.

$$4.033\,10 \text{ u} - 4.002\,602 \text{ u} = 0.030\,498 \text{ u}$$

$$0.030\,498 \text{ u}\left(\frac{1.66 \times 10^{-27} \text{ kg}}{1 \text{ u}}\right) = 5.062\,668 \times 10^{-29} \text{ kg}$$

2. Which nuclide of the three that you worked with in this exercise had the greatest mass defect? Which had the least?

   U-232; He-4

D Calculate the binding energy for each nuclide. Show your work for He-4 below. Record the binding energies in Table 1.

$$E = mc^2$$

$$= (5.062\,668 \times 10^{-28} \text{ kg})(3.00 \times 10^8 \text{ m/s})^2$$

$$= 4.556\,401\,2 \times 10^{-11} \text{ J}$$

3. Which nuclide had the greatest binding energy? Which had the least?

   U-232; He-4

4. Do your answers to Questions 2 and 3 make sense?

   Yes. Since the calculation of binding energy is the mass defect multiplied by a constant (c^2), the nuclide with the greatest mass defect should have the greatest binding energy.

E Calculate the binding energy per nucleon. Show your work for He-4 below. Record the binding energies per nucleon in Table 1.

$$\frac{4.556\,401\,2 \times\ 10^{-11}\ \text{J}}{4\ \text{nucleons}}, \text{ or } 1.139 \times 10^{-11}\ \text{J/nucleon}$$

5. Which atom had the greatest binding energy per nucleon? Which had the least?

Fe-56; He-4

6. Do your answers to Questions 3 and 5 make sense?

See TE margin for answer.

F Create a scatterplot of the four binding energies per nucleon plotted with mass numbers. Connect them with a smooth curve.

7. Describe any general observations that you have about your graph.

Answers will vary. Students should mention that the curve starts low and peaks and then drops again, though not to the same low as at the beginning.

8. Using your curve, estimate the mass number with the highest binding energy per nucleon.

Students should estimate that the binding energy per nucleon is the highest near mass number 56 (iron).

9. Compare your graph with the graph on page 542 of your textbook. What do you notice?

There are many more data points, but the basic shape of the curve should be the same.

10. If a nuclide were broken apart into its component particles, would the process release or absorb energy? Why?

The process would absorb energy because the nuclides produced by such a process would have less binding energy per particle than the original nuclides. Therefore, they would be less stable. Or it would require energy to overcome the binding energy holding the nuclear particles together.

Question 6 Answer

Answer will vary. Some students will have expected that U-232 would still be the greatest. Others may recall from class that iron is the most stable isotope, which would imply that it has the highest binding energy per nucleon.

Going Further

11. A 20-megaton hydrogen bomb releases energy equivalent to that of the explosion of 20 million tons of TNT, a powerful chemical explosive. Use the fact that 1 ton of TNT releases 4.184×10^9 J of energy to calculate how much matter is converted into energy in the explosion.

What we know:

$$m_{TNT} = 2.0 \times 10^7 \text{ tons TNT} \left(\frac{4.184 \times 10^9 \text{ J}}{1 \text{ ton TNT}} \right) = 8.368 \times 10^{16} \text{ J}$$

Unknown: $m_{converted}$

Write the formula and solve for the unknown.

$$E = mc^2$$

$$\frac{E}{c^2} = \frac{mc^2}{c^2}$$

$$m = \frac{E}{c^2}$$

Evaluate.

$$m_{converted} = \frac{8.368 \times 10^{16} \text{ J}}{(3.00 \times 10^8 \text{ m/s})^2} = 0.93 \text{ kg of matter}$$

12. Would it be wrong to use a 20 megaton hydrogen bomb during war? Why or why not?

See TE margin for answer.

____

____

____

____

____

TABLE 1

	Na-23	He-4	Fe-56	U-232
Mass of Particles (u)	23.1908	4.0331	56.4651	233.9402
Mass of Nuclide (u)	22.9898	4.00260	55.9349	232.0372
Mass Defect (u)	0.2010	0.0305	0.5302	1.9030
Mass Defect (kg)	3.336×10^{-28}	5.06×10^{-29}	8.801×10^{-28}	3.1590×10^{-27}
Binding Energy (J)	3.003×10^{-11}	4.56×10^{-12}	7.921×10^{-11}	2.8431×10^{-10}
Binding Energy per Nucleon (J/nucleon)	1.305×10^{-12}	1.14×10^{-12}	1.414×10^{-12}	1.2255×10^{-12}

THE ETHICS OF WAR

Discuss the answer to Question 12 in class. Help students state their arguments and defend their positions. Lead them to the conclusion that whatever position they take, they do so on the basis of their presupposed moral beliefs. Science can tell us how a thermonuclear explosion works and how to create one, but it cannot tell us whether to use a nuclear device.

Question 12 Answer

Answers will vary. Some students might argue that to kill someone with a huge bomb is no different than killing them with a single bullet. If the latter is allowed, so should the former be. Others might point out that the number of noncombatants killed by a huge bomb is much larger and uncontrollable and thus a huge bomb should not be allowed.

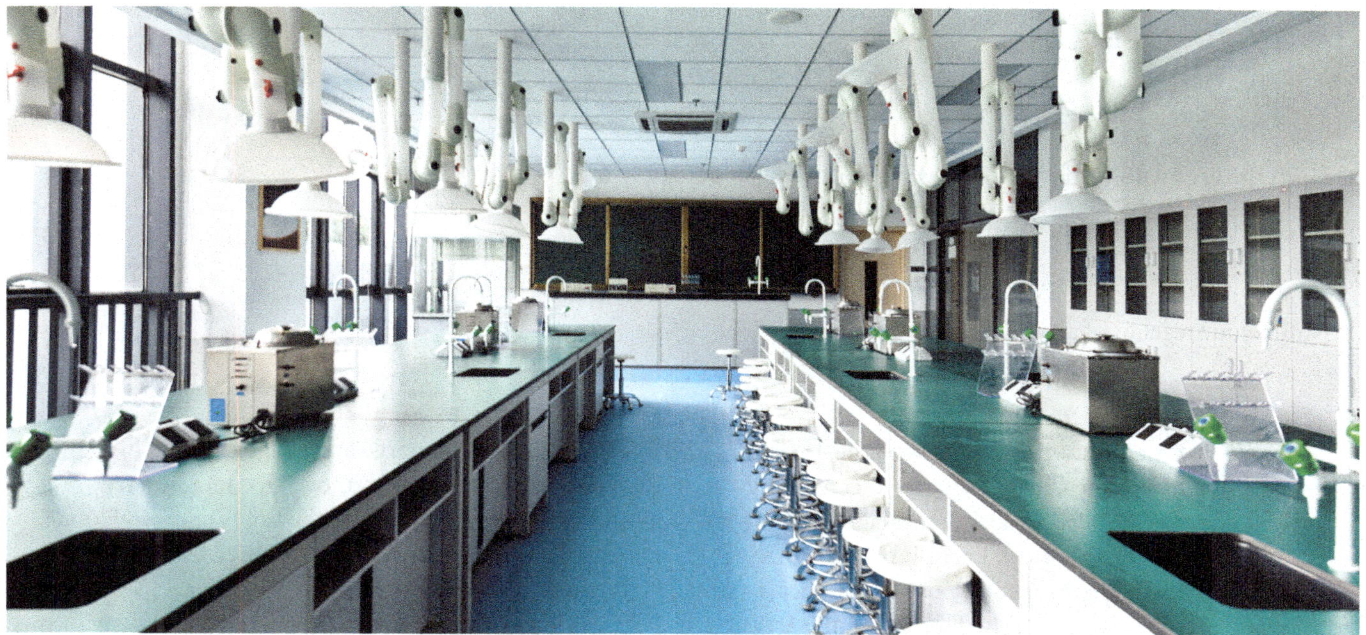

Laboratory Rules

1. Never perform an unauthorized experiment or change any assigned experiment without your teacher's permission.

2. Avoid playful, distracting, or boisterous behavior.

3. Never work alone. Students must not conduct lab activities without supervision.

4. Work at your own lab station.

5. Always work in a well-ventilated area. Use the fume hood when working with toxic vapors. Never put your head in the vent hood.

6. Always wear safety goggles when working with chemicals, glassware, projectiles, and other materials or objects that are potentially hazardous to the eyes.

7. Wear protective clothing and gloves when working with corrosive or staining chemicals.

8. While working in the laboratory, tie back long hair and avoid wearing loose clothing such as scarves or ties.

9. Never taste any chemical, eat or drink out of laboratory glassware, or eat or drink in the laboratory.

10. Always use the appropriate instruments for cutting and handle them carefully. Always cut away from yourself.

11. Thoroughly wash your hands with soap after handling any chemicals.

12. To smell a substance, gently fan its vapor toward you.

13. Never leave a flame or heat source unattended. Keep combustible materials away from heat sources.

14. When diluting acid solutions, always add the acid to water slowly. ***Never add water to an acid!***

15. When heating a test tube, point the open end away from you and others. ***Never heat a closed or stoppered container!***

16. Dispose of waste as instructed by your teacher.

17. Do not return unused chemicals to a container. Dispose of them properly.

18. Notify the teacher of any injuries, spills, or breakages.

19. Know the locations of the fire extinguisher, safety shower, eyewash station, fire blanket, first-aid kit, and Safety Data Sheets (SDS).

Appendix A
LABORATORY SAFETY AND FIRST AID RULES

Basic First Aid

Burns

Flush the area with cold water for several minutes. Do not apply ice.

Chemical Spills

Notify your teacher of all chemical spills.

ON A LABORATORY DESK

- If the material is not particularly volatile, toxic, or flammable, your teacher may have you clean the spill. For liquids, use an absorbent material that will soak up the chemical. For solids, use the designated dustpan and brush. Dispose of chemicals and cleaning materials properly. Then clean the area with soap and water.

- If the material is volatile, toxic, or flammable, and not a large spill, ask your teacher for help. If it is a large spill, you may need to evacuate the laboratory.

- If a highly reactive material, such as hydrochloric acid, is spilled, your teacher will clean it up.

ON A PERSON

- If the spill covers a large area, begin rinsing in the chemical shower, then remove all contaminated clothing and remain under the safety shower. Flood the affected body area for 15 minutes. Obtain medical help immediately.

- If the spill covers a small area, immediately flush the affected area with cold water for several minutes. Then clean the area with soap and water. Get medical attention.

- If the chemical splashes in your eyes, immediately wash them in the nearest eyewash fountain for at least 15–20 minutes. Get medical attention.

- If the spill is an acid, rinse the area with sodium bicarbonate or sodium carbonate solution; if it is a base, use citric or ascorbic acid solution.

Fire

- For any fire other than a contained fire, do *not* attempt to put it out on your own. *Understand that some fires can't be put out with water.*

- Smother a small fire in a container by covering it.

- If a person's clothes are on fire, remember to stop, drop, and roll—roll the person on the floor and use a fire blanket to extinguish the flames. The safety shower may also be used. ***Do not use a fire extinguisher***.

Swallowed Chemicals

Determine the specific substance ingested and follow the instructions on the SDS.
Contact the Poison Control Center in your area immediately.

Cuts

If the wound is superficial, clean it with a disinfectant, apply triple antibiotic ointment, and cover with a bandage. If it is a deep cut, seek immediate medical attention.

Appendix B
LABORATORY EQUIPMENT

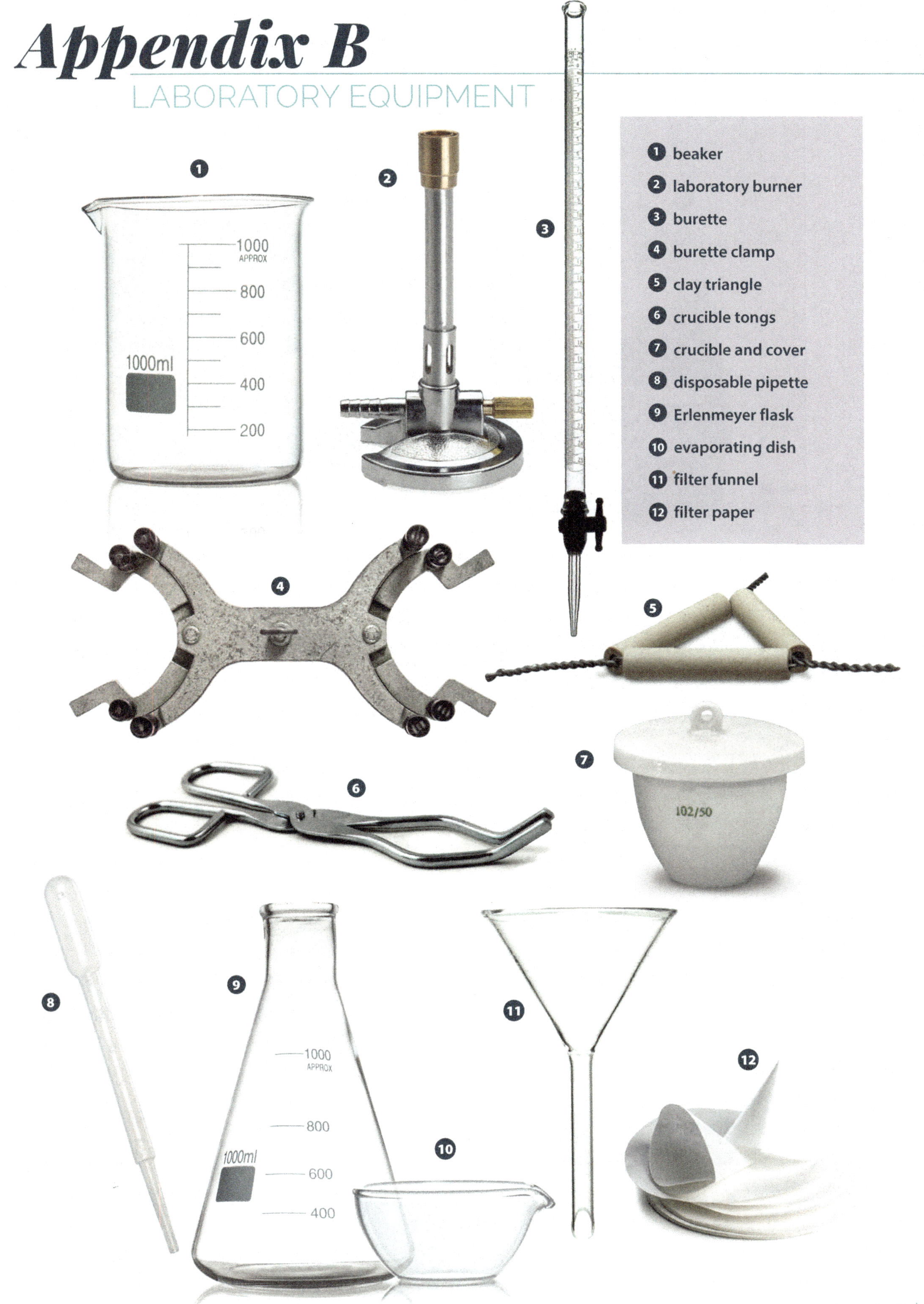

1. beaker
2. laboratory burner
3. burette
4. burette clamp
5. clay triangle
6. crucible tongs
7. crucible and cover
8. disposable pipette
9. Erlenmeyer flask
10. evaporating dish
11. filter funnel
12. filter paper

LABORATORY EQUIPMENT

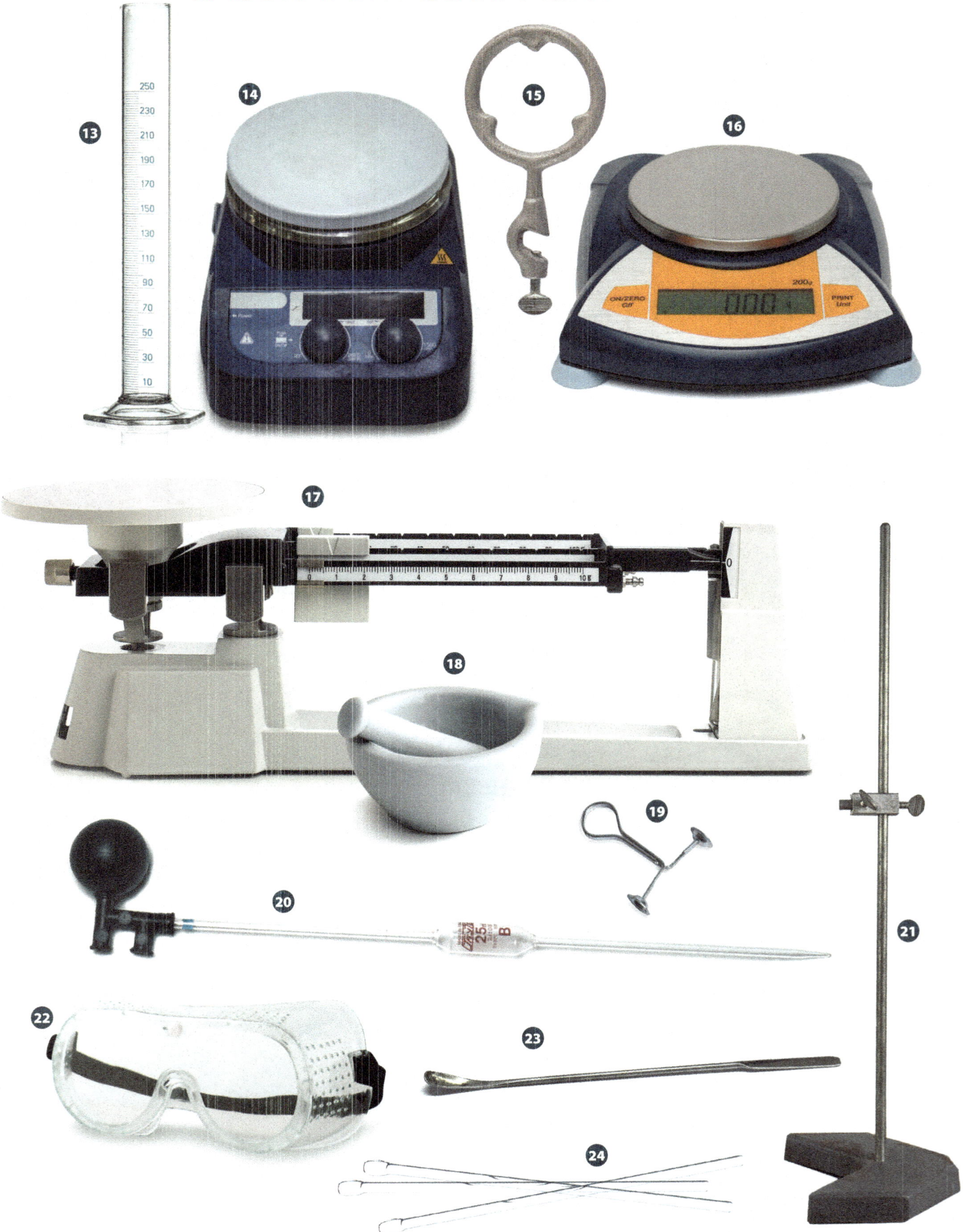

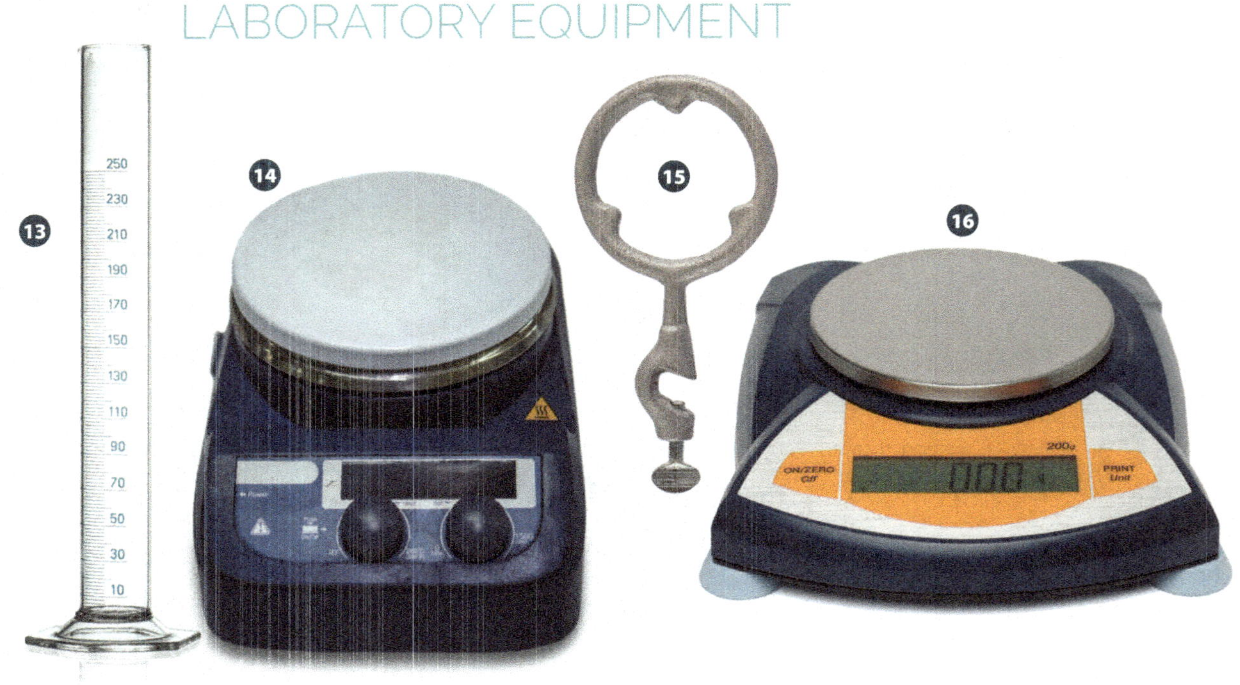

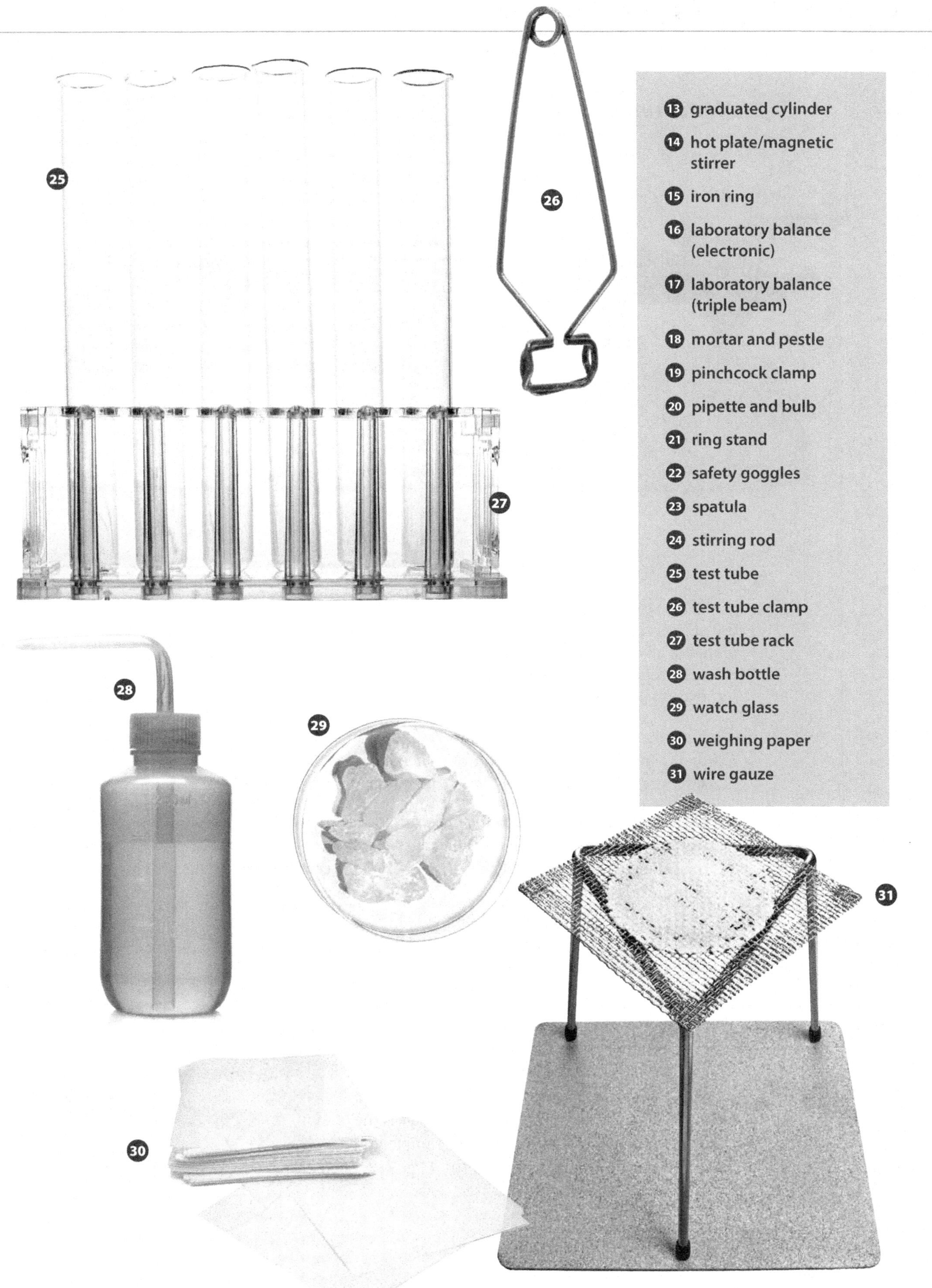

13 graduated cylinder
14 hot plate/magnetic stirrer
15 iron ring
16 laboratory balance (electronic)
17 laboratory balance (triple beam)
18 mortar and pestle
19 pinchcock clamp
20 pipette and bulb
21 ring stand
22 safety goggles
23 spatula
24 stirring rod
25 test tube
26 test tube clamp
27 test tube rack
28 wash bottle
29 watch glass
30 weighing paper
31 wire gauze

Appendix C

Measuring Mass

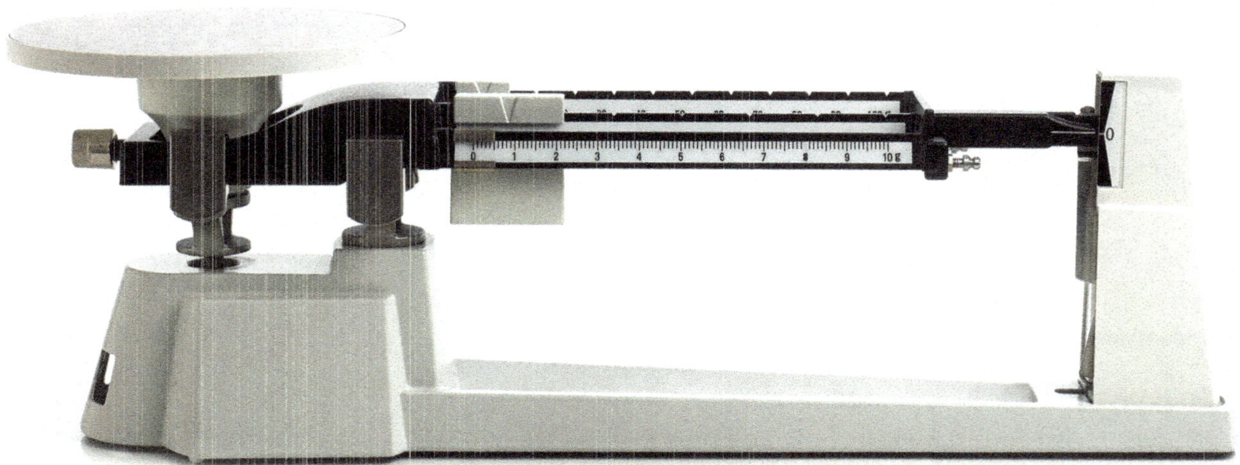

Using Mechanical Balances

The mass of a substance can be determined in the laboratory with the use of a mechanical balance. Several kinds of mechanical balances are common, but all of them operate on the same principles. To use a mechanical balance properly, follow the steps given below.

1. Place the balance on a smooth, level surface.

2. Keep the balance pan(s) clean and dry. Never put chemicals directly on the metal surface of the pan(s). When massing chemicals, place substances on a sheet of weighing paper, watch glass, or in a beaker.

3. Check the rest point of the empty balance. To do this, remove all masses from the pans and slide all movable masses to their zero positions. If the balance beam swings back and forth, note the central point of the swing. You do not have to wait until the beam stops swinging completely. If the central point lies more than two divisions from the marked zero point, ask your teacher to adjust the balance. *Do not adjust the balance yourself!*

4. Place the substance to be massed on the pan and adjust the sliding masses. Move the largest masses first, and then make final adjustments with the smaller masses. If the beam has *detents* for the sliding masses, make sure that the masses rest in the detent. The sum of all the readings is the mass of the object. The mass of the sample in the image at right would be read as 101.43 g.

Using an Electronic Balance

Electronic balances are generally faster and easier to use than their mechanical counterparts. To use an electronic balance properly, follow the instructions given below.

1. Place the balance on a smooth, level surface.

2. Keep the balance pan(s) clean and dry. Never put chemicals directly on the metal surface of the pan(s). When massing chemicals, place substances on a sheet of weighing paper, on a watch glass, or in a beaker.

3. Turn the balance on. Place the container or weighing paper that will hold the substance to be massed on the scale and make sure that there is a reading of 0 by pushing the **Tare** button.

4. Place the substance on the container or weighing paper. Add the desired substance until you have reached the appropriate mass.

Using a Laboratory Burner

Laboratory burners are a common source of heat in laboratories. They are popular because they give a hot flame and burn clean, readily available natural gas or butane. Laboratory burners work well because they mix gas with the correct amount of air to produce the most heat. If air is not mixed with the gas before it burns, not all the gas will burn, and the flame will not be as hot. If too much air is mixed with the gas, the air will extinguish the flame.

If the burner doesn't have its own gas supply, connect it to the desk gas line with a rubber hose. Open the main gas valve and light the burner with a match or burner lighter. Adjustments to the flame should be made with the needle valve and air valve (labeled in the image shown at right).

If the burner lights but the flame immediately goes out, try increasing the gas flow at the needle valve. A yellow flame signifies that insufficient air is mixing with the gas. A flame that makes a noise like a roaring wind means that too much air is entering the barrel. This extra air may cool the flame or extinguish it entirely. To get the best flame possible, rotate the barrel until the flame is entirely blue and two distinct zones appear. Place objects to be heated at the tip of the inner blue zone for quick heating. Sometimes the flame strikes back; that is, it enters the barrel and comes out the bottom. If this happens, do not panic. Turn off the gas supply and re-adjust the burner so that less air enters the barrel.

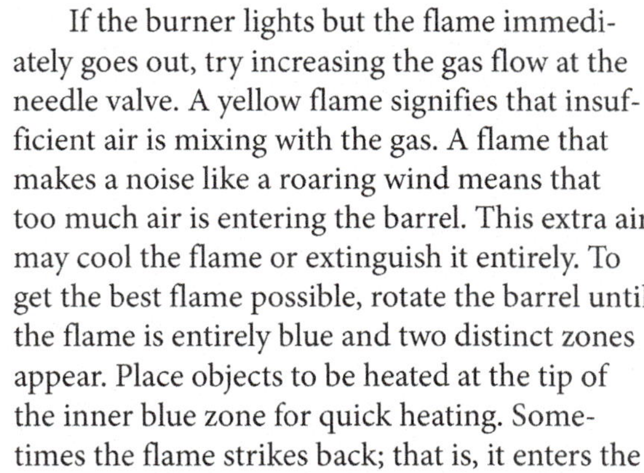

Appendix C

Measuring Chemicals

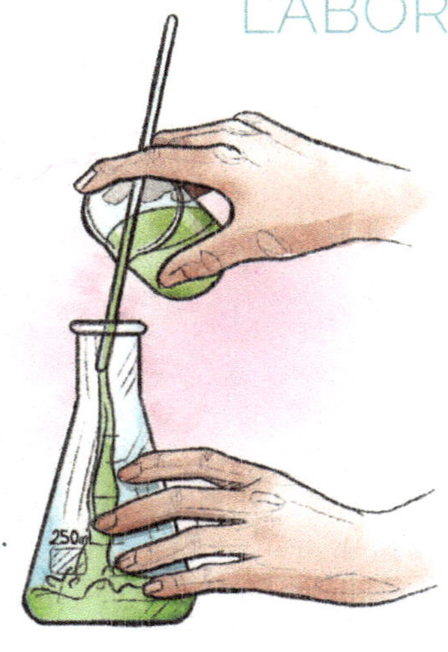

Proper technique for handling liquids is essential if you are to remain safe, keep reagents pure, and obtain accurate measurements. For increased safety, do not splash or splatter liquids when pouring. Pour them slowly down the insides of test tubes, graduated cylinders, and beakers. If anything is spilled, wipe it up promptly. To keep liquids from running down the outside of the container from which you are pouring, pour the liquid down a stirring rod.

In order to keep the liquid chemicals pure, keep stirring rods out of the stock supply. Do not let the stoppers and lids become contaminated while you are pouring. Instead, hold the stopper between your fingers. If you must put a lid down, keep the inside surface from touching the surface of the laboratory table as shown in the image at right.

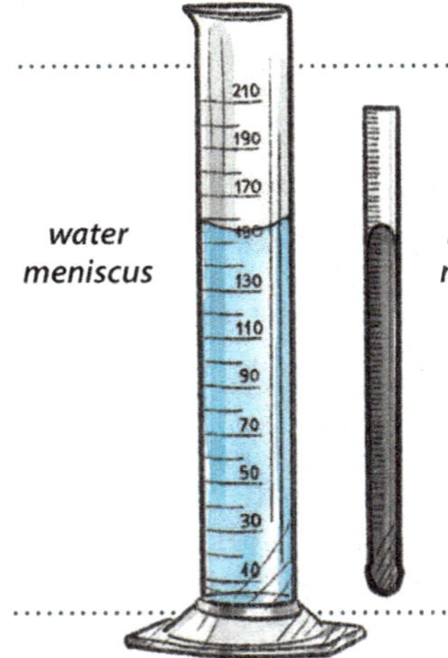

Accurate measurements of liquids can be made in burettes, graduated cylinders, and volumetric flasks. You should always measure volumes in these pieces of glassware unless you need only a rough approximation. If the liquid surface curves downward, look at the bottom of the curved surface—the *meniscus* (see left). In some cases the meniscus will curve upward; if that is the case, you should read the volume at its highest point.

To measure out solids, **1** scoop out a little of the sample onto weighing paper with a spatula, **2** gently tap the spatula until the desired amount falls off, and **3** cup the paper to pour the sample into a test tube.

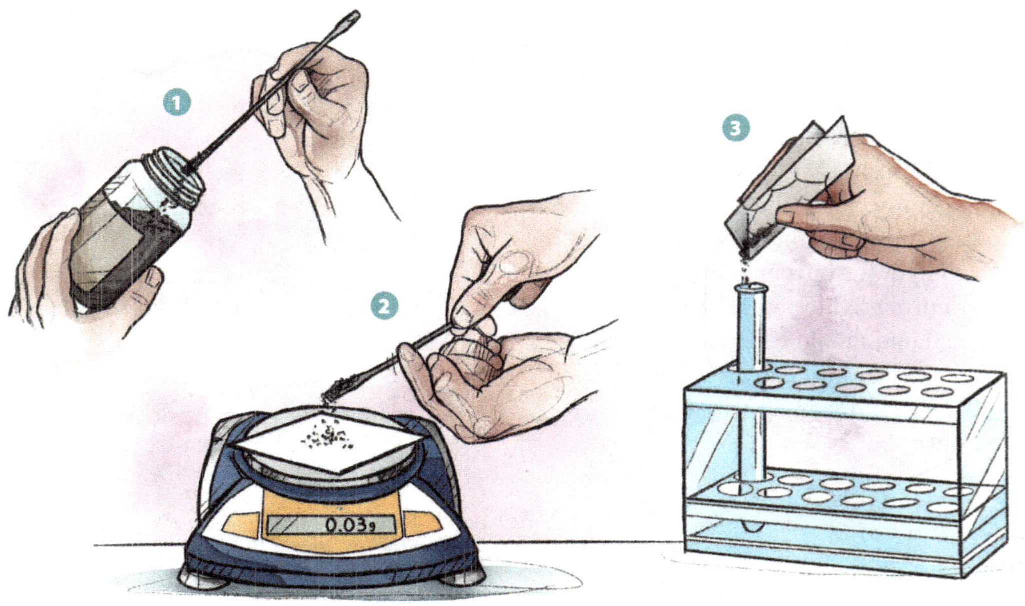

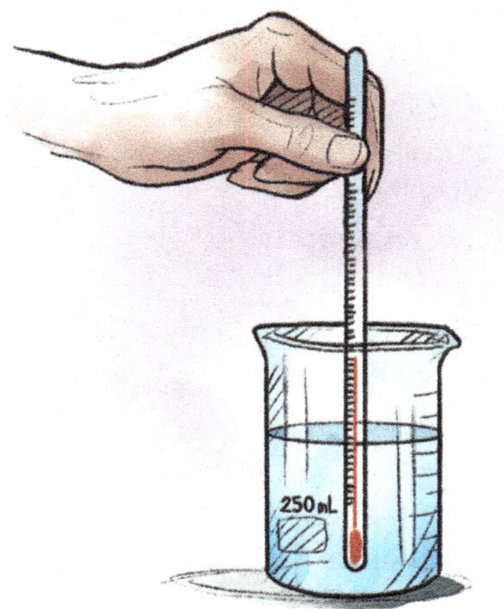

Using a Thermometer

When using a thermometer in lab activities, make sure that you are using one that has the proper temperature range for the experiment that you will be doing. Always hold a thermometer when it is in a container. *Never leave a thermometer standing upright in a container—the container is likely to get knocked over.*

When taking a reading, position the thermometer bulb just above the bottom of the container. If the bulb touches the container, your readings will be inaccurate.

If a thermometer breaks, alert your teacher and do not touch the inner contents. Some thermometers contain mercury. The spilled mercury may look fascinating, but it is toxic and can be absorbed through the skin.

Separating Liquids and Solids

Several experiments require that you separate a solid from a liquid. The most common method of separation—filtering—involves passing the solution through a fine sieve, such as filter paper. The paper allows the liquid and dissolved particles to pass through but catches undissolved particles.

The filter paper must be folded to fit the funnel. Fold it in half and then tear off the corner as shown below. Fold it again at an angle slightly less than 90° to the first fold. Open the paper to form a cone—half of the cone should have three layers of paper, and the other half should have one. Place the cone in a funnel and wet the paper with a few drops of distilled water to hold it in place. Seal the edge of the paper against the edge of the funnel so that none of the solution can go down the spout without going through the paper.

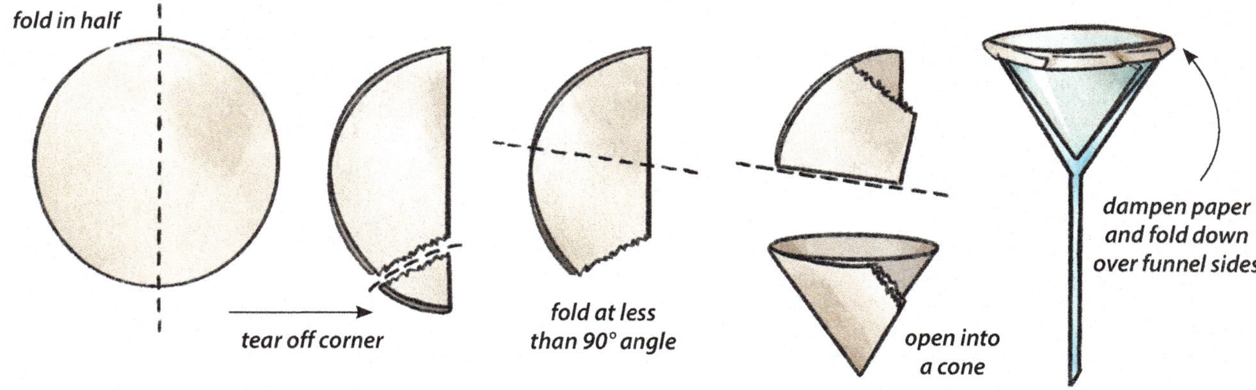

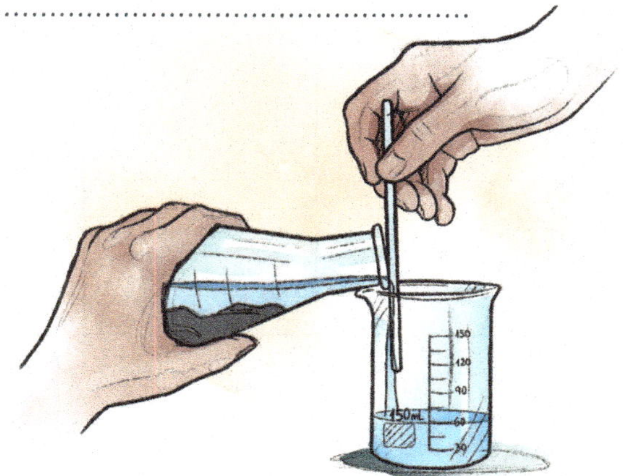

Decanting is a quick method that is often acceptable for separating a liquid from a solid. To decant, allow the solid to settle and then gently pour the liquid off the top of the residue (see left). Avoid causing turbulence that could mix the solid with the liquid. Sometimes the solid residue left in the container is rinsed off with distilled water and decanted a second time to make sure that all the liquid is separated from the precipitate.

Constructing Graphs

When data is recorded in tables, it is difficult to see the relationship that exists between sets of numbers. To make trends and patterns easier to see, you will often put your data on a graph.

In experiments that search for a cause-effect relationship between two variables, you will cause one variable (the independent variable) to change and observe the effect on the second (the dependent variable). If you were to investigate how the solubility of NH_4Cl changes with temperature, temperature would be the independent variable and solubility would be the dependent variable. Traditionally, the independent variable is plotted on the x-axis of the graph, and the dependent variable is plotted on the y-axis.

As you construct your graph, choose an appropriate scale. Don't make the graph so small that the data cannot be clearly seen or so large that the graph will not fit on a single sheet of paper. Pick a scale that will conveniently include the entire range of each variable. Keep in mind that the scales on each axis do not have to be the same. For instance, the scale on the x-axis might be 5 °C for every division, while the scale on the y-axis could be 2 g for every division. Your scale should be easy to subdivide. Subdivisions of 1, 2, 5, and 10 are the most convenient.

Once you have decided which variable will be plotted on which axis and the scales that will be used, neatly label the name of each quantity and the numbers on each axis. The title of the graph should be printed at the top of the graph. If more than one line will be sketched on the same graph, include a legend that identifies each line. Plot each of your data points by making small dots. Follow the specific lab guidelines for the proper way to handle these data points. You will usually draw a smooth line through the data points (see below).

TABLE 1

Observed Solubilities of NH_4Cl

Temperature (°C)	Solubility (g/100 mL H_2O)
10	33
20	37
30	41
40	45
50	50
60	56

Solubility of NH_4Cl. Notice how all the points are connected.

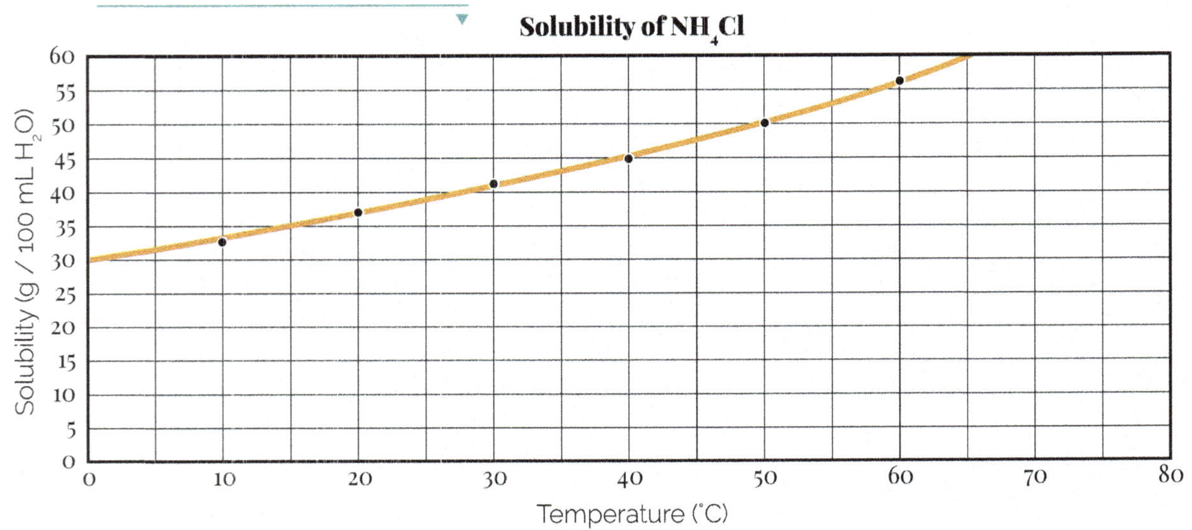

In some cases, you will want to draw a straight line even though your data points do not fall precisely in a line. If this occurs, draw a line that shows the general relationship. Be sure to make the line go through the average values of the plotted points. In the graph (below, left), line A is incorrect because it lies above the cluster of points near the bottom of the graph and below the cluster of points at the top. Line B (below, right) shows the correct method of fitting a straight line to a series of points.

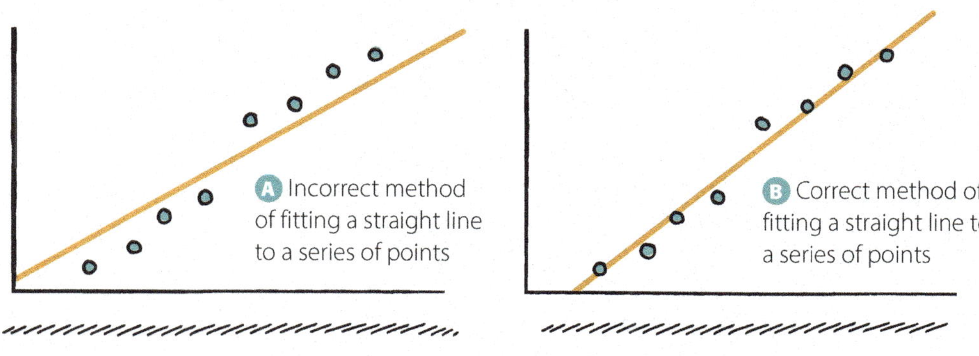

Interpreting Graphs

The shape of a graphed line tells much about the relationship between the variables. When data appears to be arranged in a straight line, the x and y variables are related in a way that can be expressed in a linear equation. A positive slope means that the y variable increases with the x variable; a negative slope indicates that the y variable decreases as the x variable increases.

Data points that curve up or down from left to right indicate that data may be best modeled by some nonlinear function. Common functions in science include quadratic, exponential, logarithmic, and inverse functions.

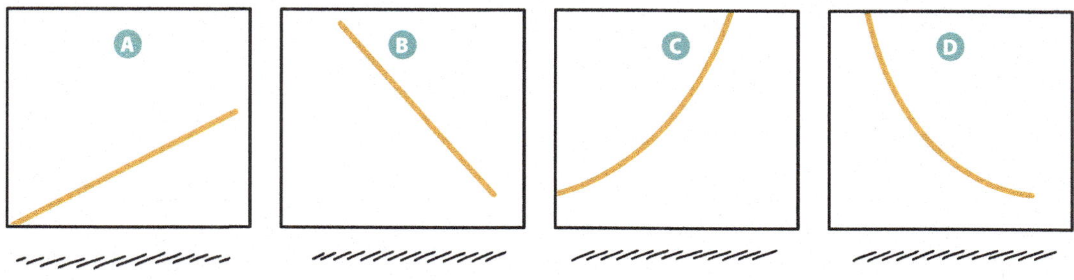

Graphs can be used to predict additional data points that have not been experimentally determined. Assuming that points between verified data points are correct because they fall on the graphed line is called *interpolation*. From the graph of NH_4Cl solubilities on the previous page, it is reasonable to assume that 43 g of NH_4Cl would dissolve at 35 °C. Extending the graphed line past the verified data points in either direction is called *extrapolation*. The graph of NH_4Cl solubilities indicates that 59 g would dissolve at 65 °C. This extrapolation is reasonable, but it may not be totally accurate.

The graphs in this lab manual can have different shapes, including linear relationships (A and B), exponential relationships (C), and inverse relationships (D).

Appendix E
WRITING FORMAL LAB REPORTS

As you have learned in your study of science, communication is an important part of the scientific process. Part of what scientists do is share what they have learned with other scientists. One example of this is the writing of formal lab reports. In a lab report, a scientist presents a problem that was investigated, describes how the investigation was conducted, and summarizes what was learned during the investigation. Many college science courses require students to present all their laboratory work in formal lab reports.

To better prepare you for college, your teacher may choose to have you submit your lab assignments in formal lab reports rather than use the fill-in-the-blank format found in your lab manual. You may be more familiar with the latter form, but there is really little to be afraid of about writing formal reports. They may be new to you, and the word "formal" may sound alarming, but with practice such reporting can become second nature.

There really isn't a standardized way of writing and compiling a formal lab report. Your teacher may require you to do all your reporting in a bound journal, or you may be asked to submit individual typed reports. Regardless of which method your teacher prefers, there are some properties of lab reports that teachers generally agree upon. Let's look at the style and mechanics of a formal lab report and the type of content that is normally included.

Style and Mechanics

The first thing to keep in mind about a formal lab report is that it is a type of *formal writing*. That means that a formal lab report is like the final draft of an essay or research paper. It should have a neat, orderly appearance. To that end, formal lab reports normally have the following elements of style and mechanics.

- Unbound lab reports should be typed in black ink. Reports in a journal should be *neatly* written.

- Lab reports should be written in complete sentences and be checked for proper spelling, punctuation, and grammar.

- Formal lab reports should be free of erasures, whiteout, or scratched-out words.

- Since objectivity is a key aspect of the scientific process, lab reports usually avoid first-person narrative and instead use an impersonal third-person form.

Example:

DON'T SAY: "I learned that 10 mL of *A* will neutralize 5 mL of *B*." (less formal)

DO SAY: "The results indicated that 10 mL of *A* will neutralize 5 mL of *B*." (more formal)

Format

Scientists (and science teachers) may disagree on the exact number of components in a formal lab report, but there is a general consensus on what is presented and the order in which it is presented. Keep in mind as you write that a formal lab report answers three basic questions.

- What problem was investigated?

- What procedure was used during the investigation?

- What was learned from the investigation?

To answer those questions, a formal lab report generally includes the following elements:

- title

- synopsis

- introduction

- list of any materials used and a description of the procedure

- data yielded by the procedure

- analysis of the data and a discussion of the findings

- sources

Let's take a closer look at each of these.

Title

A formal lab report title should succinctly describe what the lab activity was about. It should be matter-of-fact rather than creative. Consider the following possible titles for a titration experiment.

Example:

INFORMAL: Drip, Drip, Drip!

FORMAL: Comparing the Strengths of Acids Using Titration

It's OK for a formal lab report title to sound dry—save the creative writing for your language arts class! Imagine yourself as another scientist who is reading your report as part of her research; she wants to know right away what your report is about without having to read deep into the main body of the text.

Appendix E

Synopsis

A synopsis is a very brief summary of the entire lab report, that is, two or three sentences. If the title of your report catches the reader's attention, the synopsis should then give the reader just enough detail for him to decide whether he wants to read further.

Example (based on a titration lab):

The suitability of titration using an aqueous solution of sodium hydroxide (NaOH) for distinguishing between acids of different strengths was tested. The results confirmed the ability of sodium hydroxide solution to distinguish between different acids on the basis of a direct relationship between the amount of sodium hydroxide required to neutralize an acid and the number of hydronium ions produced by the acid in an aqueous solution.

Introduction

Your introduction should expand on your synopsis by giving more details about the experiment. An introduction should include the following elements.

- a description of the purpose of the experiment

- the hypothesis that will be tested

- a summary of the methods that will be used to test the hypothesis (Be brief. A more detailed description should be reserved for the Materials and Methods section that follows.)

- a brief description of the results of the procedure

- a summary of the conclusions that you have drawn on the basis of the results

Materials and Methods

The goal of this section of your lab report is to describe your investigation in enough detail to allow another person to assess its validity or even replicate it. That is why there are two parts to this section: materials *and* methods. You should first list your *materials*, including names and quantities where appropriate (e.g., for chemicals). Your list should be just that—a list. Think about how you normally see such lists in a student lab manual.

Next, describe your *methods*. You do not need to give instructions for how to do common laboratory tasks, such as using an electronic balance. Nor do you need to explain how to do common laboratory procedures if your audience is expected to be familiar with them. For example, in the titration example used above it would not be necessary to explain how a titration is done; your audience probably already knows how to do this. What your audience *would* need to know in order to replicate your experiment is the chemicals that you used along with how much and in what concentration. As you write this section, ask yourself whether someone reading your descriptions would be able to replicate your experiment exactly. If the answer is no, then you need to be more specific. It is often helpful to include a diagram of any experimental setup used.

Data

Your data section should be reserved for raw data only. This is where you report the results of your tests. Your data may be *qualitative* (i.e., based on observations) or quantitative (i.e., based on measurements). Often, a good way to report quantitative data is to use a data table. The format of data tables will vary depending on the data that you collected. An excellent technique, especially if you are using spreadsheets, is to record the units of measure in the column header and record only numerical data in the cells. This also allows you to use the computational features of the software. All quantitative data should be reported with the appropriate number of significant figures to indicate the precision of the measurements.

Analysis and Discussion

This is where the fun part begins! Now you get to interpret the data that you presented in the previous section. First, describe how the data was analyzed and show any formulas or equations that were used to do so. Sample calculations should be included. Calculated values should be reported with the proper number of significant figures. Descriptions of the results should include appropriate statistical parameters, such as the mean, median, and range of values. Where appropriate, use graphs to show trends in the data. All graphs should be properly scaled and include titles and axes labels with units.

Once you have presented the analysis of your data, you may then address the all-important question: What does it mean? Think back to your stated hypothesis—did your data and analysis support your hypothesis? Were your results in line with what you expected? How did your results compare with those of others in your class? Be sure to support any claims you make with the relevant data and analyses. Your reader—your teacher —expects to see logical arguments that are consistent with your experimental results.

Your discussion needs to be frank and honest. Don't try to make your data say something that it isn't actually saying. If your data does not support your hypothesis, then say so! Don't say that it does just because you think that might be the answer that your teacher is expecting. An important aspect of nurturing science skills is learning how to discuss inconclusive or contrary data. Examine what may have gone wrong during the experiment, make suggestions for improving future investigations, or propose a modified hypothesis for future testing.

It is also appropriate during your discussion to suggest possible practical applications of your work or to put forward additional related questions that might be investigated in the future. In doing so, you might be laying the groundwork for a future scientist. By this you can see how the process of science has links both to those who have investigated a question in the past and to those who may investigate it in the future.

Sources Cited

Much like any formal research paper, a lab report should cite sources within the body of the report and include a bibliography at the end. There are different formats for doing this. Your teacher will advise you on what format to use.

Amount required is specified for each lab group.

Items in italics may be optional depending on student approach in inquiry labs.

Item	Size	# Req.	Lab	Remarks
2-hydroxybenzoic acid ($C_7H_6O_3$)			20A	
acetic acid ($HC_2H_3O_2$)	0.1 M	10 mL	18A	
acetone		10 mL	7B	
		20 mL	13B	
alligator clip lead			19A	
ammonium chloride (NH_4Cl)		21.00 g	14A	
apple		1	21B	
apron, laboratory			1A, 2B, 6A, 10A, 10B, 11A, 11B, 12A, 12B, 13B, 14A, 14B, 15A, 15B, 16A, 16B, 17A, 17B, 18A, 18B, 19A, 19B, 20A, 20B, 21A, 21B	1 for each student
bag, plastic sandwich		1	2A	
baking powder		5 tbsp	1A	Ingredient quantities assume that each lab group makes all three recipes.
balance, laboratory		1	2A, 3B, 4, 10A, 11A, 11B, 12B, 14A, 14B, 15A, 15B, 18B, 19B, 20B, 21A	Lab 3B: *possible equipment depending on student approach*
ball, foam	2.5 cm	29	8A	Modeling clay or gumdrops are acceptable substitutes.
	7.5 cm	9		Modeling clay or marshmallows are acceptable substitutes.
banana		1	21B	
bar of soap		1	20B	
barometer		1	12B	
battery	9 V	1	7B	For conductivity tester; can substitute probeware for conductivity meter.
beaker	50 mL	7	14B	Small plastic cups are acceptable substitutions.
		1	19A, 19B, 20B	
		3	17B, 20A	
		11	21B	

Item	Size	# Req.	Lab	Remarks
beaker	100 mL	1	5B, 14B	
		5	13B	
		6	16A	
		7	16B	
	150 mL	1	2A, 10A, 19B	Lab 2A: A measuring cup is an acceptable substitution.
		2	18B	
	200 mL	3	16A	
	250 mL	1	2A, 10A, 14A, 15A, 20A, 20B	Lab 2A: A measuring cup is an acceptable substitution.
	400 mL	1	15A	
	600 mL	1	12B, 15B, 21B	
	1000 mL	1	12A	
beverages			14B	selection of sugar-containing beverages (If using soda, it should be flat.)
biscuit cutter			1A	
biuret solution		20 mL	20B	
boat, weighing		1	14B	Weighing paper is an acceptable substitution.
bottle, wash		1	14B	with distilled water
brown lunch bag		1	20B	
burette	50 mL or 1000 mL	2	18B	
	50 mL	1	19B	
burner, laboratory			2B, 5A, 10A, 11A, 11B, 12A, 15A	
butter		16 tbsp	1A	
capillary tubes, melting point		2	12A	
carbonless paper targets		2	5A	
cardboard	10 × 10 cm	1	15A, 15B	
chemical bottle with GHS label			1B	
chloride salts			5B	TBD by teacher.
chocolate-covered candies		50	4	Of two varieties; ratio TBD by teacher.
clamp, burette		1	18B, 19B	
clamp, test tube		1	14A	
clay triangle		1	2A, 10A, 11A	

Appendix F

Item	Size	# Req.	Lab	Remarks
coil of solder		1	3B	
colored pencils	set	1	5B	
compound name and formula cards	set	1	9	
compound names and formulas worksheet		1	9	
computer with internet access			13A	
computer with spread-sheet software		1	6B	Graph paper can be substituted.
conductivity tester		1	6A	*possible equipment depending on student approach*
cookie sheet		1 to 3	1A	
cooking oil			20B	
copper wool		1 g	10A	
cork		6	20B	to stopper test tubes
cornstarch solution		10 mL	20B	
cream of tartar		1/2 tsp	1A	
crucible and cover		1	11A	
crucible tongs		1	11A, 11B	
cup, foam	6–8 oz	2	15A, 15B	
cup, plastic	small	7	6A	
diffraction grating spectroscope		1	5B	
digital camera, smart-phone, or camcorder			1B	
dish, weighing		1	2A	Weighing paper is an acceptable substitution.
disposable pipettes		7	20B	
drawing compass		1	5A	
egg		1	1A	
egg white		10 mL	20B	
element samples		7	6A	TBD by teacher.
Erlenmeyer flask	125 mL	1	19B	
Erlenmeyer flask	250 mL	1	2A, 17A	
Erlenmeyer flask	250 mL	2	18B	
Erlenmeyer flask	500 mL	1	12B	
ethanoic acid (CH_3COOH)		5 mL	20A	

Item	Size	# Req.	Lab	Remarks
ethanol (CH$_3$CH$_2$OH)		20 mL	13B	
		5 mL	20A	
evaporating dish		1	7B, 11B	
fig		1	21B	
filter paper		4 pcs.	13B	
		1 pc.	20B	
filtering funnel		1	19B, 20B	
flour		6 cups	1A	
fluorescent light source		1	5B	
food coloring		few drops	13B	any color
			14B	red, blue, green, and yellow
food for testing			20B	TBD by teacher.
freezer		1	21B	
funnel		1	18B	
funnel, filtering		1	2A, 10A	
gelatin			21B	prepared
glass and rubber tubing			12B	
gloves, nitrile			2B, 6A, 10A, 10B, 11A, 11B, 12A, 12B, 13B, 14A, 15B, 16A, 16B, 17A, 18A, 18B, 19A, 19B, 20A, 20B, 21A	1 pair for each student
gloves, puncture-proof			15A, 15B	1 pair for each student
glucose solution		10 mL	20B	
glucose test strips		8	20B	
goggles			1A, 2A, 2B, 3B, 6A, 10A, 10B, 11A, 11B, 12A, 12B, 13B, 14A, 14B, 15A, 15B, 16A, 16B, 17A, 17B, 18A, 18B, 19A, 19B, 20A, 20B, 21B	1 pair for each student
graduated cylinder	10 mL	1	2A, 2B, 6A, 14A, 18A, 19B, 20B	Lab 6A: *possible equipment depending on student approach*
		2	16B	
		3	17B	
	25 mL	1	3B, 12B	Lab 3B: *possible equipment depending on student approach*
		2	10A	
		3	16A, 17B	
		4	16B	
	100 mL	1	12B, 15A, 15B, 18A, 19B, 20B	

Item	Size	# Req.	Lab	Remarks
grape		1	21B	
hammer	small	1	6A	*possible equipment depending on student approach*
hot plate		1	2A, 14A	Lab 2A: A stove is an acceptable substitution.
hydrochloric acid (HCl)	0.10 M	80 mL	16B	
		15 mL	18A	
	1.00 M	35 mL	6A	*possible equipment depending on student approach*
		75 mL	15B	
		175 mL	16A	
	6 M	7.5 mL	2B, 11B	
hydrogen peroxide (H_2O_2), 3.00%		15.0 mL	12B	
ice cubes			12A	
incandescent light source		1	5B	
iodine solution		20 mL	20B	
iron (Fe) filings		pea-sized sample	2B	
iron and sulfur mixture		pea-sized sample	2B	
iron(II) sulfate hepta-hydrate ($FeSO_4 \cdot 7H_2O$)		0.7 g	19B	
iron(II) sulfide (FeS)		pea-sized sample	2B	
iron(III) chloride ($FeCl_3$)	0.25 M	5 mL	17A	
kiwi		1	21B	
labeling tape or grease pencil			13B, 16A, 16B, 17A, 17B, 18A, 20A, 20B, 21B	
laboratory safety equipment			1B	Various; use what you have on hand (e.g., eyewash station, shower, fire blanket, fire extinguisher, first aid kit).
large bowl			1A	
LED light source		1	5B, 7B	Lab 7B: for conductivity tester
lighter, burner		1	2B, 7B, 10A, 11A, 12A, 15A	
liquid detergent			21A	
magnesium ribbon	30 cm	1	11A	
		0.10–0.15 g	15B	
	10 cm	1	19A	

Item	Size	# Req.	Lab	Remarks
magnesium sulfate hepta-hydrate ($MgSO_4 \cdot 7H_2O$)		3.0 g	21A	
magnet, bar		1	2A, 2B	
magnifying glass		1	2B, 13A	
manganese(IV) oxide (MnO_2)		1 g	12B	
marble		4	13B	
marble or ball bearing		1	5A	
marker		1	14B	
measuring cups			1A	
measuring spoons			1A	
mechanical pencil lead		4	19A	
metal shot		50–70 g	15A	
meter stick		1	3A, 5A	
methanol (CH_3OH)		5 mL	20A	
methyl orange solution			18A	
methyl red solution			18A	
metric ruler		1	21A	
microwave oven		1	20A, 21B	
milk		1.5 cups	1A	
mineral oil		20 mL	13B	
mineral samples	set	1	13A	calcite, pyrite, galena, quartz, graphite, silver (Other minerals may be considered.)
MolyMod modeling set		1	7A	Can substitute other molecular modeling sets.
mortar and pestle		1	21A	
multimeter, digital		1	19A	
nitrate salts			5B	TBD by teacher.
papaya		1	21B	
paper "isotopes"	3 cm x 3 cm	300	22A	100 each of three different colors, all marked with an "X" on one side
paper, filter		1	2A, 10A	Lab 2A: A coffee filter is an acceptable substitution.
paper, weighing		8	2B, 6A, 12B, 14A	Lab 6A: *possible equipment depending on student approach*
pastry blender			1A	
pencil, grease		1	14A	or masking tape

Item	Size	# Req.	Lab	Remarks
pennies		15	3B	*possible equipment depending on student approach*
penny		1	3B	
		4	13B	
pH meter		1	18A	optional
phenolphthalein solution			18B	
pinchcock clamp		1	12B	
pineapple		1 ea.	21B	fresh, canned, and frozen (but thawed)
pipette		1	10A, 18B	
		2	17A	
		4	13B	
		3	20A	
pipette bulb		1	14B	
pipette, disposable		1	16B	
pipette, transfer		1	11A, 11B	
		3	18A	
pipette, volumetric	10 mL	1	14B	
potassium bromate ($KBrO_3$)	0.040 M	80 mL	16B	
potassium hydrogen phthalate ($KHC_8H_4O_4$)		3 g	18B	
potassium iodide (KI)	0.010 M	80 mL	16B	
potassium nitrate (KNO_3), solid		3–4 g	15B	
potassium permanganate ($KMnO_4$)	0.020 M	60 mL	19B	
potassium thiocyanate (KSCN)	0.25 M	5 mL	17A	
presoaked wooden splints			5B	Number TBD by teacher.
probeware with temperature probe		1	13B	
propan-2-ol ($CH_3CHOHCH_3$)		5 mL	20A	
reaction plate			10B	
reference sources for nuclide masses or internet access			22B	
resistor	1 kΩ	1	7B	for conductivity tester
ring stand		1	18B, 19B	

Item	Size	# Req.	Lab	Remarks
ring stand and ring		1	2A, 7B, 10A, 11A, 11B, 15A	
ring stand with iron ring		1	21A	
rolling pin			1A	
rubber stoppers	1-hole, 2-hole	1 ea.	12B	
		10	21A	
ruler, metric		1	3A, 12A	
salt		2 tsp	1A	
sand, salt, iron mixture		2–3 g	2A	
sandpaper		1 pc.	11A, 19A	Lab 11A: 150 grit or finer
scissors		1	3A	
SDS			1B	
shampoo			21A	
shortening		1/2 cup	1A	
silver nitrate (AgNO$_3$)		0.43 g	17B	
small watch glass		1	11B	
sodium bicarbonate (NaHCO$_3$) solution		5 mL	20A	
sodium chloride (NaCl)		0.15 g	17B	
sodium hydrogen carbonate (NaHCO$_3$)		3 g	11B	
sodium hydroxide (NaOH)	3 M	20 mL	10A	
	0.4 M	150 mL	18B	
sodium sulfide (Na$_2$S)	0.1 M	25 mL	17B	
sodium thiosulfate (Na$_2$S$_2$O$_3$)	0.20 M	100 mL	16A	
	0.10 M	25 mL	16A	
	0.000 25 M	80 mL	16B	
spatula		1	1A, 2A, 2B, 6A, 11B, 12B, 14A, 21A	Lab 2A: A spoon is an acceptable substitution. Lab 6A: *possible equipment depending on student approach*
split rubber stopper, 1-hole	#4	1	15A, 15B	
spoon, deflagration		1	2B	
starch solution		5 mL	16B	
stirring rod		1	14B, 16A, 17B, 21A	

Item	Size	# Req.	Lab	Remarks
stirring rod, glass		1	10A,12A, 14A	
stopwatch		1	13B, 16A, 16B	
sugar		7 tsp	1A	
sugar		25 g	14B	
sulfur (S) powder or granules		pea-sized sample	2B	
sulfuric acid (H_2SO_4)	6 M	1 mL	19B	
sulfuric acid (H_2SO_4)	1 M	20 mL	10A	
sulfuric acid, concentrated (H_2SO_4)		5 mL	20A	as concentrated as possible
tape			19A	
tape, electrical			7B	for conductivity tester
tape, masking			12A, 14A	Lab 14A: or grease pencil
tape, paper		1	3A	
test tube		3	2B, 20A	
test tube		14	6A	*possible equipment depending on student approach*
test tube		6	7B, 21A	
test tube		1	11B, 12B, 15A	Lab 12B: large Lab 15A: medium
test tube		7	13B	Three small test tubes; size remaining four test tubes and marbles so that marbles fit snugly in tubes.
test tube	medium	4	14A	
test tube		2	17A, 17B	
test tube		21	18A	
test tube		5	21A	
test tube brush		1	21A	
test tube clamp		1	15A, 20A	
test tube rack		1	2B, 6A, 7B, 11B, 12B, 14A, 17A, 17B, 20A, 21A	Lab 6A: *possible equipment depending on student approach*
test tube rack		2	18A	
testing solutions	10 mL each		10B	aluminum nitrate, ammonium nitrate, barium nitrate, calcium nitrate, copper(II) nitrate, iron(III) nitrate, silver nitrate, zinc nitrate, sodium carbonate, sodium chloride, sodium hydroxide, sodium iodide, sodium phosphate, sodium sulfate
textbook		1	3A	
thermometer, laboratory		1	12A, 12B, 14A, 15A, 15B, 20A	
thymol blue solution			18A	
tongs, crucible		1	10A, 12A	

Item	Size	# Req.	Lab	Remarks
toothpick		36	8A	
toppings for biscuits (optional)			1A	
trisodium phosphate (Na_3PO_4)		2.0 g	21A	
unknown substances		3	7B	TBD by teacher; see note on page 59.
vegetable oil			12A	
vegetable oil		10 mL	21A	
vinegar, white (5% acetic acid)		40 mL	19A	
wash bottle		1	7B, 11B, 18B, 19B	Labs 7B, 19B: with distilled water
wash bottle		1	6A	*for distilled water; possible equipment depending on student approach*
watch glass	150 mm	2	2A	
watch glass		1	12A, 13B	
water, distilled		10 mL	19B, 21A	
water, distilled		35 mL	6A	*possible equipment depending on student approach*
water, distilled		40 mL	14A, 18B	
water, distilled		50 mL	17B	
water, distilled		60 mL	18A	
water, distilled		75 mL	20A	
water, distilled		80 mL	16B	
water, distilled		100 mL	16A, 17A	
water, distilled		250 mL	14B	
water, distilled, room-temperature		50 mL	15A, 15B	
weak acid, unknown		10 mL	18A	
weighing dish		1	15A, 15B, 19B	
weighing dish		2	17B, 21A	
weighing dishes		3	21A	
weighing paper		1	7B	
white vinegar		50 mL	18B	
wire, copper insulated		100 cm	7B	for conductivity tester
wire gauze		1 pc.	2A, 7B, 10A, 11A, 11B, 15A	
zinc strip or galvanized nail		10 cm	19A	

Amount required is specified for each lab group.

Items in italics may be optional depending on student approach in inquiry labs.

Lab	Item	Size	# Req.	Remarks
1A	apron, laboratory			1 for each student
	baking powder		5 tbsp	Ingredient quantities assume that each lab group makes all three recipes.
	biscuit cutter			
	butter		16 tbsp	
	cookie sheet		1 to 3	
	cream of tartar		1/2 tsp	
	egg		1	
	flour		6 cups	
	goggles			1 pair for each student
	large bowl			
	measuring cups			
	measuring spoons			
	milk		1.5 cups	
	pastry blender			
	rolling pin			
	salt		2 tsp	
	shortening		1/2 cup	
	spatula		1	
	sugar		7 tsp	
	toppings for biscuits (optional)			
1B	chemical bottle with GHS label			
	digital camera, smart-phone, or camcorder			
	laboratory safety equipment			Various; use what you have on hand (e.g., eyewash station, shower, fire blanket, fire extinguisher, first aid kit).
	SDS			
2A	bag, plastic sandwich		1	
	balance, laboratory		1	
	beaker	150 mL	1	A measuring cup is an acceptable substitution.
	beaker	250 mL	1	
	clay triangle		1	
	dish, weighing		1	Weighing paper is an acceptable substitution.
	Erlenmeyer flask	250 mL	1	
	funnel, filtering		1	
	goggles			1 pair for each student
	graduated cylinder	10 mL	1	
	hot plate		1	A stove is an acceptable substitution.

Lab	Item	Size	# Req.	Remarks
	magnet, bar		1	
	paper, filter		1	A coffee filter is an acceptable substitution.
	ring stand and ring		1	
2A	sand, salt, iron mixture		2–3 g	
	spatula		1	A spoon is an acceptable substitution.
	watch glass	150 mm	2	
	wire gauze		1 pc.	
	apron, laboratory			1 for each student
	burner, laboratory			
	gloves, nitrile			1 pair for each student
	goggles			1 pair for each student
	graduated cylinder	10 mL	1	
	hydrochloric acid (HCl)	6 M	7.5 mL	
	iron (Fe) filings		pea-sized sample	
	iron and sulfur mixture		pea-sized sample	
2B	iron(II) sulfide (FeS)		pea-sized sample	
	lighter, burner			
	magnet, bar		1	
	magnifying glass		1	
	paper, weighing		8	
	spatula		1	
	spoon, deflagration		1	
	sulfur (S) powder or granules		pea-sized sample	
	test tube rack		1	
	test tube		3	
	meter stick		1	
	ruler, metric		1	
3A	scissors		1	
	tape, paper		1	
	textbook		1	
	coil of solder		1	
	goggles			1 pair for each student
3B	penny		1	
	balance, laboratory		1	*possible equipment depending on student approach*
	graduated cylinder	25 mL	1	*possible equipment depending on student approach*
	pennies		15	*possible equipment depending on student approach*
4	balance, laboratory		1	
	chocolate-covered candies		50	Of two varieties; ratio TBD by teacher.

Appendix F
EQUIPMENT AND MATERIALS LIST (BY LAB)

Lab	Item	Size	# Req.	Remarks
5A	carbonless paper targets		2	
	drawing compass		1	
5A	marble or ball bearing		1	
	meter stick		1	
5B	beaker	100 mL	1	
	burner, laboratory		1	
	chloride salts			TBD by teacher.
	colored pencils	set	1	
	diffraction grating spectroscope		1	
	fluorescent light source		1	
	incandescent light source		1	
	LED light source		1	
	nitrate salts			TBD by teacher.
	presoaked wooden splints			Number TBD by teacher.
6A	apron, laboratory			1 for each student
	cups, plastic	small	7	
	element samples		7	TBD by teacher.
	gloves, nitrile			1 pair for each student
	goggles			1 pair for each student
	conductivity tester		1	*possible equipment depending on student approach*
	graduated cylinder	10 mL	1	*possible equipment depending on student approach*
	hammer	small	1	*possible equipment depending on student approach*
	hydrochloric acid	1.00 M	35 mL	*possible equipment depending on student approach*
	paper, weighing		7	*possible equipment depending on student approach*
	spatula		1	*possible equipment depending on student approach*
	test tube rack		1	*possible equipment depending on student approach*
	test tube		14	*possible equipment depending on student approach*
	wash bottle		1	*for distilled water; possible equipment depending on student approach*
	water, distilled		35 mL	*possible equipment depending on student approach*
6B	computer with spread-sheet software		1	Graph paper can be substituted.
7A	MolyMod modeling set		1	can substitute other molecular modeling sets
7B	acetone		10 mL	
	battery	9 V	1	For conductivity tester; can substitute probeware for conductivity meter.
	evaporating dish		1	
	LED light source		1	for conductivity tester
	lighter, burner			
	resistor	1 kΩ	1	for conductivity tester
	ring stand and ring		1	

Lab	Item	Size	# Req.	Remarks
7B	tape, electrical			for conductivity tester
	test tube		6	
	test tube rack		1	
	unknown substances		3	TBD by teacher; see note on page 59.
	wash bottle		1	with distilled water
	weighing paper		1	
	wire, copper insulated		100 cm	for conductivity tester
	wire gauze		1 pc.	
8A	ball, foam	2.5 cm	29	Modeling clay or gumdrops are acceptable substitutes.
		7.5 cm	9	Modeling clay or marshmallows are acceptable substitutes.
	toothpick		36	
8B	none			
9	compound name and formula cards	set	1	
	compound names and formulas worksheet		1	
10A	apron, laboratory			1 for each student
	balance, laboratory			
	beaker	150 mL	1	
		250 mL	1	
	burner, laboratory		1	
	clay triangle		1	
	copper wool		1 g	
	funnel, filtering		1	
	gloves, nitrile			1 pair for each student
	goggles			1 pair for each student
	graduated cylinder	25 mL	2	
	lighter, burner		1	
	paper, filter		1	
	pipette		1	
	ring stand and ring		1	
	sodium hydroxide	3 M	20 mL	
	stirring rod, glass		1	
	sulfuric acid, 1 M		20 mL	
	tongs, crucible		1	
	wire gauze		1 pc.	
10B	apron, laboratory			1 for each student
	gloves, nitrile			1 pair for each student
	goggles			1 pair for each student
	reaction plate			

Appendix F

Lab	Item	Size	# Req.	Remarks
10B	testing solutions	10 mL each		aluminum nitrate, ammonium nitrate, barium nitrate, calcium nitrate, copper(II) nitrate, iron(III) nitrate, silver nitrate, zinc nitrate, sodium carbonate, sodium chloride, sodium hydroxide, sodium iodide, sodium phosphate, sodium sulfate
11A	apron, laboratory			1 for each student
	balance, laboratory		1	
	burner, laboratory		1	
	clay triangle		1	
	crucible and cover		1	
	crucible tongs		1	
	goggles			1 pair for each student
	lighter, burner		1	
	magnesium ribbon	30 cm	1	
	gloves, nitrile			1 pair for each student
	pipette, transfer		1	
	ring stand and ring		1	
	sandpaper	150 grit or finer	1	
	wire gauze		1 pc.	
11B	apron, laboratory			1 for each student
	balance, laboratory		1	
	burner, laboratory		1	
	crucible tongs		1	
	evaporating dish		1	
	goggles			1 pair for each student
	hydrochloric acid (HCl)	6 M	6 mL	
	gloves, nitrile			1 pair for each student
	pipette, transfer		1	
	ring stand and ring		1	
	small watch glass		1	
	sodium hydrogen carbonate ($NaHCO_3$)		3 g	
	spatula		1	
	test tube		1	
	test tube rack		1	
	wash bottle		1	
	wire gauze		1 pc.	
12A	apron, laboratory			1 for each student
	beaker	1000 mL	1	
	burner, laboratory		1	
	capillary tubes, melting point		2	

Lab	Item	Size	# Req.	Remarks
12A	gloves, nitrile			1 pair for each student
	goggles			1 pair for each student
	ice cubes			
	lighter, burner		1	
	ruler, metric		1	
	stirring rod, glass		1	
	tape, masking			
	thermometer, laboratory		1	
	tongs, crucible		1	
	vegetable oil			
	watch glass		1	
12B	apron, laboratory			1 for each student
	balance, laboratory		1	
	barometer		1	
	beaker	600 mL	1	
	Erlenmeyer flask	500 mL	1	
	glass and rubber tubing			
	gloves, nitrile			1 pair for each student
	goggles			1 pair for each student
	graduated cylinder	25 mL	1	
		100 mL	1	
	hydrogen peroxide (H_2O_2), 3.00%		15.0 mL	
	manganese(IV) oxide (MnO_2)		1 g	
	paper, weighing		1	
	pinchcock clamp		1	
	test tube rack		1	
	rubber stoppers	1-hole, 2-hole	1 ea.	
	spatula		1	
	test tube	large	1	
	thermometer, laboratory		1	
13A	computer with internet access			
	magnifying glass		1	
	mineral samples	set	1	calcite, pyrite, galena, quartz, graphite, silver (Other minerals may be considered.)
13B	acetone		20 mL	
	apron, laboratory			1 for each student
	beaker	100 mL	5	
	ethanol (CH_3CH_2OH)		20 mL	

Appendix F
EQUIPMENT AND MATERIALS LIST (BY LAB)

Lab	Item	Size	# Req.	Remarks
13B	filter paper		4 pcs.	
	food coloring		few drops	any color
	gloves, nitrile			1 pair for each student
	goggles			1 pair for each student
	labeling tape or grease pencil		1	
	marble		4	
	mineral oil		20 mL	
	penny		4	
	pipette		4	
	probeware with temperature probe		1	
	stopwatch		1	
	test tube		7	Three small test tubes; size remaining four test tubes and marbles so that marbles fit snugly in tubes.
	watch glass		1	
14A	ammonium chloride (NH_4Cl)		21.00 g	
	apron, laboratory			1 for each student
	balance, laboratory		1	
	beaker	250 mL	1	
	clamp, test tube		1	
	gloves, nitrile			1 pair for each student
	goggles			1 pair for each student
	graduated cylinder	10 mL	1	
	hot plate		1	
	paper, weighing		1	
	pencil, grease		1	or masking tape
	test tube rack		1	
	spatula		1	
	stirring rod, glass		1	
	tape, masking		1	or grease pencil
	test tube	medium	4	
	thermometer, laboratory		1	
	water, distilled		40 mL	
14B	apron, laboratory			1 for each student
	balance, laboratory		1	
	beaker	50 mL	7	Small plastic cups are acceptable substitutions.
	beaker	100 mL	1	
	beverages			selection of sugar-containing beverages (If using soda, it should be flat.)
	boat, weighing		1	Weighing paper is an acceptable substitution.

Lab	Item	Size	# Req.	Remarks
	bottle, wash		1	with distilled water
	food coloring			red, blue, green, and yellow
	goggles			1 pair for each student
	marker		1	
14B	pipette bulb		1	
	pipette, volumetric	10 mL	1	
	stirring rod		1	
	sugar		25 g	
	water, distilled		250 mL	
	apron, laboratory			1 for each student
	balance, laboratory		1	
	beaker	250 mL	1	
		400 mL	1	
	burner, laboratory		1	
	cardboard	10 × 10 cm	1	
	cup, foam	6–8 oz	2	
	gloves, puncture-proof	pair	1	
	goggles			1 pair for each student
	graduated cylinder	100 mL	1	
15A	lighter, burner		1	
	metal shot		50–70 g	
	ring stand and ring		1	
	split rubber stopper, 1-hole	#4	1	
	test tube	medium	1	
	test tube clamp		1	
	thermometer, laboratory		1	
	water, distilled, room-temperature		50 mL	
	weighing dish		1	
	wire gauze		1 pc.	
	apron, laboratory			1 for each student
	balance, laboratory		1	
	beaker	600 mL	1	
	cardboard	10 × 10 cm	1	
	cup, foam	6–8 oz	2	
15B	gloves, puncture-proof	pair	1	
	goggles			1 pair for each student
	graduated cylinder	100 mL	1	
	hydrochloric acid (HCl)	1.00 M	75 mL	
	magnesium ribbon		0.10–0.15 g	
	gloves, nitrile			1 pair for each student

Lab	Item	Size	# Req.	Remarks
15B	potassium nitrate (KNO_3), solid		3–4 g	
	split rubber stopper, 1-hole	#4	1	
	thermometer, laboratory		1	
	water, distilled, room-temperature		50 mL	
	weighing dish		1	
16A	apron, laboratory			1 for each student
	beaker	100 mL	6	
	beaker	200 mL	3	
	gloves, nitrile			1 pair for each student
	goggles			1 pair for each student
	graduated cylinder	25 mL	3	
	hydrochloric acid (HCl)	1.00 M	175 mL	
	labeling tape or grease pencil		1	
	sodium thiosulfate ($Na_2S_2O_3$)	0.10 M	25 mL	
	sodium thiosulfate ($Na_2S_2O_3$)	0.20 M	100 mL	
	stirring rod		1	
	stopwatch		1	
	water, distilled		100 mL	
16B	apron, laboratory			1 for each student
	beaker	100 mL	7	
	gloves, nitrile			1 pair for each student
	goggles			1 pair for each student
	graduated cylinder	10 mL	2	
	graduated cylinder	25 mL	4	
	hydrochloric acid (HCl)	0.10 M	80 mL	
	labeling tape or grease pencil		1	
	pipette, disposable		1	
	potassium bromate ($KBrO_3$)	0.040 M	80 mL	
	potassium iodide (KI)	0.010 M	80 mL	
	sodium thiosulfate ($Na_2S_2O_3$)	0.00025 M	80 mL	
	starch solution		5 mL	
	stopwatch		1	
	water, distilled		80 mL	
17A	apron, laboratory			1 for each student
	Erlenmeyer flask	250 mL	1	
	gloves, nitrile			1 pair for each student
	goggles			1 pair for each student

Lab	Item	Size	# Req.	Remarks
17A	iron(III) chloride ($FeCl_3$)	0.25 M	5 mL	
	labeling tape or grease pencil		1	
	pipette		2	
	potassium thiocyanate (KSCN), 0.25 M	0.25 M	5 mL	
	test tube		2	
	test tube rack		1	
	water, distilled		100 mL	
17B	apron, laboratory			1 for each student
	beaker	50 mL	3	
	goggles			1 pair for each student
	graduated cylinder	10 mL	3	
		25 mL	3	
	labeling tape or grease pencil		1	
	silver nitrate ($AgNO_3$)		0.43 g	
	sodium chloride (NaCl)		0.15 g	
	sodium sulfide (Na_2S)	0.1 M	25 mL	
	stirring rod		1	
	test tube		2	
	test tube rack		1	
	water, distilled		50 mL	
	weighing dish		2	
18A	acetic acid ($HC_2H_3O_2$)	0.1 M	10 mL	
	apron, laboratory			1 for each student
	gloves, nitrile			1 pair for each student
	goggles			1 pair for each student
	graduated cylinder	10 mL	1	
		100 mL	1	
	hydrochloric acid (HCl)	0.10 M	15 mL	
	labeling tape or grease pencil		1	
	methyl orange solution			
	methyl red solution			
	pH meter		1	optional
	pipette, transfer		3	
	test tube		21	
	test tube rack		2	
	thymol blue solution			
	water, distilled		60 mL	
	weak acid, unknown		10 mL	

Appendix F

Lab	Item	Size	# Req.	Remarks
18B	apron, laboratory			1 for each student
	balance, laboratory		1	
	beaker	150 mL	2	
	burette	50 mL or 1000 mL	2	
	clamp, burette		1	
	Erlenmeyer flask	250 mL	2	
	funnel		1	
	gloves, nitrile			1 pair for each student
	goggles			1 pair for each student
	phenolphthalein solution			
	pipette		1	
	potassium hydrogen phthalate ($KHC_8H_4O_4$)		3 g	
	ring stand		1	
	sodium hydroxide (NaOH)	0.4 M	150 mL	
	wash bottle		1	
	water, distilled		40 mL	
	white vinegar		50 mL	
19A	alligator clip lead		2	
	apron, laboratory			1 for each student
	beaker	50 mL	1	
	gloves, nitrile			1 pair for each student
	goggles			1 pair for each student
	magnesium ribbon	10 cm	1	
	mechanical pencil lead		4	
	multimeter, digital		1	
	sandpaper			
	tape			
	vinegar, white (5% acetic acid)		40 mL	
	zinc strip or galvanized nail	10 cm	1	
19B	apron, laboratory			1 for each student
	balance, laboratory		1	
	beaker	50 mL	1	
	beaker	150 mL	1	
	burette	50 mL	1	
	clamp, burette		1	
	Erlenmeyer flask	125 mL	1	
	filtering funnel		1	
	gloves, nitrile			1 pair for each student

Lab	Item	Size	# Req.	Remarks
	goggles			1 pair for each student
	graduated cylinder	10 mL	1	
		100 mL	1	
	iron(II) sulfate hepta-hydrate ($FeSO_4 \cdot 7H_2O$)		0.7 g	
19B	potassium permanganate ($KMnO_4$)	0.020 M	60 mL	
	ring stand		1	
	sulfuric acid (H_2SO_4)	6 M	1 mL	
	wash bottle		1	with distilled water
	water, distilled		10 mL	
	weighing dish		1	
	2-hydroxybenzoic acid ($C_7H_6O_3$)			
	apron, laboratory			1 for each student
	beaker	50 mL	3	
		250 mL	1	
	ethanoic acid (CH_3COOH)		5 mL	
	ethanol (CH_3CH_2OH)		5 mL	
	gloves, nitrile			1 pair for each student
	goggles			1 pair for each student
	labeling tape or grease pencil		1	
	thermometer, laboratory		1	
20A	methanol (CH_3OH)		5 mL	
	microwave oven		1	
	pipette		3	
	propan-2-ol ($CH_3CHOHCH_3$)		5 mL	
	sodium bicarbonate ($NaHCO_3$) solution		5 mL	
	sulfuric acid, concentrated (H_2SO_4)		5 mL	
	test tube		3	
	test tube clamp		1	
	test tube rack		1	
	water, distilled		75 mL	
	apron, laboratory			1 for each student
	balance, laboratory		1	
	bar of soap		1	
20B	beaker	50 mL	1	
	beaker	250 mL	1	
	biuret solution		20 mL	
	brown lunch bag		1	

EQUIPMENT AND MATERIALS LIST (BY LAB)

Lab	Item	Size	# Req.	Remarks
20B	cooking oil			
	cork		6	to stopper test tubes
	cornstarch solution		10 mL	
	disposable pipettes		7	
	egg white		10 mL	
	filter paper		1 pc.	
	filtering funnel		1	
	food for testing			
	gloves, nitrile			1 pair for each student
	glucose solution		10 mL	
	glucose test strips		8	
	goggles			1 pair for each student
	graduated cylinder	10 mL	1	
		100 mL	1	
	iodine solution		20 mL	
	labeling tape or grease pencil		1	
21A	apron, laboratory			1 for each student
	balance, laboratory		1	
	liquid detergent			
	magnesium sulfate heptahydrate ($MgSO_4 \cdot 7H_2O$)		3.0 g	
	metric ruler		1	
	mortar and pestle		1	
	gloves, nitrile			1 pair for each student
	ring stand with iron ring		1	
	rubber stoppers		10	
	shampoo			
	spatula		1	
	stirring rod		1	
	test tube		5	
	test tube brush		1	
	test tube rack		1	
	trisodium phosphate (Na_3PO_4)		2.0 g	
	vegetable oil		10 mL	
	water, distilled		10 mL	
	weighing dish		2	
	weighing dishes		3	

Lab	Item	Size	# Req.	Remarks
21B	apple		1	
	apron, laboratory			1 for each student
	banana		1	
	beaker	50 mL	11	
		600 mL	1	
	fig		1	
	freezer		1	
	gelatin			prepared
	goggles			1 pair for each student
	grape		1	
	kiwi		1	
	labeling tape or grease pencil		1	
	microwave oven		1	
	papaya		1	
	pineapple		1 ea.	fresh, canned, and frozen (but thawed)
22A	paper "isotopes"	3 cm x 3 cm	300	100 each of three different colors, all marked with an "X" on one side
22B	reference sources for nuclide masses or internet access			

Appendix G

LABORATORY EXAM

You, a Student Chemist!

This laboratory exam is designed to show your teacher the skills and thought processes that you have developed in the chemistry laboratory setting. This is your opportunity to be a student chemist!

As you get ready to attack the following procedures, think about the guidelines below.

1. Follow safety protocols.

2. Use significant figures when reporting data and calculating values.

3. Focus on using procedures that will give you maximum accuracy, such as doing multiple trials of a procedure.

4. If you run out of time to complete a section, at least record what you did get done.

5. Never leave chemicals or an experiment overnight without checking first with your teacher.

Procedure

SEPARATING MIXTURES

Your goal is to separate a mixture of sand, salt, and iron. Here are some smaller goals to focus on as you get started.

1. Find the percent composition for each substance in the mixture.

2. Determine your accuracy in percent recovered and by comparing your experimental percent compositions with those values from your teacher.

3. Concisely and completely outline the process that you used to determine the mass-to-mass ratio of salt to sand.

DETERMINING CONCENTRATION

Determine the amount concentration of a sample of HCl as accurately as possible.

1. Gather supplies to determine this amount concentration. You may consider reacting the HCl with something else that you can measure to determine the concentration of HCl.

2. Record your data and show sample calculations.

CANDLE CHEMISTRY

Explore the chemistry of candles, meeting the goals below.

1. Demonstrate that candles use oxygen and release water and carbon dioxide.

2. Determine the physical state of the wax when combustion takes place.

3. Experimentally determine the calories released per gram of candle consumed as the candle burns.

4. Write a concise description of the procedures used to support your conclusions. Include any data or calculations.

MATCHSTICK ROCKETS

Build a matchstick rocket. As solid fuel turns to gas, the volume of the match head increases dramatically. If the match head is contained, the pressure produced can be directed to propel the match a significant distance. Use these ideas to make your rocket travel as far as possible. You must build your rocket using the following requirements.

1. Use only one match per rocket; no other propellants are allowed.

2. The whole rocket must experience movement; fragments from rockets that disintegrate on the launching pad will not be considered.

3. Distance of travel is measured from the launch site to the farthest portion of the rocket once it has come to rest.

Appendix G
LABORATORY EXAM: TEACHER GUIDE

You, a Student Chemist!

» Demonstrate creative problem-solving skills in a laboratory setting.

» Perform laboratory techniques with accuracy and precision.

» Solve problems using scientific inquiry and scientific tools.

This Laboratory Exam consists of a number of separate activities. The intent of each is to demonstrate a particular set of laboratory skills. Students will need to think back over all the skills they have learned during this course, decide on which ones are suited to completing the task, and then carry out the procedure.

Consider allowing students to pick one or two sections of the Laboratory Exam to complete. You may choose to have them do all the sections.

You may want to assign point values to each step or part of the procedure. Make sure that each different procedure has an equal point value if you opt to give your students a choice in the activities they complete. If you choose to use all these procedures, see the Laboratory Exam grading rubric on page 274.

Give your students resources to solve their problems. You may choose to assign point deductions for failing to comply with safety regulations or for "buying" help from the teacher to point them in the right direction to solve their problems.

Also, provide your students with guidelines for their final report, such as whether you would like specific sections on discussion, data, analysis, and results. Tell them whether you would like their reports typed out and whether they need a bibliography.

It may be helpful to give students guidelines on how they should work together as a team and how they should report their contributions to the assignment.

You may also choose to do a preliminary grading of a lab group's results, giving them the opportunity to retry the experiment if they get poor results.

For the matchstick rocket procedure, consider having a demonstration to see which rocket travels the farthest. An option would be to give bonus points to students whose rockets travel farther. This project may be an individual one.

Procedure

SEPARATING MIXTURES

This activity is similar to Lab 2A. If you did Lab 2A, you can still have your students do this activity. Challenge them with how much better they can do with another year of laboratory experience behind them.

You can make the mixture or have the students make a mixture. If the students make the mixture, you may need to modify Goal 2 below.

1. Students will need to isolate each component of the mixture. They can then find the mass fraction of each component compared with their recovered mass.

2. They should determine the percent error of the mass recovered compared with the initial mass of their sample. If you made the initial mixture, then they should also calculate the percent errors of each component that they recovered.

3. This part of the process is assessing both their organization of a process and their ability to communicate.

DETERMINING CONCENTRATION

Determine the amount concentration of a sample of HCl as accurately as possible. To make 100 mL of a 1.20 M HCL solution, pour 70 mL of distilled water in a volumetric flask. Then add 10.0 mL of 12 M HCl to the flask. Gently swirl the flask until thoroughly mixed and then dilute with distilled water to 100 mL.

1. Students can react the HCl with sodium carbonate or a metal. (Zinc or magnesium would work well.) Hydrogen gas is released to remind students about ignition sources. Students can also do a titration of the HCl with sodium hydroxide. A 1 M NaOH solution would work well (4.0 g NaOH dissolved in distilled water to make a 100 mL solution).

2. This goal will demonstrate students' ability to do stoichiometric calculations.

CANDLE CHEMISTRY

This activity is a good opportunity for your students to work with gases and the gas laws. This will be challenging for them since they must determine how they can demonstrate each of the following aspects of the reaction.

1. To demonstrate that oxygen is consumed, students can burn two candles in similar environments: one being normal air and one being oxygen-rich. To demonstrate that burning candles produce carbon dioxide, students can burn a candle in a container and then pass the product gases through lime water or a solution with bromothymol blue. They will need to research and experiment to show the expected color change that is associated with the presence of CO_2. To show water as a product, students can condense water from the resultant gases.

2. Most students will recognize that the wax is not solid. They will then have to determine whether there is any evidence that the wax is still liquid, or a gas, or perhaps even a plasma.

3. This portion can be accomplished using calorimetry.

4. This part of the process is assessing both their organization of a process and their ability to communicate, including mathematical reasoning.

MATCHSTICK ROCKETS

Matchstick rockets are simple to build. The simplest design is a wooden matchstick with the match head wrapped in aluminum foil. Students will need to build a launch stand also. To launch the rocket, heat the foil with another match or a candle. When the match head ignites, the expanding gases trapped within the foil produce the "thrust" for the matchstick rocket. *Have water available for any potential fires*.

Appendix G

LABORATORY EXAM GRADING RUBRIC

Point values have been left blank for you to determine.

	Excellent (___ pts)	Good (___ pts)	Fair (___ pts)	Needs Work (___ pts)
Separating Mixtures				
Mass Ratio of Salt-Sand Mixture	100%–85% accuracy	84%–70% accuracy	69%–50% accuracy	less than 50% accuracy
Percent Recovered	100%–85% accuracy	84%–70% accuracy	69%–50% accuracy	less than 50% accuracy
Process Description	thorough and orderly description	good and well-ordered description; may lack minor details	average description and organization; lacks detail	little or no description
Determining Concentration				
Amount Concentration Results	100%–85% accuracy	84%–70% accuracy	69%–50% accuracy	less than 50% accuracy
Data and Calculations	thorough and orderly	good and well-ordered; may contain minor errors	average quality and organization; may be incomplete and/or include errors	little or no data and calculations; may include major errors
Candle Chemistry				
Description of Role of Oxygen and Water	Detailed description demonstrates thorough understanding.	Description demonstrates good understanding but may lack detail or contain minor conceptual errors.	Description demonstrates average understanding but lacks detail and/or contains conceptual errors.	Description demonstrates little or no understanding; may be incomplete and/or contain major conceptual errors.
Description of Role of Carbon Dioxide and Wax	Detailed description demonstrates thorough understanding.	Description demonstrates good understanding but may lack detail or contain minor conceptual errors.	Description demonstrates average understanding but lacks detail and/or contains conceptual errors.	Description demonstrates little or no understanding; may be incomplete and/or contain major conceptual errors.
Energy Calculations	thorough and orderly	good and well-ordered; may contain minor errors	average quality and organization; may be incomplete and/or include errors	little or no data and calculations; may include major errors
Process Description and Data on Energy Released per Gram of Candle Consumed	Detailed description demonstrates thorough understanding.	Description demonstrates good understanding but may lack detail or contain minor conceptual errors.	Description demonstrates average understanding but lacks detail and/or contains conceptual errors.	Description demonstrates little or no understanding; may be incomplete and/or contain major conceptual errors.

Matchstick Rockets*				
Movement of Matchstick Rocket	Rocket travels at least 1 m from launch pad.	Rocket moves from launch pad.	Rocket fragments, but fragments move from launch pad.	Rocket ignites but does not move.
Miscellaneous				
Use of Laboratory Time	Excellent	Good	Fair	Poor
Lab Report	Detailed, well-written, and properly formatted report demonstrates thorough knowledge of all tasks and their related chemistry concepts.	Properly formatted report demonstrates good knowledge of all tasks and their related chemistry concepts but may lack detail and/or contain minor conceptual errors.	Report demonstrates some knowledge of tasks and their related chemistry concepts but lacks detail and contains conceptual errors.	Report demonstrates little or no knowledge of tasks and their related chemistry concepts; may be incomplete and/or contain major conceptual errors.

Consider giving bonus points for first and second place distances for matchstick rockets.

Photo Credits

Key: (t) top; (c) center; (b) bottom; (l) left; (r) right; (bg) background

COVER

Nick Poon/Moment/Getty Images

FRONT MATTER

i real444/iStock/Getty Images Plus/Getty Images; **ii** ilbusca /DigitalVision Vectors/Getty Images; **v** knowlesgallery /iStock/Getty Images; **vi-vii** Portra/Digital Vision/Getty Images; **vii**r ilbusca/Digital Vision Vectors/Getty Images; **viii**t Rat0007/iStock/Getty Images; **viii**b Erik Isakson/Getty Images; **ix**l Liudmyla Liudmyla/iStock/Getty Images; **ix**r serezniy/iStock/Getty Images; **xi** Jon Feingersh Photography Inc/DigitalVision/Getty Images; **xii**t Hispanolistic /E+/Getty Images; **xii**b Dan Kitwood/Getty Images News /Getty Images; **xiii** Hispanolistic/E+/Getty Images; **xv** (glass-ware set) choness/iStock/Getty images; **xv** (evaporating dish) rrocio/iStock/Getty Images; **xv** (tongs) Betka82 /iStock/Getty Images; **xv** (crucible) Rabbitmindphoto /Shutterstock.com; **xv** (ring stand) nipastock/iStock/Getty Images; **xv** (wash bottle) © Turtle Rock Scientific/SCIENCE SOURCE; **xvi** sanjeri/E+/Getty Images; **xvii**r ErikaMitchell /iStock/Getty Images; **xvii**c Liudmyla Liudmyla/iStock /Getty Images; **xvii**l miflippo/iStock/Getty Images; **xvii**bg Migrenart/iStock/Getty Images Plus/Getty Images

CHAPTER 1

1 Anjelika Gretskaia/Moment/Getty Images; **2** Eskay Lim /EyeEm/Getty Images; **3** whitemay/Digital Vision Vectors /Getty Images; **5** David Tadevosian/Shutterstock.com; **6** JohnnyGreig/E+/Getty Images; **7**l Migrenart/iStock/Getty Images Plus/Getty Images; **7**r ErikaMitchell/iStock/Getty Images

CHAPTER 2

9 Hulton Archive/Getty Images; **15** ilbusca/Digital Vision Vectors/Getty Images

CHAPTER 3

19 ilbusca/Digital Vision Vectors/Getty Images; **23** dlerick /iStock/Getty Images

CHAPTER 4

25 FactoryTh/iStock/Getty Images; **28** stockcreations /Shutterstock.com

CHAPTER 5

33 Rawpixel/iStock/Getty Images; **37** Haitong Yu/Moment /Getty Images; **40** © NOAO/AURA/NSF/SCIENCE SOURCE

CHAPTER 6

45 Dan Reynolds Photography/Moment/Getty Images; **47** Spiral Periodic Table by Robert W Harrison/Wikimedia Commons/ CC By-SA 3.0

CHAPTER 7

57 Archive Photos/Moviepix/Getty Images

CHAPTER 8

63 Salvator Barki/Moment/Getty Images; **69** zf L/Moment /Getty Images

CHAPTER 9

77 Artranq/iStock/Getty Images

CHAPTER 10

85 Jack Andersen/Photodisc/Getty Images; **89**l DedMityay /iStock/Getty Images; **89**r Andrew Brookes/Cultura/Getty Images

CHAPTER 11

93 AROON PHUKEED/Moment/Getty Images; **99** ilbusca /Digital Vision Vectors/Getty Images

CHAPTER 12

107 by wildestanimal/Moment/Getty Images; **111** Stephen Frink Collection / Alamy Stock Photo

Periodic Table
OF THE ELEMENTS

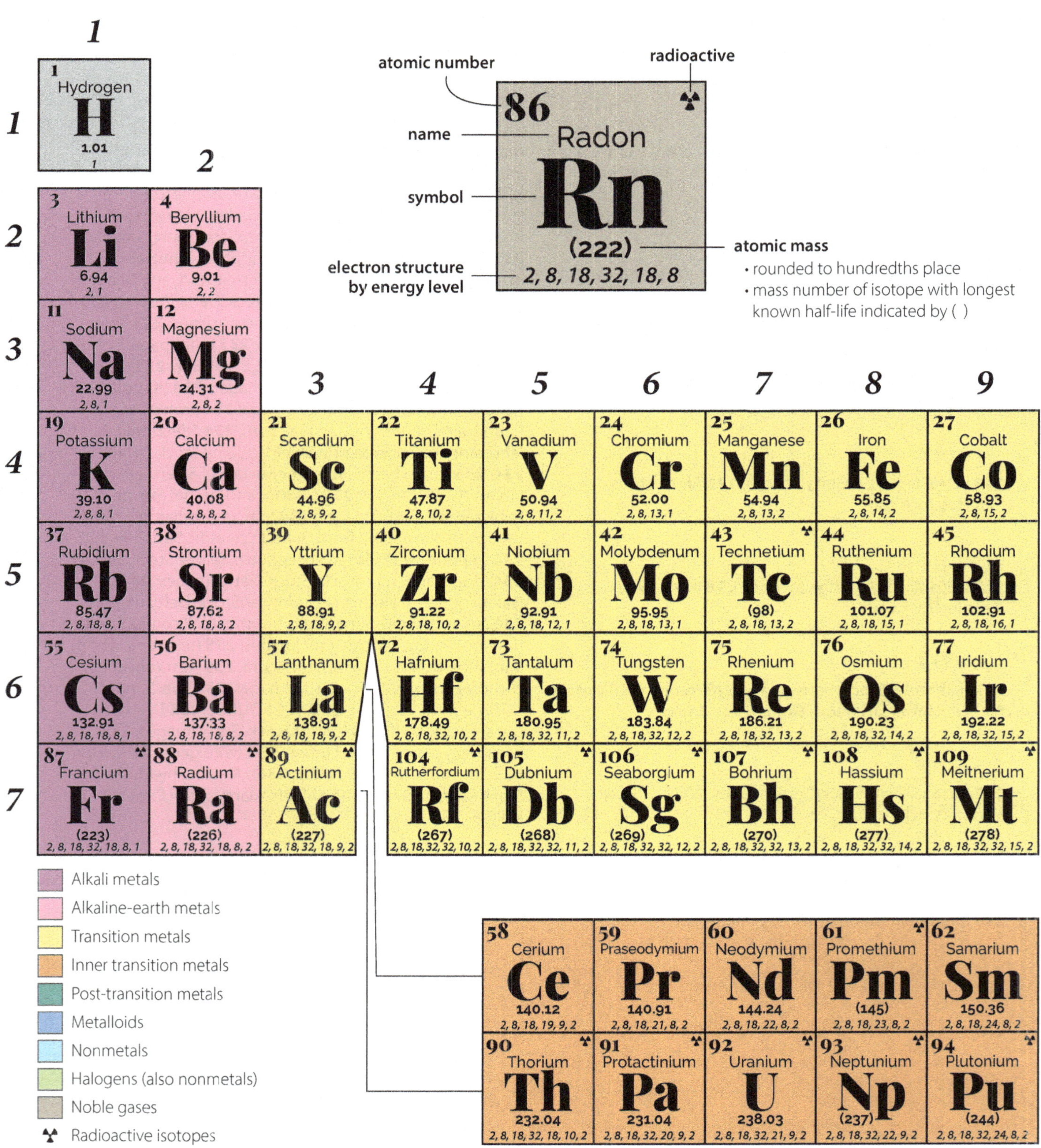

Periodic Table of the Elements

10	11	12	13	14	15	16	17	18
								He 2 Helium 4.00 (2)
			B 5 Boron 10.81 (2, 3)	**C** 6 Carbon 12.01 (2, 4)	**N** 7 Nitrogen 14.01 (2, 5)	**O** 8 Oxygen 16.00 (2, 6)	**F** 9 Fluorine 19.00 (2, 7)	**Ne** 10 Neon 20.18 (2, 8)
			Al 13 Aluminum 26.98 (2, 8, 3)	**Si** 14 Silicon 28.09 (2, 8, 4)	**P** 15 Phosphorus 30.97 (2, 8, 5)	**S** 16 Sulfur 32.06 (2, 8, 6)	**Cl** 17 Chlorine 35.45 (2, 8, 7)	**Ar** 18 Argon 39.95 (2, 8, 8)
Ni 28 Nickel 58.69 (2, 8, 17, 1)	**Cu** 29 Copper 63.55 (2, 8, 18, 1)	**Zn** 30 Zinc 65.38 (2, 8, 18, 2)	**Ga** 31 Gallium 69.72 (2, 8, 18, 3)	**Ge** 32 Germanium 72.63 (2, 8, 18, 4)	**As** 33 Arsenic 74.92 (2, 8, 18, 5)	**Se** 34 Selenium 78.97 (2, 8, 18, 6)	**Br** 35 Bromine 79.90 (2, 8, 18, 7)	**Kr** 36 Krypton 83.80 (2, 8, 18, 8)
Pd 46 Palladium 106.42 (2, 8, 18, 18)	**Ag** 47 Silver 107.87 (2, 8, 18, 18, 1)	**Cd** 48 Cadmium 112.41 (2, 8, 18, 18, 2)	**In** 49 Indium 114.82 (2, 8, 18, 18, 3)	**Sn** 50 Tin 118.71 (2, 8, 18, 18, 4)	**Sb** 51 Antimony 121.76 (2, 8, 18, 18, 5)	**Te** 52 Tellurium 127.60 (2, 8, 18, 18, 6)	**I** 53 Iodine 126.90 (2, 8, 18, 18, 7)	**Xe** 54 Xenon 131.29 (2, 8, 18, 18, 8)
Pt 78 Platinum 195.08 (2, 8, 18, 32, 17, 1)	**Au** 79 Gold 196.97 (2, 8, 18, 32, 18, 1)	**Hg** 80 Mercury 200.59 (2, 8, 18, 32, 18, 2)	**Tl** 81 Thallium 204.38 (2, 8, 18, 32, 18, 3)	**Pb** 82 Lead 207.2 (2, 8, 18, 32, 18, 4)	**Bi** 83 Bismuth 208.98 (2, 8, 18, 32, 18, 5)	**Po** 84 Polonium (209) (2, 8, 18, 32, 18, 6)	**At** 85 Astatine (210) (2, 8, 18, 32, 18, 7)	**Rn** 86 Radon (222) (2, 8, 18, 32, 18, 8)
Ds 110 Darmstadtium (281) (2, 8, 18, 32, 32, 17, 1)	**Rg** 111 Roentgenium (282) (2, 8, 18, 32, 32, 17, 2)	**Cn** 112 Copernicium (285) (2, 8, 18, 32, 32, 18, 2)	**Nh** 113 Nihonium (286) (2, 8, 18, 32, 32, 18, 3)	**Fl** 114 Flerovium (289) (2, 8, 18, 32, 32, 18, 4)	**Mc** 115 Moscovium (290) (2, 8, 18, 32, 32, 18, 5)	**Lv** 116 Livermorium (293) (2, 8, 18, 32, 32, 18, 6)	**Ts** 117 Tennessine (294) (2, 8, 18, 32, 32, 18, 7)	**Og** 118 Oganesson (294) (2, 8, 18, 32, 32, 18, 8)

Eu 63 Europium 151.96 (2, 8, 18, 25, 8, 2)	**Gd** 64 Gadolinium 157.25 (2, 8, 18, 25, 9, 2)	**Tb** 65 Terbium 158.93 (2, 8, 18, 27, 8, 2)	**Dy** 66 Dysprosium 162.50 (2, 8, 18, 28, 8, 2)	**Ho** 67 Holmium 164.93 (2, 8, 18, 29, 8, 2)	**Er** 68 Erbium 167.26 (2, 8, 18, 30, 8, 2)	**Tm** 69 Thulium 168.93 (2, 8, 18, 31, 8, 2)	**Yb** 70 Ytterbium 173.05 (2, 8, 18, 32, 8, 2)	**Lu** 71 Lutetium 174.97 (2, 8, 18, 32, 9, 2)
Am 95 Americium (243) (2, 8, 18, 32, 25, 8, 2)	**Cm** 96 Curium (247) (2, 8, 18, 32, 25, 9, 2)	**Bk** 97 Berkelium (247) (2, 8, 18, 32, 27, 8, 2)	**Cf** 98 Californium (251) (2, 8, 18, 32, 28, 8, 2)	**Es** 99 Einsteinium (252) (2, 8, 18, 32, 29, 8, 2)	**Fm** 100 Fermium (257) (2, 8, 18, 32, 30, 8, 2)	**Md** 101 Mendelevium (258) (2, 8, 18, 32, 31, 8, 2)	**No** 102 Nobelium (259) (2, 8, 18, 32, 32, 8, 2)	**Lr** 103 Lawrencium (266) (2, 8, 18, 32, 32, 8, 3)